Todor Doychev

The dynamics of finite-size settling particles

Dissertationsreihe am Institut für Hydromechanik
Karlsruher Institut für Technologie (KIT)

Heft 2015/1

The dynamics of finite-size settling particles

by
Todor Doychev

KIT Scientific Publishing

The dynamics of finite-size settling particles

Zur Erlangung des akademischen Grades eines

DOKTOR-INGENIEURS

von der Fakultät für

Bauingenieur-, Geo- und Umweltwissenschaften

des Karlsruher Instituts für Technologie (KIT)

genehmigte

DISSERTATION

von

Dipl.-Math. techn. Todor Doychev

geboren in Targovishte

Tag der mündlichen Prüfung: 09. Mai 2014

Referent: Prof. Dr. Markus Uhlmann
Korreferent: Prof. Dr. Sivaramakrishnan Balachandar

Karlsruhe (2014)

To Sveta

Abstract

This thesis contributes to the fundamental understanding of the physical mechanisms that take place in pseudo-turbulent particulate flows. Particulate two-phase flows are of particular interest in great number of environmental, natural and technical processes. Different flow configurations have been considered in the present work: (i) the settling of a single particle in a uniform flow, (ii) the flow through an assembly of fixed randomly distributed particles and (iii) the settling of many particles in an initially quiescent flow field. The particles under consideration are rigid, heavy, finite-size spherical particles with intermediate Galileo (Reynolds) numbers of the order of $\mathcal{O}(100)$. All considered flows are unbounded in all three spatial directions, mono-dispersed and in the dilute regime. High fidelity data sets are generated by means of direct numerical simulation employing an immersed boundary method for the representation of the fluid-solid interface. The results provide detailed picture of the particle and fluid motion over long time intervals. The present flow configurations allow the study of different effects on the particle and fluid motion, such as collective and mobility effects, effects of the solid volume fraction and the Galileo number.

The results evidence that depending on the particle settling regime (i.e. Galileo number and particle-to-fluid density ratio) the particles may exhibit strong inhomogeneous spatial distribution. It is found that the particles are preferentially located in regions with downward fluid motion. The particles inside clusters experience smaller drag coefficient values than the average and larger settling velocities than the average. The flow in all flow cases is observed to exhibit characteristic features of pseudo-turbulence, i.e. fluctuations in the flow field are generated solely by the relative motion of the particles with respect to the fluid. The particle-induced flow field is found to be highly anisotropic with dominant vertical components. The magnitudes of the fluid velocity fluctuations are found out to be comparable to those of fully turbulent particulate flows reported in previous studies. In agreement with previous studies, the energy spectrum of the flow field is observed to experience -3 decay power law with the wave number. Conditional average of the flow field revealed that in the particle assemblies the mean particle wake is significantly attenuated (when compared to the wake of an isolated particle at the same Galileo numbers) by the motion of the particles and the hydrodynamic interactions in the inter-wake regions of the flow. The simulations indicate that, in the present flow configurations, the collective and mobility effects play significant role for the particle and fluid motion.

Zusammenfassung

Partikelströmungen sind ein oft auftretendes Phänomen in der Umwelt, in der Natur sowie in vielen technischen Prozessen. Das grundlegende Verständnis der verschiedenen Mechanismen, die in solchen Strömungen vorkommen, sind von besonderem Interesse für viele Ingenieursanwendungen. Daher ist eine zuverlässige und korrekte Vorhersage dieser Strömungsvorgänge von großer Bedeutung. Trotz der Bedeutung und der breiten Palette von Anwendungen von Partikelströmungen wurden die grundlegenden Mechanismen in solchen Strömungen bis heute noch nicht vollständig erforscht und verstanden. Ein möglicher Grund dafür ist der äußerst komplexe Charakter dieser Strömungsvorgänge. Leider ist die theoretische Beschreibung der physikalischen Prozesse nur für sehr begrenzte Strömungsbedingungen anwendbar. Die meisten der derzeit verwendeten Methoden basieren auf empirischen Annahmen für die Bestimung von Größen wie die mittlere Sedimentationsgeschwindigkeit der Partikel, die Widerstandskoeffizienten, die räumliche Verteilung der Partikeln sowie die Amplituden der Geschwindigkeitsfluktuationen. Die Genauigkeit der Ansätze ist jedoch oftmals unzureichend, da in der Regel eine große Anzahl von Annahmen und Vereinfachungen der Fragestellungen benötigt wird. Daher wird der Anwedungsbereich solcher Methoden stark eingeschränkt.

Die vorliegende Arbeit befasst sich mit Fragestellungen wie der Dynamik von schweren sedimentierenden Partikeln, den Eigenschaften des durch die Partikeln induzierten Strömungsfeldes sowie den daraus resultierenden Zwei-Wege-Kopplungsmechanismen. Außerdem trägt diese Arbeit zum grundlegenden Verständnis der Mechanismen in solchen Fragestellungen bei, indem sie hoch genaue Daten bereitstellt und auswertet. Die Ergebnisse beinhalten die Daten von verschiedenen direkten numerischen Simulationen von Strömungen mit einer großen Anzahl von schweren Partikeln endlicher Größe, die unter Einwirkung der Schwerkraft sedementieren. Die Problemstellungen in dieser Arbeit sind vereinfacht, indem die betrachteten Strömungsgebiete nicht durch Wände beschränkt sind. Dies erlaubt uns, sich ausschließlich auf die grundlegenden physikalischen Prozesse zu konzentrieren.

Die Oberfläche der Partikel wird dabei mit Hilfe der Immersed Boundary Method diskretisiert. Die verschiedenen Ströungskonfigurationen bestehen im Einzelnen aus: (i) der Sedimentation eines einzelnen Partikels in einer gleichförmigen Strömung, (ii) der Strömung durch eine zufällig verteilte feste Anordnung von Partikeln und (iii) der Sedimentation von vielen Partikeln in einem (zunächst) ruhenden Fluid. Das Hauptaugenmerk liegt dabei auf den Schwankungen der Strömungsgrößen in der durch die Partikel induzierten Strömung sowie auf der Charakterisierung der Partikelgeschwindigkeiten, der hydrodynamischen Kräfte, des Nachlaufs der Partikel sowie der räumlichen Verteilung der Partikel. Die Ergebnisse der drei Strömungskonfigurationen werden im Folgenden kurz erläutert.

Sedimentation eines Partikels in einer gleichförmigen Strömung.
Die Sedimentation eines einzelnen schweren kugelförmigen Partikels unter dem Einfluss der Schwerkraft im zunächst ruhenden Fluid wurde mittels direkter numerischer Simulationen untersucht. Dabei wurde die Sedimentation in zwei verschiedenen Regimen betrachtet. Im ersten Regime sedimentiert das Partikel auf einer stationären axialsymmetrischen Trjektorie mit Galileo Zahl Ga $= 121$. In dem zweiten Regime sedimentiert das Partikel wieder auf einer stationären Bahn, jedoch ist die Partikeltrajektorie unter einem bestimmten Winkel mit der vertikalen Achse. Die Galileo Zahl berechnet sich in dieser Konfiguration zu 178. Der Dichtenquotient zwischen dem Partikel und dem Fluid beträgt in den zwei Simulationen $\rho_p/\rho_f = 1.5$. Besonderes Augenmerk wurde auf die Bestimmung der entsprechenden räumlichen und zeitlichen Auflösung der Bewegung der Partikel gelegt.

Der Schwerpunkt liegt hier auf der Charakterisierung der Partikelsedimentationsgeschwindigkeit, der hydrodynamischen Kräfte und des Partikelnachlaufs. Die Ergebnisse wurden mit den verfügbaren Daten aus der Literatur erfolgreich verglichen und wurden in den folgenden Simulationen als Referenzdaten herangezogen. Die gute Übereinstimmung der Ergebnisse mit den Vergleichsdaten deutet darauf hin, dass die charakteristischen Strömungseigenschaften mit hoher Genauigkeit reproduziert wurden.

Strömung durch eine feste Anordnung von zufällig verteilten Partikeln.
Es wurde eine Strömung durch eine zufällig verteilte feste Anordnung von Partikeln mit zwei verschiedenen Parametersätzen simuliert. Die Größe der einzelnen Partikel sowie die zeitliche und die räumliche Diskretisierung der Strömung wurden wie in den vorher beschriebenen Simulationen einzelner Partikel gewählt. Die Anzahl der Partikel wurde so gewählt, dass die Strömung als verdünnt angesehen werden kann. Die Strömung wurde mit einer konstanten mittleren Geschwindigkeit angetrieben, mit der die Partikel-Reynolds Zahl sich zu $Re_p = 141$ and $Re_p = 245$ berechnet.

Es wurde beobachtet, daß in beiden Fällen die Fluktuationen der Fluidgeschwindigkeit stark anisotrop sind, mit einer dominanten vertikalen Geschwindigkeitskomponente. Dies wurde auf die Wirkung des Partikelnachlaufs zurückgeführt. Dabei ist die Höhe der Anisotropie in guter Übereinstimmung mit den Beobachtungen in der Literatur. Der Verlauf der berechneten Energiespektren in Abhängigkeit von der Wellenzahl zeigt eine Steigung von -3, was eine charakteristische Eigenschaft von pseudo-turbulenten Strömnugen ist. Weiterhin wurde beobachtet, dass die Amplituden der Fluktuationen um den räumlichen Mittelwert höher sind als die um den zeitlichen Mittelwert. Eine weitere Erkenntnis ist, dass der mittlere Widerstandskoeffizient der Partikel sehr nah an den Widerstandskoeffizienten eines einzelnen Partikels ist. Der Widerstandskoeffizient von den Partikeln, die sich im Nachlauf eines anderen Partikels befinden, ist signifikant kleiner als der Koeffizient des führenden Partikels. Die Größe des mittleren Partikelnachlaufs wurde im Vergleich zum Nachlauf eines isolierten Parikels verkleinert.

Sedimentation von vielen Partikeln in einem (anfangs) ruhenden Fluid.
Es wurden drei direkte numerische Simulationen von vielen Partikeln, die in einem anfangs ruhenden Fluid sedimentieren, durchgeführt. Die drei Strömungsfälle sind eindeutig durch Parameterkonfigurationen, bestehend aus dem Dichtequotienten ρ_p/ρ_f, der Galileo Zahl Ga und des Volumenanteils der Partikel Φ_s, bestimmt. Für die erste Simulation lautet die Parameterkonfiguration ($\rho_p/\rho_f = 1.5, \text{Ga} = 121, \Phi_s = 0.005$), für die zweite ($\rho_p/\rho_f = 1.5, \text{Ga} = 178, \Phi_s = 0.005$)

und für die dritte ($\rho_p/\rho_f = 1.5, \mathrm{Ga} = 178, \Phi_s = 0.0005$). Die räumliche Anordnung der Partikel wurde eingehend studiert. Dabei wurde beobachtet, dass die Partikel im zweiten und dritten Strömungsfall inhomogen verteilt sind. Sie bilden Agglomerationen in der vertikalen Richtung. In der ersten Simulation jedoch wurden keine Agglomerationen gefunden. Weiterhin wurde die mittlere relative Sedimentationsgeschwindigkeit der Partikel, bezogen auf den globalen Mittelwert der Fluidgeschwindigkeit, ausführlich analysiert. Die mittlere relative Sedimentationsgeschwindigkeit im zweiten und dritten Fall war viel höher als die eines isolierten Partikels. Im ersten Fall andererseits unterschied sich diese Geschwindigkeit zu der eines isolierten Partikels nur gering. Wenn aber die mittlere relative Sedimentationsgeschwindigkeit auf die lokale Fluidgeschwindigkeit in der Nähe der Partikel bezogen wird, fällt dieser Unterschied in allen drei Fällen gering aus. Als Konsequenz befinden sich die Partikel im zweiten und dritten Strömungsfall überwiegend in den Bereichen mit abwärts gerichteter Strömung. Weiterhin wurde herausgefunden, dass die Fluktuationen der lokalen Fluidgeschwindigkeit in der Nähe der Partikel maßgeblich zu den Fluktuationen der Partikelgeschwindigkeit beiträgt. Die Partikelgeschwindigkeit, die hydrodynamischen Kräfte und die Fluidgeschwindigkeit sind hoch anisotrop. Die Amplituden der Fluktuationen der Fluidgeschwindigkeit in dem ersten Fall konnten zufriedenstellend mit einem empirischen Modell reproduziert werden. In den anderen zwei Fällen, in denen Partikelansammlungen beobachtet wurden, führte das empirische Modell allerdings zu keinem Erfolg. Weiterhin wurde eine Energiebilanz für die turbulente kinetische Energie für das Strömungsfeld aufgestellt und beobachtet, dass die potentielle Enegie und die Dissipationsrate die zwei dominantesten Beiträge sind und sich gegenseitig ausbalancieren. Der mittlere Partikelnachlauf wurde signifikant beeinträchtigt durch die Partikelbewegung und die Agglomeration. Die Statistik der Strömung steht in allgemeiner Übereinstimmung mit den Ergebnissen aus der Literatur.

Acknowledgments

This work was supported by the German Research Foundation (DFG) under the project UH $242/1 - 1$ for which I am grateful. The simulations were performed partially at Jülich Supercomputing Center (JSC), at the Leibniz Supercomputing Center (LRZ) of the Bavarian Academy of Science and Humanities and at the Steinbuch Center for Computing (SCC) of Karlsruhe Institute of Technology. The computer resources, technical expertise and assistance provided by these centers are gratefully acknowledged.

I would like to express my deepest gratitude to my supervisor and advisor Prof. Markus Uhlmann for his excellent supervision, scientific guidance, continuous support and patience. His selfless help and encouragement were great motivation for my research and writing of this thesis. Without his dedication and commitment to my research I would have not been able to carry out the research and finish this thesis in the present form.

I would like to sincerely thank Prof. Sivaramakrishnan Balachandar for agreeing to participate in my PhD committee and for his disposition as the second referee of this thesis. I would like to thank also Prof. Jan Dušek, Prof. Willy Dörfler and Prof. Thomas Seelig for their interest in my research and disposition as members of my PhD committee.

I am particularly grateful to my fellow doctoral student Aman for the countless hours of stimulating discussions and the time spent for proofreading of this thesis. I am also thankful to my fellow lab-mates and the staff I was able to meet during my time at the Institute for Hydromechanics: Manolo, Clemens, Conny, Agathe, Yoshi, Angelika, Heidi, Marco, Fred for their support, feedback, friendship and for providing me an excellent working atmosphere.

Last but not least, I would like to thank to my family for their unconditional support and encouragement during the course of this thesis. In particular, I would like to thank my wife, Svetlana, for her love and all the sacrifices she made supporting me and cheering me up during this journey.

Contents

Abstract i

Zusammenfassung iii

Acknowledgments vii

Nomenclature xiii

I Introduction to and Fundamentals of particulate flows 1

1 Introduction 3
1.1 Motivation . . . 3
1.2 Particulate flows: State of the art - challenges and open questions . . . 5
1.2.1 Particle settling velocity and hydrodynamic force . . . 5
1.2.2 Turbulence modification by the dispersed phase . . . 8
1.2.3 Spatial distribution of the dispersed phase . . . 11
1.3 Objectives of the thesis . . . 14
1.4 Outline of the thesis . . . 14

2 Particulate flows - Fundamentals 17
2.1 Physical background . . . 17
2.1.1 Basic assumptions . . . 17
2.1.2 Turbulent flows . . . 18
2.1.3 Relevant parameters . . . 21
2.1.4 Classification of particulate flows . . . 25
2.2 Governing equations . . . 28
2.2.1 Navier-Stokes equations . . . 28
2.2.2 Equations of the dispersed phase . . . 30
2.2.3 Coupling of fluid and particle equations . . . 32
2.3 Numerical model . . . 32
2.3.1 Direct numerical simulation of the carrier phase . . . 32
2.3.2 Immersed boundary method . . . 34
2.3.3 Details of the implementation . . . 35

3 Settling of a single particle in an ambient fluid 37
3.1 Introduction . . . 37
3.2 Overview of the regimes of particle motion . . . 38
3.3 Numerical set-up . . . 40
3.4 Results . . . 43
3.4.1 Particle motion . . . 43
3.4.2 Particle wakes . . . 47
3.5 Conclusion . . . 54

4 Flow through a stationary array of randomly distributed spheres 57
4.1 Introduction . . . 57
4.2 Computational set-up . . . 59
4.3 Results . . . 61
4.3.1 Flow field induced by the particles . . . 61
4.3.2 Energy spectrum and two-point correlations of the flow field . . . 67
4.3.3 Particle hydrodynamic force . . . 71
4.3.4 Particle wakes . . . 76
4.4 Conclusion . . . 82

II Settling of many finite-size particles in an initially quiescent flow field 85

5 Introduction and computational set-up 87
5.1 Flow configuration, initial- and boundary conditions . . . 90
5.2 Temporal evolution of the system . . . 92
5.3 Conclusion . . . 96

6 Spatial structure of the dispersed phase 99
6.1 Qualitative cluster analysis . . . 99
6.2 Voronoï Analysis . . . 101
6.3 Conditionally averaged particle concentration . . . 107
6.4 Effect of the solid volume fraction . . . 110
6.5 Conclusion . . . 113

7 Particle motion 117
7.1 Mean particle settling velocity and hydrodynamic force . . . 117
7.1.1 Fluid velocity seen by the particle . . . 117
7.1.2 Particle average settling velocity . . . 120
7.1.3 Particle Hydrodynamic force . . . 126
7.2 Fluctuating quantities . . . 135
7.2.1 Probability distribution functions . . . 141
7.2.2 Lagrangian statistics . . . 149
7.3 Conclusion . . . 157

8 Flow field induced by the particles **161**
8.1 Qualitative flow description 161
8.2 Statistics of the fluctuating field 163
8.3 Energy budget 173
8.4 Energy spectrum and two-point correlations of the flow field 175
8.5 Particle wakes 179
8.6 Conclusion 185

9 Summary, conclusions and recommendations for future works **189**
9.1 Settling of a single particle in an ambient fluid 190
9.2 Flow through a stationary array of randomly distributed spheres 190
9.3 Settling of many particles in an initially quiescent flow field 191
9.4 Recommendation for future work 192

Appendices **195**

A Averaging procedures **197**
A.1 Phase averaging 197
A.2 Particle-conditioned averaging 198

B Flow field velocity decomposition **201**

C Two-point correlation and one dimensional spectrum of the fluid velocity field **203**

D Derivation of the Energy equation for two-phase particulate flows **205**
D.1 Notation 205
D.2 Derivation of the kinetic energy equation 207
D.3 Energy equation for particulate flows 208
D.4 Turbulent kinetic energy budget 210
D.5 Extended derivation 212

Bibliography **215**

List of Tables **231**

List of Figures **233**

Nomenclature

Mathematical symbols

D/Dt	material derivative, viz. eq. D.15
$\cdot$	dot product of two vectors
$\times$	cross product of two vectors
∇	Nabla operator, i.e. in Cartesian coordinates, $\nabla = (\partial/\partial x, \partial/\partial y, \partial/\partial z)$
∇^2	Laplace operator, i.e. in Cartesian coordinates, $\nabla^2 = (\partial^2/\partial x^2, \partial^2/\partial y^2, \partial^2/\partial z^2)$
$\langle\cdot\rangle_p$	solid-phase averaging
$\langle\cdot\rangle_t$	time averaging
$\langle\cdot\rangle_z$	spatial averaging of a fluid quantity in the vertical direction z
$\langle\cdot\rangle_{\Omega_f}$	fluid-phase averaging
$\langle\cdot\rangle_{\Omega_f,t}$	averaging in time and fluid-phase averaging
$\langle\cdot\rangle_{p,t}$	averaging in time and solid-phase averaging
ϕ'_f	fluctuations of some flow field quantity with respect to the instantaneous space phase averaged flow field (cf. eq. 4.1)
ϕ''_f	fluctuations of some flow field quantity with respect to the time averaged flow field (cf. eq. 4.2)
ϕ'''_f	fluctuations of some flow field quantity with respect to the space and time averaged flow field (cf. eq. 4.3)

Greek symbols

α	angle btw. the particle velocity vector and the vertical axis of the Cartesian coordinate system
$\delta_{u_{p,i}}$	particle velocity increment
Δt	time step used for the temporal discritization
$\Delta x, \Delta y, \Delta z$	width of the computational grid in x-, y- and z-direction

ε	rate of energy transfer from the large scales
η	Kolmogorov length scale
θ	wake momentum diameter, i.e. $(C_d d_p^2/8)^{1/2}$
κ	wave number vector
λ	Taylor micro scale
Λ	length scale for large structures (integral length scale)
λ_x, λ_z	ave length in the horizontal (x) and vertical (z) direction
λ_2	second largest eigenvalue of the tensor $\mathbf{S}^2 + \mathbf{\Omega}^2$
μ	dynamic viscosity of the fluid, i.e. $\mu = \nu\rho_f$
ν	kinematic viscosity of the fluid
π	transcendental number
ρ_f	fluid density
ρ_p	particle density
$\sigma_{\mathcal{V}}$	standard deviation of the normalized Voronoï volumes
σ	stress tensor
Σ	used to indicate summation
τ	separation time
τ_η	Kolmogorov time scale
τ_f	some relevant fluid time scale
τ_g	gravitational time scale
τ_Λ	time scale for large structures (integral time scale)
τ_ν	viscous time scale
τ_p	Stokes time scale or particle relaxation time
τ_{sep}	separation time scale
$\boldsymbol{\tau}$	viscous stress tensor, i.e. $\boldsymbol{\tau} = \mu(\partial_j u_i + \partial_i u_j)$
ϕ	some relevant flow field variable
Φ_s	solid volume fraction, viz. eq. 2.8
Φ_f	indicator function for the fluid-phase, viz. eq. A.1
Φ_p	indicator function for the solid-phase, viz. eq. A.2

ϕ_{ref}	some relevant reference quantity
ϕ_s^{cond}	particle conditioned local volume fraction
Ψ_b	buoyancy force, viz. eq. 8.7
Ψ_p	force term due to two-way coupling, viz. eq. 8.8
ω_p	angular velocity of the particle
$\mathbf{\Omega}$	anti-symmetric part of the velocity gradient tensor
Ω_A	three dimensional averaging domain
Ω_p	orientation of the particle with respect to coordinate axis
∂V_f	boundary of the domain occupied by the fluid

Roman symbols

a_g	effective gravitational acceleration
A_V	aspect ratio defined as the ratio of the largest horizontal extension to the largest vertical extension
$\mathbf{a}_p$	particle acceleration
A_p	projected cross-sectional area of the particles
C_d	particle drag coefficient, viz. eq. 3.7
C_d^{Clift}	particle drag coefficient according to the relation in Clift et al. (1978)
C_d^i, $C_d^{S_i}$	particle drag coefficients according to the definitions in eq. 7.10 - 7.11
C_d^S	particle drag coefficient as defined in eq. 7.8
C_d^{SN}	particle drag coefficient according to Schiller and Naumann (1933)
$C_d^{\pm}$	particle drag coefficients defined as in eq. 7.12
C_u, C_w	constants used for the modeling of the flow field fluctuations (viz. equations 8.3 and 8.4)
$d_{i,j}$	distance between the centers of mass of the i-th and j-th particle
d_{min}	average distance to the nearest neighbor as defined in eq. 6.3
d_p	particle diameter
$\mathbf{e}$	unit normal vector defined in the Cartesian coordinate system $\mathbf{e} = (e_x, e_y, e_z)^T$
$\hat{\mathbf{e}}$	unit normal vector for the coordinate system defined in equations 3.8 - 3.10. $\hat{\mathbf{e}} = (\hat{e}_x, \hat{e}_y, \hat{e}_z)^T$

$\mathcal{E}$	energy spectrum, (cf. eq. 2.6)
$E_{\alpha\alpha}$	one dimensional spectra of the flow field
f_{ij}	residual hydrodynamic force between the i-th and j-th particle
f_V	body force
$\mathbf{F}_{coll}$	forces due to particle-particle collisions
$\mathbf{F}_{\nu}$	volume forces exerted by the fluid upon the particle
$\mathbf{F}_{\sigma}$	hydrodynamic forces exerted by the fluid upon the particle
$\mathbf{g}$	gravity
Ga	Galileo number
h_{wz}	half-width of the (mean) particle wake
$\mathbf{I}$	identity tensor
I_r	relative turbulence intensity, viz. eq. 7.5
I_{rV}	relative turbulence intensity, viz. eq. 7.6
$\mathbf{J}_p$	moment of inertia tensor
$\tilde{k}_f$	turbulent kinetic energy in the (mean) particle wake
$K(\phi)$	kurtosis of some quantity ϕ
l_e	recirculation length of the (mean) particle wake
l_p	distance from the center of the particle to a surface element ds on the surface of the particle
$\mathbf{L}$	period of the domain $\mathbf{L} = (L_x, L_y, L_z)$
$\mathcal{L}$	some relevant fluid length scale
m_p	mass of particle
$\mathbf{n}$	normal unit vector
N_l	number of Lagrangian marker points used for the discretization of the fluid-particle interface
n_p	particle number density
N_p	number of particles in the computational domain
N_{proc}	number of processors used for the simulation
$N_{f,t}$	number of available snapshots of the flow field

$N_{p,t}$	number of available snapshots of the particles
$\mathbf{p}$	pressure
Re	Reynolds number (cf. eq. 2.2)
Re_{∞}^{clift}	particle Reynolds number based on w_{∞}^{clift} and particle diameter d_p
r_p	particle radius
Re_p	particle Reynolds number
$R_{\alpha\alpha}$	two-point correlation function of the flow field
Re_{λ}	Taylor micro scale Reynolds number
Re_{Λ}	Reynolds number based on u_{Λ} and Λ
$R_{\mathcal{S}}$	radius of a spherical shell $\mathcal{S}$ centered around a reference particle
$Re_p^{\mathcal{S}}$	Reynolds number based on the velocity $w_s^{\mathcal{S}}$ (cf. eq. 7.3)
Re_{∞}	Reynolds number based on the particle terminal settling velocity w_{∞}
$\mathbf{S}$	symmetric part of the velocity gradient tensor
$\mathcal{S}$	spherical shell centered around a reference particle
St	Stokes number
$S(\phi)$	skewness of some quantity ϕ
t	time
$\mathbf{T}$	torque of the particle
T_I	characteristic time for the particle response to flow changes (viz. eq. 7.24)
T_{int}	integral time scale, viz. eq. 7.23
$\mathcal{U}$	some relevant fluid velocity scale
u_d	velocity deficit in the (mean) particle wake
u_{d0}	velocity deficit in the (mean) particle wake on the vertical axis passing through the center of the particle
$\mathbf{u}_f$	fluid velocity vector, i.e. $\mathbf{u}_f = (u_f, v_f, w_f)^T$
$\mathbf{u}_g$	gravitational velocity scale
$u_f^{\mathcal{S}}$	instantaneous particle relative velocity w.r.t. to the fluid velocity $\mathbf{u}_f^{\mathcal{S}}$ as defined in eq. 7.1
$\mathbf{u}_{p,rel}$	particle relative velocity with respect to the fluid-phase-averaged velocity

$u^{S}_{p,rel}$	fluid velocity defined on the shell spherical shell $\mathcal{S}$ centered around a reference particle
u^{S}_{s}	instantaneous mean settling velocity of the particles as defined in eq. 7.2
u_{η}	Kolmogorov velocity scale
u_{Λ}	velocity scale for large structures
u_{∞}	particle terminal velocity
$\mathbf{U}_{\infty}$	free stream incoming velocity
w^{clift}_{∞}	terminal settling velocity
$w_{p,\infty}$	particle terminal settling velocity
w_s	terminal settling velocity of a single particle in an initially quiescent fluid
w_{∞}	terminal relative velocity of the particles
V_f	volume occupied by the fluid
$\mathcal{V}_i$	Voronoï cell associated to the i-th particle
V_i	Voronoï volume of the i-th particle
V_p	volume occupied by all particles
V_{Ω}	volume occupied by the two phases
$\mathbf{x}$	position vector, i.e. in Cartesian coordinates $\mathbf{x} = (x, y, z)^T$
$\mathbf{x}_p$	position of the particle with respect to the center of mass, i.e. $\mathbf{x}_p = (x_p, y_p, z_p)^T$
$\tilde{\mathbf{x}}$	position vector in a coordinate system centered at a reference particle aligned with the Cartesian coordinate system, i.e. $\tilde{\mathbf{x}} = (\tilde{x}, \tilde{y}, \tilde{z})^T$
$\hat{\mathbf{x}}$	position vector in a coordinate system centered at a reference particle, as defined in equations 3.8 - 3.10, i.e. $\hat{\mathbf{x}} = (\hat{x}, \hat{y}, \hat{z})^T$
Y^2	particle dispersion, viz. eq. 7.22

Abbreviations

CFL	Courant-Friedrichs-Lewy number
CPU	central processing unit
DNS	direct numerical simulation
HPC	high performance computing
IBM	immersed boundary method

JSC	Jülich supercomputing center
LES	large eddy simulation
LRZ	Leibniz-Rechenzentrum München
$RANS$	Reynolds averaged Navier-Stokes equations
SCC	Steinbuch center for computing Karlsruhe
$p.d.f.$	probability distribution function
$r.m.s.$	root mean square

Part I

Introduction to and Fundamentals of particulate flows

Chapter 1

Introduction

1.1 Motivation

Particle-laden flows are involved in a large number of environmental, natural and technical processes. The fundamental understanding of the mechanisms that take place in such flows and thereby the correct prediction of such flows is of great importance, since they could have high impact on the everyday life of people and the ecosystem. For example volcanic eruptions could release high particulate concentrations in the atmosphere where particles are ejected upwards at high speeds and disperse in large scales causing serious difficulties for air traffic and in particularly serious cases could influence the global climate. Therefore the understanding and prediction of the particle concentration and transport for such events is crucial for assessing the impact on the environment and taking adequate precaution steps. Another example is the deposition of sand particles in the atmosphere by strong winds. For example massive Sahara plumes of dust, also known as the Saharan Air Layer play an important role in decreasing the probability of "cyclogenesis" or hurricane formation. The cloud and rain formation is another important example for particulate flows in the atmosphere. In this case the particles are considered to be liquid droplets. The understanding of the controlling processes for the *cloud* and *rain* formation is of great importance for the weather prediction and climatology. Especially in the context of the recent studies concerning the world climate change it is considered one of the most urgent scientific problems. The main mechanisms responsible for the rain and hailstone formation are the condensation of vapor and the coalescence of water droplets, since the droplets need to reach sufficient size in order to fall. In this context the droplet collisions play an important role for the growth process and are of particular interest for the meteorologists.

An important natural phenomenon involving fluid-particle interactions is *sediment transport* in rivers and coastal areas. The sediments for example could consist of natural materials, pollutants, heavy metals and others. Sediment transport could lead for example to silting of river beds or coastal land erosion which could have devastating consequences for adjoining structures. In sedimentation processes the gravity acts mostly orthogonally to the mean flow. This in turn increases the role of the lift force on the particles, which is responsible for the deposition of the particles and is one of the forces causing erosion by lifting-up of particles.

The emission of small micron particles as result of industrial processes in urban environment shows another area in which multiphase (particulate) flows have direct impact on our everyday life. The dispersion process of such particle *pollutants* is of great interest for the proper design of cities, where effects such as "street canyons" between buildings has to be considered. This has immediate effect upon the human health and as such leads to another important dispersed multiphase problem, viz. the motion and the deposition of micro particles in the human respiratory system.

Besides their environmental impact, particulate flows are also important in many engineering applications. For example in the energy producing systems, *fluidized beds* are typical example of particulate multiphase flows. In such reactors the particles of solid fuel are suspended by hydrodynamic force of upward gas flows. In this case the drag force on the particles which keeps them suspended and the particle interaction are of particular interest, since their knowledge will allow better control of the combustion process. In nuclear reactors the use of bubbles was studied as control parameter of temperature and flow rates. For example it was observed that in wall bounded flows the bubbles tend to accumulate in vicinity of the wall, which could be used to carry vapor away from hot surfaces.

Despite the importance and the wide range of applications of particulate flows, the fundamental mechanisms involved in such flows are far from completely understood. One possible reason for this is the complex character of the flow under consideration. In dispersed particulate flows usually the number of particles is overwhelming. Additionally, the flow field introduces hydrodynamic force and torque on the particles, which can lead to complex interaction between the particles and the flow field. Moreover, difficulties arise due to the large range of scales at which the particle-fluid and particle-particle interaction take place. In vast number of applications the particles are well separated from each other and they have to be considered separately. In such cases the small scale interaction of the individual particles with the background flow has to be considered. On the other hand, it is well known that when the density of the particles is different from that of the fluid they have inertia and do not follow the flow field exactly. As a consequence, the particles will sample the flow field inhomogeneously. Such regions with high particle concentration could be orders of magnitude larger than the particles. As a consequence of the inhomogeneities in the particle spatial distribution, the particle collision frequencies could be greatly affected. Thus, such phenomena should be considered in applications where the collision rate of the particles is important, e.g. particle rain formation, fluidized bed reactors. When the particle velocity is sufficiently large for particle wakes to develop, the interaction between the particles and the fluid becomes more complicated. Particle wakes could lead either to enhancement or attenuation of the turbulence in the flow field and in some cases to complex particle-particle interactions. The mutual interaction of the wakes of multiple particles could cause complex velocity fluctuations beyond that of classical turbulence. In case of absence of classical turbulence in the background flow this phenomenon has been known as *pseudo turbulence*.

The richness of applications and complexity of dispersed multiphase flows has caused a lot of difficulties in the process of their study over the years. The present work will try to address some of the topics involved in the dynamics of particles settling under the influence of gravity, as well as the characteristics of the flow field induced by the settling particles and its impact upon the dispersed phase. Furthermore, this study aims at improving the understanding of the fundamental processes that take place in such flows and also aims at providing a deep and new insight into this processes by trying to isolate the effects of different key parameters. In order to do so, the problem is simplified by eliminating influences due to wall effects, while allowing to focus solely on the fundamental physics involved in the particle-particle and particle-fluid interactions. The problems are tackled by gradually increasing the complexity of the particulate systems in question. First, the forces and dynamics of single particle settling in an ambient fluid are characterized. Then a random arrangement of fixed particles is considered. Here the main focus is on the fluctuations in the flow field due to the particle wakes and on the hydrodynamic forces induced by the flow field. The complexity is further increased by allowing the particles to move and interact with each other freely. The study of this set-up will represent the major part of this work. The emphasis of this part is on the characterization of the particle velocities, hydrodynamic forces and spatial distribution of the dispersed phase, as well as the flow field induced by the particles.

The present work employs direct numerical simulation with immersed boundary method for the representation of the fluid-solid interface. This allows the generation of high-fidelity data for the flow configurations described above and we believe that this data set will contribute to the research of particulate flows. In the following the challenges and the open questions which were identified in previous works will be discussed by giving a short literature review of the most relevant studies.

1.2 Particulate flows: State of the art - challenges and open questions

1.2.1 Particle settling velocity and hydrodynamic force

One of the most important and most studied aspects of the dynamics of particles immersed in a (turbulent) flow is the hydrodynamic force exerted by the fluid upon the particles and have always appealed to the interest of the scientific community. One of the earliest contribution to the understanding of the physics involved in such problems is the discovery of the Archimedes' principle, which determines the hydrostatic force acting on an immersed object. Later Galileo (1612) published short treatise on the subject where he emphasized the implications of the principle on the dynamics of submerged or partially submerged bodies in water. But, the fundamentals of the classical mechanics and fluid mechanics were first laid after the publication of *Principia Mathematica* (1687) by Newton (1687). He devoted his second book entirely to the examination of fluid flows and immersed objects. Since then large number of studies were devoted to the characterization and determination of the hydrodynamic forces exerted by the fluid upon the particles.

In the early days of the scientific inquiry, the systematic study of the hydrodynamic force was mostly accessible through theoretical works. Unfortunately, an universal theoretical definition and description of the hydrodynamic forces was only feasible for flows with very small particle velocities (creeping flow). In the following century significant advances were made in the theoretical description of the hydrodynamic force for creeping flows. The works of Poisson (1831), Stokes (1845), Basset (1888), Oseen (1911, 1913), Faxén (1922), Gatignol (1983) and Maxey and Riley (1983) contributed significantly to the knowledge and understanding of the hydrodynamic force in such flows. In his pioneering work Poisson (1831) was the first who solved the potential flow equation around a sphere. Stokes (1845) described the motion of a rigid-sphere moving with a very low velocity in a viscous flow. His work is considered to be a cornerstone in the research of particulate flows and the hydrodynamic drag force and drag coefficient in such configurations still bears his name. Oseen (1911, 1913) extended the Stokes' drag by using a correct asymptotic analysis, which is called Oseen's drag coefficient. Faxén (1922) introduced the so called Faxén terms which account for non-uniformity in the flow. In a theoretical work Gatignol (1983) used a Faxén formulae for the estimation of the drag and the torque of a solid particle moving in an unsteady non-uniform Stokes flow. Finally Maxey and Riley (1983) defined a Lagrangian equation of motion of a solid sphere under creeping flow conditions, which considers unsteady, non-uniform fluid motion and body forces.

On the other hand, for flows with finite or large particle velocities it was not possible to obtain an analytical expression for the hydrodynamic force. Thus, most of the works were initially experimental and later with the rise of the computing, numerical. The drag component of the hydrodynamic force was identified as one of the most important hydrodynamic parameters, especially for flows with large particle velocity. The drag force is often expressed in form of the particle drag coefficient. In experiments the drag coefficient is usually computed based on the particle relative velocity and a balance between the gravity and the drag force. The drag coefficient of stationary isolated sphere subjected to uniform flow was extensively studied and a large number of experimental data as well as empirical correlations for the drag coefficient and the particle relative velocity is available in the literature. A comprehensive review of the experimental data can be found in Clift et al. (1978) and in a more recent work of Brown and Lawler (2003). As noted by Lovalenti and Brady (1993), the generation of vorticity on the particles and its advection in the particle wake needs to be considered in the determination of the drag coefficient. The various wake structures that may exist behind the particle could lead to unsteadiness of the hydrodynamic force and also to the onset of lift force and torque on the particle. For example Bouchet et al. (2006) showed that each change of the wake structure is accompanied with change of the drag with overall qualitative effect of the wake instabilities being an increase of the drag. With advance of the measuring and computational techniques and capabilities several works investigated the wake structures behind a fixed particle (Natarajan and Acrivos, 1993; Johnson and Patel, 1999; Tomboulides and Orszag, 2000; Ghidersa and Dušek, 2000) and behind a mobile particle (Jenny et al., 2004; Veldhuis et al., 2005, 2009; Horowitz and Williamson, 2010) in a uniform flow. A recent review on the wake structures and the resulting changes in the particle trajectories and drag coefficient can be found in Ern et al. (2012).

The studies mentioned above have focused on particles subjected to uniform flows. It is well known, that turbulence can have significant effect on the hydrodynamic force exerted upon the particles. Past experimental measurements on the effect of the carrier-phase turbulence on the force on a particle have been so far inconclusive. Some authors reporting increase (Uhlherr and Sinclair, 1970; Brucato et al., 1998), or decrease (Rudolff and Bachalo, 1988) and others observed insignificant changes (Wu and Faeth, 1994, 1995) of the drag coefficient. The disagreement of the available data is an evidence of the challenges the turbulent particulate flow present. A discussion on the subject can be found in the book of Michaelides (2006, chapter §7.1). Experimental results can be found for example in Tsuji et al. (1991) and more recently in Doroodchi et al. (2008). A rigorous numerical study of the influence of free stream turbulence on the particle drag and lift force of an isolated particle can be found in Bagchi and Balachandar (2003). They considered a particle with size larger than the smallest flow structures subjected to a frozen isotropic turbulent flow. The authors considered different free stream turbulence intensities and different particle Reynolds numbers. However, they did not observed a substantial effect of the free stream turbulence on the mean drag, but noted that with increasing particle size the accuracy of the prediction of the instantaneous drag decreased. Later Burton and Eaton (2005) studied numerically the interaction between a fixed large particle and decaying homogeneous isotropic turbulence. The authors compared their results to those predicted by the Lagrangian equation of motion and noted considerable differences. More recently Kim and Balachandar (2011) performed similar studies as in Bagchi and Balachandar (2003). The authors compared the hydrodynamic forces obtained from the simulation with those in the absence of ambient turbulence. They showed that non-uniformity of the flow can play an important role in the determination of the particle forces and that the influence on the particle drag force was similar to the Faxén correction. Further, the lift force of the particles was shown to be influenced by the surrounding turbulence and the self-induced wake structure.

The complexity increases, when multiple particles are considered. One of the still not fully understood and unresolved issues considering such applications is the force exerted by the fluid, on an assembly of particles. For stationary particles, in an early work Kaneda (1986) studied the drag on a sparse random array of fixed spheres in the creeping flow regime. Recently Hill et al. (2001b,a) and Tenneti et al. (2011) performed numerical experiments on the subject and extended the study by considering fluid inertia and high particle volume concentration. From the simulations the authors derived empirical correlations for the average drag force and the Reynolds number and showed that the drag force has a more complex dependance on the Reynolds number. A review on the topic can be also found in Koch and Hill (2001, chapter §6).

For mobile particles on the other hand, it is well accepted that as soon as the settling of multiple particles is considered and the mean distance between the particles is small enough for hydrodynamic interaction between the particles to take place, the average settling velocity of the particles will most probably deviate from the terminal settling velocity of a single particle in the same conditions. For such flows the dynamical interplay between the particle hydrodynamic forces and the particle settling velocity is of particular interest. Several studies were devoted to the correct prediction of the particle settling velocity for the ensemble settling of multiple particles. In the Stokes regime, Batchelor (1972) theoretically showed that the average particle velocity depends linearly on the particle volume concentration. In his study he considered only randomly dispersed suspension of spheres.

On the other hand, when inertial effects are no longer negligible, there is no theoretical approach to estimate the effect of the particle volume concentration on the average particle settling velocity. A widely used empirical correlation was derived by Richardson and Zaki (1954) for flows with high particle volume concentrations. Later Smith (1998) proposed a model for the settling velocity for particles which are randomly distributed in space. The common observation was that the average settling velocity decreased as the particle volume concentration was increased. In this context the expression *hindered settling velocity* is commonly used and several studies were dedicated to the subject, e.g. Parthasarathy and Faeth (1990b); Felice et al. (1995); Felice (1999); Yin and Koch (2007). The general accepted mechanism responsible for the decrease of the average particle settling velocity is that an upward flux of fluid is generated in order to compensate for the downward flux of fluid generated by the settling of the particles, which would lead to the net reduction of the average settling velocity.

When the fluid phase exhibits a free stream turbulence which is independently generated, a number of issues are still not well understood, for example whether turbulence would increase or decrease the average settling velocity or whether the average drag coefficient is affected. For small particles in the Stokes regime the effect of several parameters on the settling velocity were thoroughly investigated (Wang and Maxey, 1993; Yang and Lei, 1998; Aliseda et al., 2002; Yang and Shy, 2003, 2005; Bosse et al., 2006; Snyder et al., 2008). On the other hand, for large particles with finite to large Reynolds numbers relatively few studies are available.

The scarcity of experimental data are most probably due to the difficulties encountered in obtaining an accurate measurements of the particle settling velocity.

From the above studies, four theories that predict the turbulence effect upon the particle settling velocity have emerged: vortex trapping (Tooby et al., 1977), non-linear drag (Wang and Maxey, 1993), loitering effect (Nielsen, 1993) and the preferential sweeping effect (Wang and Maxey, 1993). The difficulty rises, since the first three effects are mainly responsible for the decrease of the settling velocity and the fourth leads to its increase. Moreover, several effects could appear simultaneously and compete with each other. Their relative importance in turbulent particulate flows is incompletely defined and depend on the controlling parameters of the flow. The effect of the non-linear drag is important for large particles, whereas the preferential sweeping effect is dominant for small particles (Wang and Maxey, 1993).

1.2.2 Turbulence modification by the dispersed phase

One of the most fundamental problems related to particulate flows is the effect of the dispersed phase on the background flow, particularly when the flow is considered to be turbulent. In a large number of studies it has been observed, that the addition of even a modest amount of solid particles can significantly affect the statistics of the carrier fluid turbulence. Some investigators reported enhancement, while others attenuation of the carrier-phase turbulence. Thus, one of the open questions in turbulent particulate flows is whether reduction or enhancement of the turbulence should be expected after particles are introduced to turbulent flows. This phenomenon has been often referred in the literature as turbulence *modification* or *modulation* by the dispersed phase. In the past a significant amount of research was devoted to the understanding of the physical mechanisms and the identification of the controlling parameters. Despite their abundance, currently there is no generally applicable theory or model available that predicts the turbulence modification. The difficulties stem most probably from the contribution of several mechanisms responsible for the turbulence modulation. For example Michaelides (2006) lists seven different mechanisms, which could have an impact on the turbulence. Moreover, multiple mechanisms can act simultaneously and to our knowledge their relative influence is still not fully understood.

One of the earliest attempts to derive an analytical model for turbulence modulation was made by Hinze (1971). His model is based on dimensionless and order of magnitude analysis. He discussed several aspects of the turbulence modulation such as the role of the particle volume concentration and that precise information of the flow is crucial. First with advances in the experimental methods in the 1980's a quantitative investigation was possible. Gore and Crowe (1989) compiled a large number of experimental data and identified the ratio of the characteristic particle length scale to the characteristic length of the flow as the controlling parameter for turbulence modulation. Later Hetsroni (1989) additionally considered particle inertia and suggested the Stokes number (the ratio of the particle-to-fluid relevant time scales) as controlling parameter. In the following years a number of phenomenological models were proposed, e.g. Yuan and Michaelides (1992); Yarin and Hetsroni (1994); Crowe (2000). The first two models consider turbulence generation in the particle wake and the third one is based on an averaging of the kinetic energy equation. A comprehensive discussion on the models can be found for example in Michaelides (2006).

Large number of numerical studies addressed the subject of turbulence modulation and contributed to the understanding of the process involved therein. Most of the studies, however, considered small particles with length scales smaller than the smallest length scale of the flow.

The majority of the numerical research focused on the modulation of homogeneous forced or decaying isotropic turbulence. Forced isotropic turbulence was investigated in Squires and Eaton (1990); Eaton (1994); Boivin et al. (1998); Sundaram and Collins (1999); Hwang and Eaton (2006), whereas decaying isotropic turbulence was studied numerically in Elghobashi and Truesdell (1993); Druzhinin and Elghobashi (1999); Ferrante and Elghobashi (2003). The general observation is that the particles modulate the turbulence by altering the inertial transfer mechanism of turbulent kinetic energy. In the absence of gravity or mean shear, the energy at the large scales is damped by the particles, whereas the energy at the small scales is enhanced. This process is usually accompanied by increase of the viscous dissipation rate and hence increase in the rate of energy transfer from the large scales. This is often referenced in the literature as the *cross-over* or *pivoting* effect. As a result, a net decrease of the total turbulent kinetic energy is observed in most of the aforementioned numerical studies. On the other hand, when particles are subjected to the influence of gravity, the particles transfer their momentum to the small turbulent scales in an anisotropic manner and energy is transferred from the direction of gravity to the other two directions. This could cause a reverse cascade, which tends to build up energy at the large scales and reducing the decay rate of turbulent energy. Depending on the Stokes number this could lead to net enhancement or attenuation of the turbulence intensity. Several experimental studies, with ambiguous outcomes, were devoted to the verification of the observations made above, e.g. Schreck and Kleis (1993); Hussainov and Kartushinsky (2000); Geiss et al. (2004); Poelma et al. (2007). For more thorough discussion on the numerical and experimental works mentioned above, the reader is referred to the review in Poelma and Ooms (2006).

Particles with size comparable to or larger than the smallest flow scales have received less attention, presumably due to the lack of a scale-separation as well as the fact that the local flow around the particle happens at finite Reynolds numbers (ratio of the inertial-to-viscous forces) lead to considerable additional complexity. When the particle size is comparable or larger to the smallest turbulence scales, the numerical approaches in the aforementioned studies could loose their validity (Eaton, 2009). Only recently direct numerical simulations of the turbulence modulation by the particles with size comparable to the smallest turbulence scales were performed, e.g. Ten Cate et al. (2004); Lucci et al. (2010); Yeo et al. (2010); Gao et al. (2013). Both studies noted that the particles attenuated the turbulence due to enhanced dissipation rate in vicinity of the particles. This effect and the turbulence attenuation by the particles was reviewed recently in Balachandar and Eaton (2010).

A more fundamental approach to the turbulence modulation, due to the vortex shedding in the particle wakes, is to alternatively study the interaction of the free-stream turbulent flow with a single particle. The effect of free-stream homogeneous turbulence on the particle wake was investigated experimentally for example in Wu and Faeth (1994, 1995). Later Bagchi and Balachandar (2004) complemented their study numerically by performing direct numerical simulation of an isolated particle in an isotropic turbulent flow. The common conclusion was that the turbulent kinetic energy in the wake of the particle was enhanced due to the presence of turbulence in the background flow, causing faster recovery than in a uniform ambient flow. The streamwise velocity fluctuations in the wake were more enhanced than the velocity fluctuations in the cross-stream direction. Similar observation were also made in Mittal (2000). Recent studies on the particle wake in turbulent environment were performed in Legendre et al. (2006) and on the wake attenuation of a sphere was performed by Risso et al. (2008); Amoura et al. (2010).

The wake generated by particles with finite Reynolds numbers can contribute significantly to turbulence production. Thus, the characterization of the mechanisms and the characteristics of the particle-induced turbulence is essential. Parthasarathy and Faeth (1990a) theoretically and experimentally investigated the continuous-phase properties of the flow induced by the sedimentation of spherical particles in a nearly stagnant water bath in the very dilute regime. In their configuration all turbulent properties were solely due to the effects of turbulence modulation by the particles. They used the principle of superposition to predict the fluid velocity fluctuations and yielded reasonable predictions for many properties of the flow. Later Mizukami et al. (1992) extended the experiments by considering the settling of particles in stagnant air. More recently Chen et al. (2000); Chen and Faeth (2000); Lee et al. (2003) considered similar experimental setup and parameter range as Parthasarathy and Faeth (1990a) and drew analogies between the turbulent inter-wake region and the grid-generated turbulence. Kajishima and Takiguchi (2002); Kajishima (2004a) studied numerically the interaction between the particles and the particle-induced turbulence and observed enhancement of turbulence due to inhomogeneous particle distribution.

The flow field induced by the rise of light particles or bubbles in an otherwise quiescent fluid was investigated extensively in the past and large body of literature on the subject exist. The flow field turbulence induced by the bubbles is often referred to as *pseudo-turbulence* to distinguish it from the classical turbulence, such as for example homogeneous isotropic turbulence. An interesting feature of the turbulent like motion observed in such flow configurations is that the particles may influence the inertial transfer of turbulent kinetic energy, hence, changing the slope/shape of the turbulent energy spectra from the well known Kolmogorov scaling exponent $-5/3$ to more steep exponent. However, the results from the literature have been so far inconclusive. For already turbulent flows, Lance and Bataille (1991) observed in an experimental study that with increasing volume concentration of the bubbles the Kolmogorov exponent was substituted by $-8/3$. In contrast the experiments of Mudde et al. (1997); Rensen et al. (2005) and the numerical study of Mazzitelli et al. (2003) found the classical $-5/3$ power law. Lance and Bataille (1991) suggested the ratio of the bubble-induced kinetic energy and the kinetic energy of the flow without bubbles as an appropriate parameter to characterize the type of the bubbly flow. Rensen et al. (2005) later called this parameter *bubblance* and used it to parametrize the relative influence of the turbulent fluctuations induced by the bubbles. A theoretical study of the velocity fluctuations induced by bubbles moving in an isotropic homogeneous turbulence based on the potential flow theory can be found in van Wijngaarden (1998). For flows which are not *a priori* turbulent, recent investigation showed that the energy spectrum of the bubble-induced liquid velocity fluctuations follows -3 power law (Bunner and Tryggvason, 2002; Martínez-Mercado et al., 2010; Riboux et al., 2010, 2013). In a theoretical work Risso (2011) obtained -3 power law for the turbulent kinetic energy spectra. The spatial distribution of the bubbles was identified as the main reason for the power law. It is interesting to note that the same decay rate could be observed in other two-phase flows regardless if particles or bubbles are considered. Recently, the experimental work of Mendez-Diaz et al. (2013) confirmed the results of Risso (2011). They concluded that the specific details of the hydrodynamic interactions among the bubbles didn't affected the way in which the pseudo-turbulent fluctuations were produced. Despite the recent effort on the subject, surprisingly little is known about the flow field generated by the settling particles and to what extent the results already obtained in the context of homogeneous bubbly flows apply to the case of heavy solid particles.

1.2.3 Spatial distribution of the dispersed phase

It has been known for a quite some time that particles suspended in turbulent flows might be dispersed inhomogeneous manner and can form regions with high particle concentrations. This phenomenon is commonly referred to as *preferential concentration* or *clustering* and correlates the instantaneous particle concentration field with the turbulent flow field. Particle clustering can have a significant effect on both the properties of the particles and the turbulent flow field. The knowledge of the mechanisms leading to clustering and the parameters controlling the process have received a lot of attention from the scientific community and significant amount of studies were devoted to the subject. Substantial amount of experimental (Maxey, 1987; Fessler et al., 1994; Eaton and Rouson, 2001; Yang and Shy, 2005; Eaton et al., 2005; Wood et al., 2005; Saw et al., 2008) and numerical (Squires and Eaton, 1991; Wang and Maxey, 1993; Rouson and Eaton, 2001; Bec et al., 2007; Dejoan and Monchaux, 2013) studies were dedicated to the subject. A review of the clustering of small particles can be found in Eaton and Fessler (1994). The preferential particle concentration was also reviewed recently by Balachandar and Eaton (2010). In all of the aforementioned investigations the Stokes number was identified as the major parameter determining the degree of preferential concentration and the strongest clustering occurred at Stokes number near unity. Moreover, the general observation is that as the Stokes number approaches zero, particles tend to follow streamlines precisely, whereas at very large values of the Stokes number the particles tend to follow ballistic trajectories. In both cases uniform distribution of the particles was observed. On the other hand for intermediate values of the Stokes number, inertia prevents the particles from following exactly the streamlines and heavy particles are centrifuged away from regions with high vorticity and leads to an accumulation of particles in the straining flow regions. For light particles on the other hand the opposite was observed. This mechanism is also called *inertial clustering*, since it is caused by the different inertia of the particles and the fluid.

A consequence of the particle clustering is that the probability of finding pairs of particles is enhanced. A measure of this probability and thus of the degree of clustering is the so called *radial distribution function* and as identified by Sundaram and Collins (1997) if evaluated at contact provides precise correction to the collision kernel. Reade and Collins (2000) demonstrated that the radial distribution function grows as a power-law of the inverse of the separation distance for separations smaller than the Kolmogorov scale. In an analytical work Chun et al. (2005) developed a theory to predict the radial distribution function in turbulent particulate flows with small Stokes numbers. Recently Saw and Salazar (2012); Saw and Shaw (2012) compared numerical and experimental results with the theoretical model of Chun et al. (2005) confirming the power-law of the radial distribution function. Salazar et al. (2008) investigated the inertial clustering in homogeneous isotropic turbulence by performing detailed comparison between numerical predictions and experimental measurements of the radial distribution function. Later De Jong et al. (2010) extended the study by investigating the temporal development of the radial distribution function.

Recently a new particle clustering mechanism, the so called *sweep-stick* mechanism, was proposed (Goto and Vassilicos, 2006, 2008). The main concept is that in contrast to low Reynolds number flows, where the clustering was attributed to the action of the smallest scale eddies, at high Reynolds number flows, the turbulence owns large range of multiscale coherent eddies, which all can affect the preferential concentration of the particles. Thus, at high Reynolds numbers the clustering is not a single scale phenomenon, but rather has a multiscale nature (Boffetta et al., 2004) and therefore clustering could differ from the high-strain regions of the flow (Goto and Vassilicos, 2008). The sweep-stick mechanism relates the particle clustering to the multiscale coherent eddies in the turbulent flow. Yoshimoto and Goto (2007) showed that the clustering of heavy particles in homogeneous turbulence have self-similar multiscale nature and that particle clustering take place even at moderate-to-large Stokes numbers. Later Coleman and Vassilicos (2009) investigated the preferential concentration in the inertial range of Stokes numbers and concluded that the sweep-stick mechanism was more prominent than the inertial clustering.

Considering particle with size similar or larger than the smallest length scales of the flow (finite-size), the experimental results of Monchaux et al. (2010) showed particle clustering behavior consistent with the sweep-stick mechanism and maximum preferential concentration for particles with Stokes numbers near unity. Furthermore, the authors also observed self similarity of the particle clusters similar to the results of Yoshimoto and Goto (2007). A review on the inertial clustering, the sweep-stick mechanism and another different clustering mechanisms can be found in Monchaux et al. (2012). The authors also provide a comprehensive review of different numerical and experimental configurations and different techniques for the cluster identifications. In the majority of the works considering small particles, the Kolmogorov time scale was usually used for the calculation of the Stokes number. Guala et al. (2008) investigated experimentally the clustering of large particles in homogeneous isotropic turbulence. Preferential concentration of the large particles was observed on scales comparable with the Taylor microscale and they suggested the replacement of the Kolmogorov time with the time scale corresponding to the Taylor microscale in the calculation of the Stokes number. They concluded that the clustering mechanism for large particles is similar as for sub-Kolmogorov size particles.

Recently, Fiabane et al. (2012) studied the effect of the Stokes number on the preferential concentration of finite-size neutrally buoyant particles. The authors did not observe any preferential concentration of the particles. They suspected that the pressure distribution at the surface of the particles was the dominant factor in the dynamics of the particles. Nishino and Matsushita (2004) performed experiments of finite-size heavy particles with high Reynolds number in homogeneous turbulence and observed anisotropic clustering of the particles. The particles formed streamwise elongated structures, *columnar like structures*. They noted that even in extremely dilute regime the distribution of the particles deviated from the homogeneous distribution. Similar columnar like structures were also observed in the numerical works of Kajishima and Takiguchi (2002); Kajishima (2004b). They considered, however, dilute pseudo-turbulent flows. The formation of the clusters was particle Reynolds number depended and the clusters persisted over long time intervals. The authors concluded that the clusters are formed due to the particle wakes.

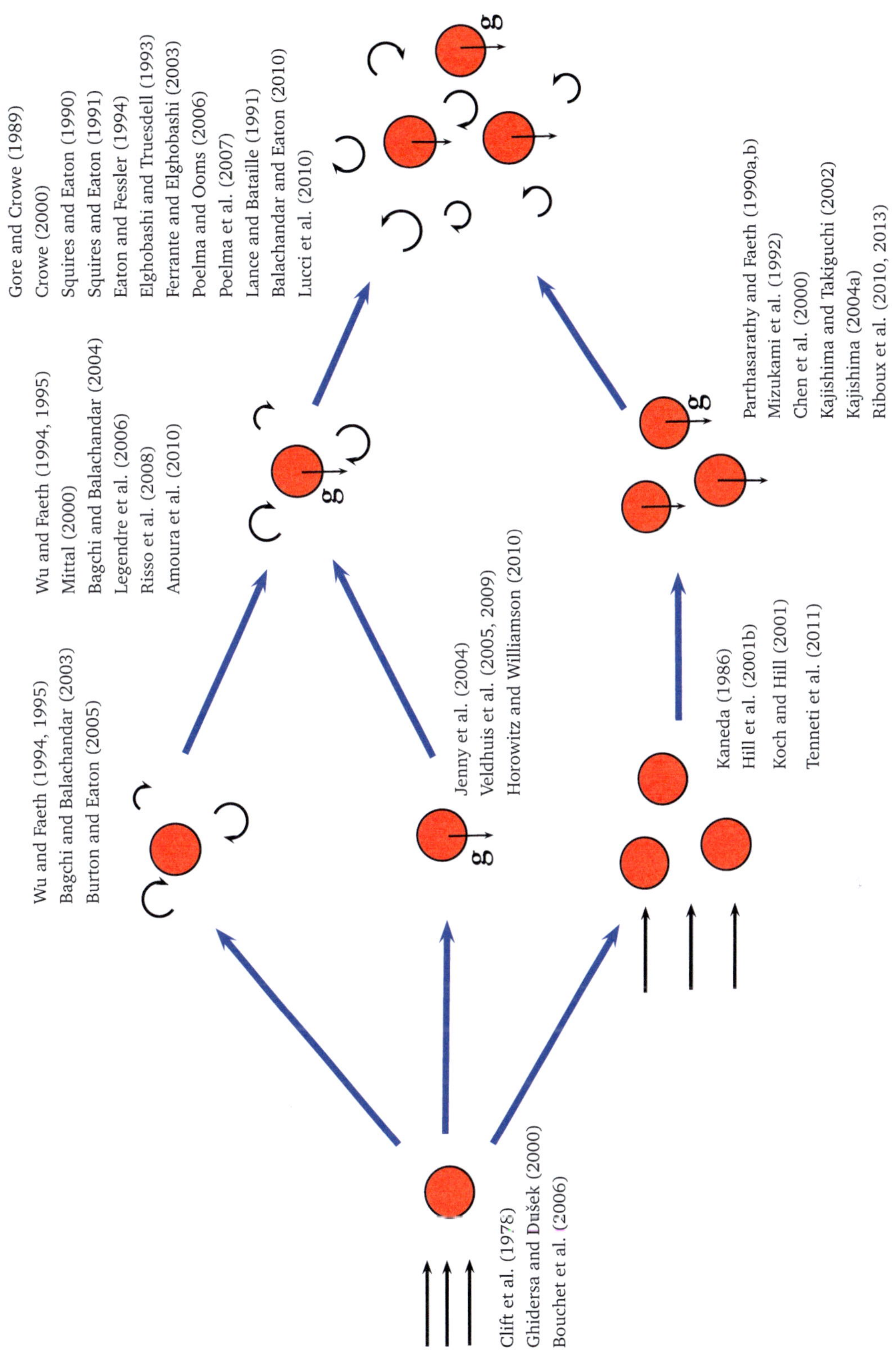

Figure 1.1: Overview of some possible flow configurations for dispersed phase flows with selected papers from sections §1.2.1 - §1.2.3. Black straight arrows indicate uniform flow. Black curved arrows indicate turbulent flows.

1.3 Objectives of the thesis

The objective of this thesis is to extend the existing knowledge base and understanding of the fundamental mechanisms and processes involved in dilute, high Reynolds number particulate flows. More specifically, a unique data set is generated by means of direct numerical simulations (DNS) with immersed boundary method for the representation of the fluid-solid interface. This facilitates the generation of high-fidelity data set with high spatial and temporal resolutions for both the fluid and the particle properties. Here we consider incompressible flow around a large number of heavy rigid spherical particles of finite-size with particle Reynolds numbers of several hundreds. The simulated system will be dilute, mono-dispersed and will include gravity. For such configurations, the interaction of the (turbulent) carrier flow and the particles can lead to a number of coupling mechanisms, such as particle clustering, modulation of the flow field due to the particles and modified particle settling velocity. The thesis aims to address the knowledge gaps related to the dynamics of such fluid-particle systems with respect to the coupling phenomena. In particular we will address the following open questions:

- Do finite-size particles exhibit clustering and what is the exact mechanism of clustering?
- What are the characteristics of the particle-induced flow field?
- How does the particle clustering affect the hydrodynamic forces and the settling rate of the particles?

In order to study the fundamentals of the coupling mechanisms, the problem is simplified by considering simple flow geometries, hence, eliminating inhomogeneities due to wall effects. We performed a series of simulations with gradually increasing the complexity of the settling of single and multiple heavy finite-size particles in an initially quiescent flow field. Additionally we simulated the flow through a random arrangement of fixed spheres. This will allow to study the different processes by isolating and assessing several different aspects, such as the effect of particle mobility, particle volume concentration and particle Reynolds numbers. The emphasis here is on providing (i) characterization of the flow field induced by the particles and turbulence modulation by the particles, (ii) detailed statistical description of the hydrodynamic forces exerted by the fluid upon the particles and its implications on the particle dynamics, (iii) characterization of the spatial distribution of the dispersed phase and detailed analysis of the statistical consequences for both the flow field and the particles. This analysis is by no means meant to be a conclusive study of dispersed two-phase flows. It is mainly meant to contribute to the understanding of the complex mechanisms of the fluid-particle interaction. In particular we anticipate that the main impact of the present results will be in the following areas: (i) providing high-quality statistics of significant physical quantities, (ii) providing deep and new insight into the dynamical processes driving fluid-particle interaction and (iii) laying the foundations for the improvement of engineering-purpose models, in particular the point-particle approach.

1.4 Outline of the thesis

The outline of the thesis is as follows. In chapter §2 an overview on the governing equations and the most common fundamental aspects of multiphase fluid dynamics is provided. Chapter §3 presents the numerical results for the single particle settling in an ambient fluid for two different Galileo numbers. Here statistics of the particle wakes, settling velocity and hydrodynamic forces are evaluated and presented. Chapter §4 provides the results of two direct numerical simulations of the flow through a random arrangement of fixed particles. The main focus in this chapter is on the fluctuations in the flow field due to the particle wakes and on the hydrodynamic forces induced by the flow field upon the particles. A detailed analysis of the mean particle wake is presented as well. In chapters §5 - §8 detailed discussion of the results from three direct numerical simulations of freely moving large number of particles is presented. The emphasis of this part is on the characterization of the particle velocities, hydrodynamic forces and spatial distribution of the dispersed phase as well as the flow field induced by the particles. The effects of particle clustering and the particle volume concentration are discussed as well. Chapter §9 provides a general conclusion and summary of the thesis, recommendations for future work and outlook are provided as well. Extended discussion and additional materials are provided in the appendices.

Chapter 2

Particulate flows - Fundamentals

The main objective of this chapter is to review in detail the state-of-the-art and the most common physical aspects involved in dispersed two-phase flows. The chapter is structured as follows. First in §2.1.1 we provide the basic assumptions which limits the physical scope of the present work. As next, a short heuristic introduction on turbulent flows is given in §2.1.2. Based on this, the most important physical parameters and the various length, velocity and time scales common to (turbulent) particulate flows are introduced in §2.1.3. This then allows us to classify the various regimes existing in particulate flows, providing more precise characterization of here considered dispersed flows (§2.1.4). After the physical introduction, the mathematical description of the physical processes is introduced in the form of governing equations in §2.2. Finally, in section §2.3 the numerical treatment of the equations introduced in section §2.2 along with the numerical methods used in the thesis are presented.

2.1 Physical background

2.1.1 Basic assumptions

Particulate flows are involved in a large range of applications and are characterized by very wide spectrum of important time and length scales. Depending on the specific parameters of the two-phase flow, the physical properties of the flow may vary significantly and this could affect the methodology and the techniques used for their investigation. In order to limit the scope of the present work, a number of assumptions need to be made. The objective of this section is to introduce the assumptions made in the present study along with few essential definitions of dispersed phase flows which apply specifically to particulate flows.

For the mathematical description of the flow, the concept of the *continuum* is important for the development of conservation equations for the carrier-phase. In the continuum approximation, individual molecules and molecular interactions are neglected and it is assumed that the fluid consists of continuous matter. The continuum hypothesis, further allows for a meaningful definition of the fluid density. This assures that at each point of the continuous fluid there is a unique value of the fluid properties, such as density, velocity and pressure, and that they vary continuously from point to point. The continuum assumption is often parametrized by the magnitude of the Knudsen number, Kn, which is the ratio of the mean-free path of molecules to some characteristic flow length scale. Commonly the smallest length scales in the flow field, which are usually called the *macroscopic* length scale, are employed. In general, the continuum approximation is justified for small values of the Knudsen number, $\mathrm{Kn} \ll 1$. For Knudsen numbers of order of unity and larger, a molecular flow regime is established and the flow can not be considered as a continuum (Pope, 2000; Currie, 2002).

In the present study the particles in general will be referred to as *dispersed-phase* and are assumed to be rigid solid unattached bodies of spherical shape immersed in a fluid for which the continuum hypothesis holds and will be referred to as *continuous-phase*. The term rigid-body implies non-deformable body with constant shape. The present work is limited to one dispersed-phase and one continuous-phase. Such flows are also often referred to as *monodispersed*. No mass and heat transfer between the phases are considered in the present work.

The density of the particles is assumed to be homogeneous across the volume occupied by the particles and to be larger than the density of the fluid. Regarding the continuous-phase, the flow is assumed to be incompressible and only Newtonian fluids are considered. The former assumption constrains constant fluid density and the later is the linear relation between the viscous stresses and the rate-of-strain tensor (Batchelor, 1967, § 3.3, p. 146)

$$\tau_{ij} = \mu \left(\frac{\partial u_{f,i}}{\partial x_j} + \frac{\partial u_{f,j}}{\partial x_i} \right), \qquad \text{for } i,j = 1,2,3 \tag{2.1}$$

In the above equation τ_{ij} is the viscous stress tensor, $\mu = \nu \rho_f$ is the dynamic viscosity of the fluid, defined as the product of the fluid density ρ_f and the kinematic viscosity of the fluid ν, $\boldsymbol{u}_f = (u_f, v_f, w_f)^T$ is the fluid velocity vector defined on a Cartesian coordinates (x, y, z). In the present work the fluid and the particle phases are distinguished by the subscripts "f" and "p" respectively. Accordingly the continuous-phase and dispersed-phase quantities are defined in a different frame of reference. The continuum density and fluid velocity fields are *Eulerian fields* and are defined in a Eulerian frame of reference, $\boldsymbol{x} = (x, y, z)^T$. The x and y axis denotes the horizontal directions perpendicular to the direction in which gravity $\boldsymbol{g}$ acts, while the z axis is aligned with the direction of gravity. The dispersed-phase related quantities on the other hand are defined in the Lagrangian frame of reference, $\boldsymbol{x}_p = (x_p, y_p, z_p)^T$. Throughout the thesis bold symbols are used to denote vector and tensor quantities and the indices x, y, z or $1, 2, 3$ denote vector components in the Eulerian frame of reference, e.g. u_z or u_3 denotes the vertical component of the vector $\boldsymbol{u}$.

2.1.2 Turbulent flows

Most flows under consideration in this thesis are assumed to be either uniform or pseudo-turbulent flows. Pseudo-turbulent flows are often defined as instantaneous flows in which the fluctuations in the continuous-phase are generated solely due to the presence of immersed bodies, i.e. in the present work rigid spherical particles. However, we will also consider multiphase-flows with instantaneous homogeneous turbulent flow field (initially isotropic), that is homogeneous turbulent flows with no external turbulence forcing except for the dispersed-phase. In order to better characterize and quantify the pseudo-turbulence, the statistical description of both the classical turbulence and the pseudo-turbulence are compared. The aim of the present section is to provide a brief description and characterization of turbulent flows from somewhat a heuristic viewpoint. This section is by no means an exhaustive qualitative description of turbulent flows and we refer the reader to the excellent textbooks of Tennekes and Lumley (1972); Hinze (1975); Frisch (1995); Pope (2000).

Turbulent flows are omnipresent in our every day life, from the smoke from the chimney, or the flow in steered cup of coffee to turbulence in the atmosphere. Despite the widespread occurrence of turbulence, the physical phenomenon is still not completely understood and is considered as one of the most challenging and perhaps frustrating topics in fluid mechanics. The problem of turbulence has been studied for long time and yet there is no deterministic (complete) and precise definition of turbulence. Many of the researchers were in favor of giving a rather intuitive characterization of turbulence, which however provides some description of the physical attributes of turbulence. For example Richardson (1922) stated:

"Big whorls have little whorls,
which feed on their velocity.
And little whorls have lesser whorls.
And so on to viscosity."

This reflects on the multiscale nature of turbulence and the *cascade* of energy, whereby energy is transferred from the large scales to successively smaller scales until it is dissipated on molecular scales. Hinze (1975) provided yet another definition which reflects on the irregular and random nature of turbulent flows:

"Turbulent fluid motion is an irregular condition of the flow
in which the various quantities show a random variation with
time and space coordinates, so that statistically distinct
average values can be discerned."

In the past the description of turbulence comprised a number of characteristic features. A few of them are (i) chaotic, random behavior; (ii) three dimensional, rotational; (iii) large range of length and time scales and (iv) diffusion and dissipation of energy (Tennekes and Lumley, 1972).

Today it is well accepted that the turbulent flows are characterized by a single non-dimensionless control parameter, the Reynolds number:

$$\mathrm{Re} = \mathcal{L}\mathcal{U}/\nu, \tag{2.2}$$

where $\mathcal{L}$ and $\mathcal{U}$ are some relevant length and velocity scale of the flow and ν is the kinematic viscosity of the fluid. The Reynolds number expresses the relative importance of inertial and viscous forces and provides a good measure of the strength of turbulence. At small Reynolds number the viscous terms dominate the fluid motion and the flow is commonly found to be laminar.

The above definition of turbulence given by Richardson (1922) revealed two dominant processes, the transfer of energy to successively smaller scales and the dissipation of energy. The energy cascade is one of the most important characteristic phenomenon of turbulent flows and large number of physical aspects relates to it. The mechanisms related to the transfer of energy from large to small scales is best studied in the context of homogeneous isotropic turbulence. The flow is said to be homogeneous, if all statistics are invariant under shift in position. As a result, the mean flow field is uniform and under appropriate choice of frame of reference,it can be taken to be zero. When the flow is also statistically invariant under rotations and reflections of the coordinate system, then it is said to be isotropic. The concept of homogeneous isotropic turbulence is extremely important in the turbulence research and most of the turbulence theories relies on it.

The theory introduced by Kolmogorov (1941) is certainly the most famous theory of isotropic turbulence. Based on the cascade theory of Richardson (1922), Kolmogorov was the first to incorporate the multiscale nature of turbulence in a phenomenological theory, which is stated in form of three hypothesis. Kolmogorov argued that the directional information of the large scales is lost as the energy passes down the cascade. As a consequence the small scale motions are assumed to be statistically isotropic and universal for high Reynolds number turbulent flows. He also identified that the rate at which energy is dissipated is extremely important parameter which controls the energy flux from large scales where energy is injected to the small scales where it is dissipated. Consequently, he hypothesized that the statistics of the small scales are determined by the rate of net energy transfer from the large scales ε and the kinematic viscosity of the fluid ν. Given this two parameters, there are unique length, velocity and time scales that can be defined:

$$\eta \equiv (\nu^3/\varepsilon)^{1/4}, \tag{2.3}$$

$$u_\eta \equiv (\varepsilon\nu)^{1/4}, \tag{2.4}$$

$$\tau_\eta \equiv (\nu/\varepsilon)^{1/2}. \tag{2.5}$$

These scales were identified by Kolmogorov and are assumed to be the smallest scales in turbulence and now they bear his name. The Reynolds number based on the Kolmogorov scales is unity, i.e. $\eta u_\eta/\nu = 1$.

The range of scales at which energy is dissipated is referred to as *dissipation range* with dominating viscous effects, whereas the range of large scales at which energy is injected is referred to as *energy-containing range*. The large fluid scales are often characterized by the so called integral length scale Λ, which is a measure for the size of the largest fluid structures. For high Reynolds numbers the separation between the energy containing range of the large scales and the dissipation range of the small scales becomes more distinctive and the intermediate range increases. The intermediate range of scales at which the effective energy transfer takes place is referred to as *inertial subrange*. As the name imply, motions in this range are determined by inertial effects with viscous effects being negligible. Kolmogorov demonstrated that the inertial subrange is universal and uniquely determined by the dissipation rate ε and is independent of the kinematic viscosity ν. From dimensional analysis, in the inertial subrange the energy spectrum $\mathcal{E}(\kappa)$ can be written as:

$$\mathcal{E}(\kappa) = C\varepsilon^{2/3}\kappa^{-5/3}, \tag{2.6}$$

where C is a universal constant and κ is the wave number. The spectrum is also known as the Kolmogorov $-5/3$ spectrum and is for high Reynolds numbers universal. A schematic diagram of the energy cascade and typical energy spectrum at very high Reynolds number are illustrated in figure 2.1.

The Kolmogorov's theory is considered to be one of the milestones in the turbulence research. A detailed discussion of the Kolmogorov's theory will go beyond the scope of the present work and the interested reader is referred to the comprehensive review in the excellent textbook of Frisch (1995).

The random nature of turbulence, which originates from the chaotic nature of the flow, poses strong limitations on the methods used for studying turbulent flows. However, the statistical properties of the flow appear to be reproducible. This has led in the past to

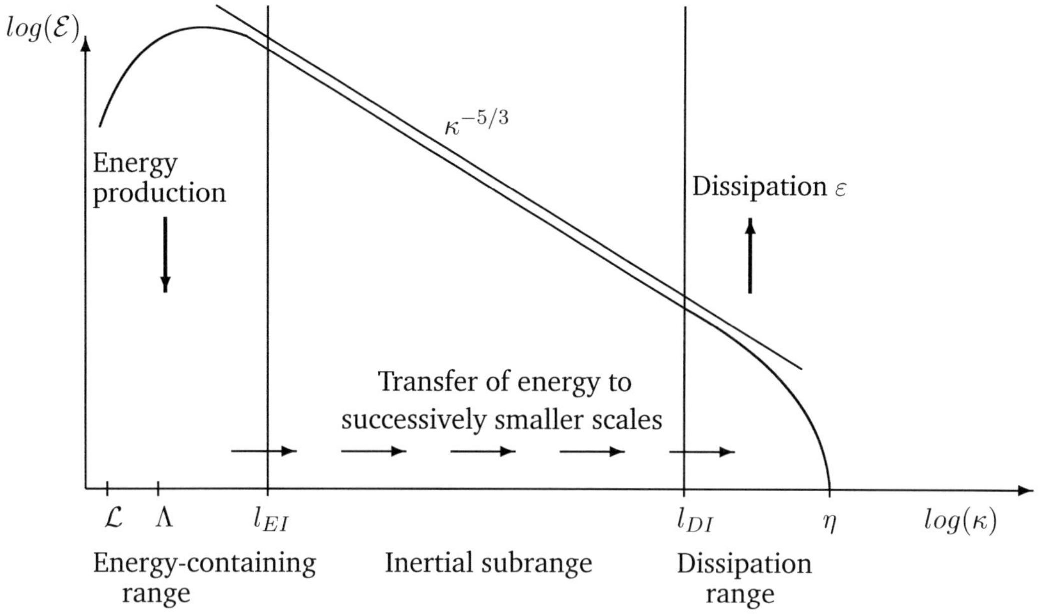

Figure 2.1: A schematic diagram of the energy cascade and energy spectra at very high Reynolds number. The diagram was inspired by Pope (2000). The wave numbers corresponding to the various length scales introduced above are depicted on the x-axis. The length scales l_{EI} und l_{DI} indicate the demarcation lines between the energy containing and dissipation ranges.

the development of rich set of statistical quantities for description and characterization of turbulence. For example Tennekes and Lumley (1972); Pope (2000) provide good entry point into the statistical description of turbulent flows. In the course of this work several statistical quantities are introduced and will be defined in the following sections on the fly as they are needed.

2.1.3 Relevant parameters

In studying two-phase flows usually one is confronted with a large set of physical properties and parameters of the flow making the characterization of (turbulent) particulate flows extremely difficult. Several parameters could lead at the same time to number of competing mechanisms with completely opposite effects on both the dispersed- and carrier-phase. Even for single-phase turbulent flows there is a wide range of physical properties that describe the flow. One reason is the separation of scales originated from the energy cascade mechanism. Combined with the stochastic nature of turbulence and physical properties of the dispersed phase as well as the random distribution of the particles, it makes the parametric description of particulate flows a formidable challenge. At this point, we will provide a summary of all relevant physical properties and scales related to particulate flows. Using these, a number of dimensionless parameters controlling the properties and regime of the two-phase flow are introduced as well. For clarity some of the specific definitions of the here presented quantities are omitted and will be introduced later in course of the work. We would like also to refer the interested reader to the appendices where detailed definitions of all parameters used in the thesis is presented.

The physical properties of the carrier-phase are the already introduced fluid density ρ_f and kinematic viscosity ν. Along with this the single-phase flows are governed by wide range of length, velocity and time scales. The smallest scales of the flow are determined by the kinematic viscosity of the fluid ν and dissipation rate ε and are called Kolmogorv's scales η, u_η, τ_η (s. section 2.1.2). The large scales inherent to the energy containing range in the energy cascade are determined by the turbulence generation mechanism and are mostly based upon the flow geometry. The large scale structures can be characterized by a length scale Λ, velocity scale u_Λ and and time scale τ_Λ. These scales are often referred to as integral length and time scales. The Reynolds number Re_Λ based on this scales is in general large in value and is comparable to the Reynolds number Re in equation 2.2, so the direct effects of viscosity are negligibly small. Most of the complexity of turbulence lie in the inertial range with scales between the dissipative and the integral length scales. A prominent representative of this length scales is the Taylor micro-scale λ which is much larger than the Kolmogorov length scale and much smaller than the integral length scale. One may define another Reynolds number based on the Taylor micro-scale, the well known Taylor micro-scale Reynolds number Re_λ which is traditionally used to characterize isotropic turbulence and is defined as:

$$\mathrm{Re}_\lambda = \frac{u'_f \lambda}{\nu}, \tag{2.7}$$

where u'_f is the characteristic velocity of the eddies with size λ.

In the present study only spherical rigid particles with constant density are considered. Thus, the presence of the particles introduces only three additional physical parameters, the particle density ρ_p, particle diameter d_p and the volume load of the particles per unit total volume Φ_s. The volume load is defined as the ratio of the volume occupied by all particles V_p and the volume occupied by the two-phases V_Ω and is often referred to as the *solid volume fraction*:

$$\Phi_s = \frac{V_p}{V_\Omega} \quad \text{with} \quad V_p = N_p \frac{1}{6}\pi d_p^3. \tag{2.8}$$

In the above equation N_p is the number of particles in the volume V_Ω. Given the solid volume fraction and the particle diameter, a number density can be calculated as:

$$n_p = \frac{6\Phi_s}{\pi d_p^3}. \tag{2.9}$$

The particles introduce also various time and velocity scales. Commonly, the motion of the particle in the flow is described by the *particle relative velocity*, defined as the difference of the particle and fluid velocities $\boldsymbol{u}_{p,rel} = \boldsymbol{u}_p - \boldsymbol{u}_f$. Here we would like to note that the choice of the fluid velocity used to calculate $\boldsymbol{u}_{p,rel}$ is not trivial, especially for particles with sizes larger than the smallest flow length scales, and will be discussed in more details in section 7.1.1. For the discussion in the current and the following sections of this chapter, we will assume single particle in an uniform flow with clearly defined incoming velocity $\boldsymbol{U}_\infty$, viz. $\boldsymbol{u}_f = \boldsymbol{U}_\infty$, except otherwise stated. Based on the particle relative velocity a particle Reynolds number can be defined:

$$\mathrm{Re}_p = \frac{|\boldsymbol{u}_{p,rel}| d_p}{\nu}. \tag{2.10}$$

The particle Reynolds number characterize the flow in vicinity of the particle and can be interpreted as the non-dimensional ratio of fluid inertial forces to viscous forces with respect to the flow around the particle. The particle Reynolds number is an important and widely used parameter for the characterization of the particle motion.

A commonly used particle velocity scale is the so called *particle terminal settling velocity*, $w_{p,\infty}$. The velocity $w_{p,\infty}$ measures the maximum settling velocity a particle can attain settling under the influence of gravity in quiescent conditions before drag balances gravitational forces. Another velocity scale, u_g, can be also obtained from the effective gravitational acceleration, $a_g = |1 - \rho_p/\rho_f||\boldsymbol{g}|$, resulting from the balance of buoyancy and the weight of the particles. This velocity scale is also known as gravitational velocity scale and is calculated as:

$$u_g = \sqrt{|1 - \rho_p/\rho_f||\boldsymbol{g}|d_p}, \tag{2.11}$$

A Reynolds number can be defined based on the gravitational velocity scale, to obtain a dimensionless parameter known as the Galileo number:

$$Ga = \frac{u_g d_p}{\nu} = \frac{\sqrt{|1 - \rho_p/\rho_f||\boldsymbol{g}|d_p^3}}{\nu} \tag{2.12}$$

Together with the density ratio of both phases ρ_p/ρ_f, the Galileo number plays an important role in the characterization of the settling regime of particles, which will be discussed in more detail in section §3.2.

Similar to the carrier-phase, a various time scales for the dispersed-phase can be derived. The time scales for the particles are important for the establishing of non-dimensional parameters to characterize the response of the particles to changes in the velocity and the acceleration of different flow scales. By far the most prominent time scale is the *Stokes time scale* or particle *relaxation time*, which relates to the time required for a particle to respond to a change in velocity. The Stokes time scale is defined as:

$$\tau_p = \frac{\rho_p}{\rho_f}\frac{d_p^2}{18\nu}. \tag{2.13}$$

As the name implies, the Stokes time scale holds for particulate flows in the Stokes regime with densities much larger than the fluid density, i.e. $Re_p < 1$ and $\rho_p/\rho_f \gg 1$. The particle relaxation time is the time required for a particle released from rest in a uniform flow with velocity $\boldsymbol{U}_\infty$ to achieve 63% of the free stream velocity $\boldsymbol{U}_\infty$. The Stokes time scale is independent of the particle relative velocity and depends entirely on the physical properties of the fluid (ρ_f, ν) and the particle (ρ_p, d_p). This is especially advantageous, since it allows an *a priori* calculation of τ_p. Even though the relation 2.13 holds for flows with low Reynolds number, it is still widely used even for flows with finite particle Reynolds number. Besides the Stokes time scale, different time scales based on the gravitational forces and a particle velocity scale can be defined. Based on the gravitational velocity and the particle terminal settling velocity, a particle gravitational time scales can be derived:

$$\tau_g = \frac{d_p}{u_g}, \tag{2.14}$$

$$\tau_g = \frac{w_{p,\infty}}{|\boldsymbol{g}||1 - \rho_p/\rho_f|}. \tag{2.15}$$

Alternatively, a purely viscous time scale can be defined as $\tau_\nu = d_p^2/\nu$.

Based on the above definitions a number of non-dimensional quantities can be formed. Some of them have been already introduced, the solid volume fraction (Φ_s), the various fluid and particle Reynolds numbers (Re, Re_λ, Re_Λ, Re_p), the Galileo number (Ga) and the particle-to-fluid density ratio (ρ_p/ρ_f). Considering the length scales of the flow field and the particles, the particle size is often compared to different length scales of the flow field. The expressions *point-particles* or *finite-size particles* are related to the ratio of the particle diameter to the Kolmogorov length scale, d_p/η. For values larger than unity the particles are considered to have finite-size and effects related to their size need to be accounted for, while for values smaller than unity the particles are considered as point source of mass. For finite-size particles, additionally, the particle size is often compared to the size of the large flow structures or the size of flow structures from the inertial range, i.e. d_p/Λ, d_p/λ. This is especially important in the context of the filtering effect of the flow by the particles and is closely related to the acceleration statistics of the particles, the preferential sampling of the flow by the particles and the sweep-stick mechanism.

To compare the response time of a particle τ_p and a typical flow time scale τ_f, the ratio of these two is introduced. This ratio is known as the particle *Stokes number*

$$\mathrm{St} = \frac{\tau_p}{\tau_f}. \tag{2.16}$$

The Stokes number is an important parameter for determining whether, and to what degree, the particle and fluid velocity field are correlated for given fluid time scale. As for the length scales, different time scales for the fluid as well for the particles can be used for the calculation of the Stokes number. In such a way, the relative importance of different flow scales corresponding to the time scale used in St can be estimated. For example the Stokes number based on the Kolmogorov time scale, $\mathrm{St}_\eta = \tau_p/\tau_\eta$ can provide a good estimation of how well the particles will follow the continuous-phase streamlines. This is for example important for the measurement techniques. For a very small values of the Stokes number ($\mathrm{St} \to 0$), the time for the particle velocity to react to the changes in the fluid velocity is negligible and the particles can be considered as tracers. As the Stokes number increases, the particles will take longer to adapt to the changes in the surrounding fluid velocity field. For Stokes number of the order of unity ($\mathrm{St} \sim 1$), the particle is susceptible for large scale motions. For very large Stokes number ($\mathrm{St} \gg 1$) the particle inertia is large and the effect of the flow field upon the particles accordingly decreases. By using different flow time scales (any of the above defined particle and fluid time scales), the effect of the flow structures related to this scales upon the particles can be studied. Finally, one of the most important application of the Stokes number is the classification or characterization of particulate flows based on its value.

Dimensional considerations

As can be seen (turbulent) particulate flows are characterized by a large set of scales and physical parameters. Depending on the particulate flow under consideration a number of non-dimensional parameters can be identified that characterize the two-phase flow. The number of parameters vary with the flow configuration. In the following, based on dimensional considerations (Buckingham, 1914), we will present a set of non-dimensional quantities, for which we believe they represent the mechanisms of the flows under consideration in this study. We start first with the simplest configuration, the settling of single particle in an otherwise ambient fluid. As next, we present the controlling parameters for similar configuration, but with large number of particles and finally, we consider the sedimentation of multiple particles in a turbulent flow field.

Given the fluid density ρ_f, the kinematic fluid viscosity ν, the vector of gravitational acceleration $\boldsymbol{g}$ on the one hand, and the particle diameter d_p and particle density ρ_p on the other hand, dimensional analysis shows that the problem of single particle settling in an otherwise quiescent fluid is governed by two non-dimensional parameters. Both of which were already introduced, the particle-to-fluid density ratio ρ_p/ρ_f and the Galileo number Ga. When considering multiple particles, one additional parameter come into play, the solid volume fraction Φ_s. When additionally the background flow is turbulent, the parameter space increases by two additional non-dimensional groups, yielding total of five parameters, e.g. the Reynolds number of the background turbulence and a particle-to-fluid length scale ratio. Commonly the Taylor micro-scale Re_λ is considered in homogeneous turbulence and for the length scale ratio, the particle diameter and the Kolmogorov length scale are considered d_p/η. These non-dimensional parameters are by far not the only possible. Depending on the application or the objectives, a range of alternative options exist, one can choose for example different length and/or time scales from the pool of scales defined above. For example one may consider the ratio of the particle diameter to the Taylor micro-scale λ or to the integral length scale Λ. Instead of the Galileo number, the particle Reynolds number can be considered. However, the parametrization of the problem by the Galileo number is in particular useful, since the determination of Ga requires only the physical properties of the particles and the fluid and can be calculated *a priori*. Analogously, a different definition of the flow Reynolds number can be used.

In the present work all flows include gravitational effects. In order to estimate this effects, the dimensionless parameter based on the ratio of a typical fluid fluctuation, such as the root-mean-square value of the fluctuating velocity u'_f, to the particle terminal velocity, $I_r = u'_f/u_\infty$ is considered. For large values of I_r the background flow is dominant and gravitational effects can be neglected, whereas for small values of I_r gravity will dominate the motion of the particle. In turbulent flows additionally the Stokes number plays an important role in determining the effect of the background flow upon the particles.

The above introduced physical properties and various scales are summarized in table 2.1. As can be seen, due to the vast number of parameters the parametrization of particulate flows is not trivial task. This probably could be one of the reasons for the difficulties in predicting turbulent particulate flows.

2.1.4 Classification of particulate flows

Dispersed multiphase-flows incorporate complex interactions between the particles and the fluid. The interaction between the two phases could lead to a number of complex mechanisms, as detailed in section §1. Depending on the flow properties, the relative influence of this mechanisms can vary and have strong impact over the equations governing the physical processes and the numerical or experimental techniques used for their study. The difficulties stem most probably from the large set of parameters present in turbulent dispersed flows, as shown in the previous section. Thus the correct estimation of the characteristics and regime of the multiphase flow is of great importance. To help better estimate the expected type of interactions, the dispersed flows are classified based on number of dimensionless quantities (section §2.1.3), which express the relative importance of different interaction mechanisms. The aim of this section is to provide a brief overview over the different classifications and related terminology, which one usually encounters in the context of dispersed flows.

Using the above introduced non-dimensional groups, dispersed flows can be classified based on the:

	continuous-phase	**dispersed-phase**
Physical properties		
density $[\mathrm{ML}^{-3}]$	ρ_f	ρ_p
kinematic viscosity $[\mathrm{L}^2\mathrm{T}^{-1}]$	ν	–
dissipation rate $[\mathrm{L}^2\mathrm{T}^{-3}]$	ε	–
Length, time and velocity scales		
length scales [L]	η, λ, Λ, $\mathcal{L}$	d_p
time scales [T]	τ_η, τ_λ, τ_Λ	τ_p, τ_g, τ_ν
velocity scales $[\mathrm{LT}^{-1}]$	$\boldsymbol{u}_\eta$, $\boldsymbol{u}_\lambda$, $\boldsymbol{u}_\Lambda$, $\boldsymbol{U}_\infty$, $\mathcal{U}$	$\boldsymbol{u}_p$, $\boldsymbol{u}_\infty$, $\boldsymbol{u}_g$
Dimensionless groups		
Reynolds number $[\sim \mathcal{U}\mathcal{L}/\nu]$	Re_λ, Re_Λ	Re_p, Ga
particle-to-fluid density ratio	ρ_p/ρ_f	
particle-to-fluid length scale ratio	d_p/η, d_p/λ, d_p/Λ	
solid volume fraction	Φ_s	
number density	n_p	
relative turbulence intensity	I_r	
Stokes number	St, St_η	

Table 2.1: Physical properties for (turbulent) particulate flow. Additionally, various length, time and velocity scales introduced in section §2.1.3 are presented. Dimensionless groups involved in particulate flows are also listed. For detailed definition of the different parameters the reader is referred to section §2.1.3.

- particle size compared to the smallest flow length scales (Kolmogorov length scale), i.e. d_p/η. When the ratio is large the particles are assumed to have *finite-size* and effects such as boundary layer effects on the surface of the particle and wake shedding need to be considered. For small values on the other hand the particles are considered as *point-particles*. In most of the point-particle flow configurations, the particles are assumed to have small Reynolds numbers ($\mathrm{Re}_p \ll 1$) and this has, for example, a great impact on the equation of motion, the forces exerted by the fluid upon the particles (e.g. analytical expression for the drag force and coefficient, the so called Stokes drag), or on the requirements of the numerical techniques. In the case of finite-size particles, there are no analytical expressions neither for the particle equation of motion nor for the hydrodynamic forces. Additionally, numerical simulations of finite-size particles impose stronger requirements on the numerical methods as well on the computational costs.
- particle-to-fluid density ratio ρ_p/ρ_f. The particles are assumed to be *light* for $\rho_p/\rho_f < 1$, *heavy* for $\rho_p/\rho_f > 1$ and *neutrally buoyant* for $\rho_p/\rho_f = 1$. Neutrally buoyant particle are of particular interest for experimentalists, since they tend to follow the flow streamlines exactly (note, this apply only for small particles). The particle trajectories for the light as well for the heavy particles deviate from that of the fluid. Light and heavy particles differentiate each other for example on the effect of turbulent eddies upon the particles. Heavy particles are centrifuged out of the turbulent vortices and accumulate in regions with hight strain rates, whereas the opposite is true for light particles. Particles with density different from that of the fluid are also known as *inertial particles*.
- solid volume fraction Φ_s. Based on Φ_s, dispersed phase flows are classified as *dilute* or *dense* flows. In dense flows the particle-particle interactions (particle collisions) dominate

the particle motion, whereas in dilute flows this effects are negligible and particles interact with each other through the hydrodynamic interactions in the flow field. With increasing Φ_s the mean inter-particle distance decreases and the likelihood of particle collisions increases respectively. When the solid volume fraction is small, the effect of the particles on the flow field is negligible. We would like to note that there is no definitive scaling parameter that defines the boundaries between dilute and dense flows. Commonly for values of Φ_s in the order of $\mathcal{O}(10^{-3})$ and smaller the flow is assumed to be dilute. The number density n_p or the mean time between particle collisions could also be used as classification parameters between dilute and dense flows.

An important concept in the analysis of dispersed flows (and multiphase flows in general) is the phase coupling. The phase coupling describes how both phases interact with each other and which interaction mechanisms are dominant in a particular configuration. In general coupling can take place through mass, momentum and energy transfer between the phases. In the present work only momentum coupling is considered. The particles considered in the present study are assumed to be rigid, thus no mass and energy exchange between the phases are present. Energy coupling is used here in the context of heat transfer and is not considered. The interested reader is referred to the textbook of Crowe et al. (2011) for more detailed discussion on mass and energy coupling. The momentum coupling is mainly due to the presence of the particles and results from the hydrodynamic force on the dispersed- and continuous-phase. As a result of this type of coupling, a variety of interactions between the phases could occur, which may lead to changes in the turbulence and net momentum of the continuous-phase. Based on the particle concentration, the coupling between the dispersed- and continuous-phase can be further differentiated on the basis of the degree of coupling, which tends to increase as Φ_s increases. By order of increasing particle concentration, dispersed flows can be classified in three coupling regimes:

- **one-way coupling:** in these flows the effect of the particles on the continuous-phase is negligible. However, the particle motion is affected by the fluid phase. In case of numerical investigation this type of coupling is particularly advantageous, since it allows the calculation of the flow-phase in an *a priori* step and the subsequently integration of the particle equation of motion. The one-way coupling is estimated to apply for particle concentration in the range of 10^{-6} to 10^{-5} (Elghobashi and Truesdell, 1993; Elghobashi, 1994).
- **two-way coupling:** at higher particle concentration the dispersed-phase could also affect the fluid flow by changing the momentum or effective density. The effect of two-way coupling is particularly important in the context of turbulence modulation (section §1), an attenuation or enhancement of the turbulent fluctuations in the flow can be observed depending on the particle characteristics. Here the particles can interact implicitly with each other through fluid-dynamic interactions, e.g. particle wake attraction. In contrast to the case of one-way coupling the fluid- and particle-phase equations should be solved simultaneously. Two-way coupling is found to be most prominent for solid volume fractions in the range of 10^{-5} to 10^{-3}.
- **four-way coupling:** at even higher concentrations, the particle-particle interactions and dynamics becomes more important and dominant. For such flows the collisions between the particles become increasingly important and need to be taken into account. This type of coupling occurs for solid volume fractions larger than 10^{-2}.

We would like to note that the definition of the boundaries between the various coupling regimes can be subjective and there is no definitive demarcation value which will divide distinctive coupling regions. More detailed discussion of the coupling effects and especially the implications on the numerical approaches can be found in Elghobashi and Truesdell (1993); Elghobashi (1994, 2006); Balachandar and Eaton (2010).

2.2 Governing equations

After defining the assumptions and thus the physical scope of the present work, we are now able to present the mathematical formulation of the physical problem in the form of governing equations. The aim of this section is to introduce the equations governing the continuous-phase as well as the dispersed-phase. We first derive the equations for flows without interfaces in relatively standard manner. The issues of boundary and initial conditions required for the solution of introduced equations are then shortly addressed. Next the equations of rigid body motion are presented. The forces acting upon a particle suspended in a viscous flow are also briefly elaborated.

2.2.1 Navier-Stokes equations

It is well accepted that the Navier-Stokes equations describe accurately the motion of a fluid. Before we proceed with the introduction of the equations, the assumptions made in the present work are for convenience shortly summarized. In the present work we assume that the continuum hypothesis holds and we deal with incompressible flows and Newtonian fluids. The following governing equations are then derived with this constraints in mind.

The governing equations are derived on the principle of mass and momentum conservation. The principle of conservation of mass states that mass is not created or destroyed. For incompressible flows the mass conservation equation reads in integral form:

$$\oint_{\partial\Omega_f} \boldsymbol{u}_f \cdot \boldsymbol{n} = 0. \tag{2.17}$$

In the above equation $\partial\Omega_f$ denotes the boundary of a general domain occupied by the fluid, $\boldsymbol{n}$ is the normal unit vector to the surface $\partial\Omega_f$ and $\cdot$ is the dot product between two vectors. The above equation states that the volume of any fluid element cannot change and is also called *continuity equation*. The differential form of the above equation can be simply written as:

$$\nabla \cdot \boldsymbol{u}_f = 0, \tag{2.18}$$

where the ∇ is the so called Nabla operator defined as $\nabla = (\frac{\partial}{\partial x}, \frac{\partial}{\partial y}, \frac{\partial}{\partial z})$. The above condition on the fluid velocity field is also known as divergence-free condition.

The equation of motion is derived by using the momentum conservation principle specifying that the rate of change of fluid momentum in the volume occupied by the fluid Ω_f is the difference in momentum across the boundary $\partial\Omega_f$ plus the net force acting on the volume:

$$\frac{\partial \boldsymbol{u}_f}{\partial t} + \nabla \cdot (\boldsymbol{u}_f \boldsymbol{u}_f) = \nabla \cdot \boldsymbol{\sigma} + \boldsymbol{f} \tag{2.19}$$

In the above equation t denotes the time, $\boldsymbol{\sigma}$ is the stress tensor and $\boldsymbol{f}$ the volume force acting on the fluid. The stress tensor $\boldsymbol{\sigma}$ may be decomposed into a pressure and viscous part and can be written as:

$$\boldsymbol{\sigma} = -\frac{1}{\rho_f}\left(p\boldsymbol{I} + \boldsymbol{\tau}\right), \tag{2.20}$$

in which $\boldsymbol{I}$ is the identity tensor, p the pressure and the viscous part $\boldsymbol{\tau}$ of the tensor is given by equation 2.1.

The set of equation 2.18 and 2.19 is known as the Navier-Stokes equations. A detailed derivation of the Navier-Stokes equations along with properties of the equations and some special cases of the equations can be found for example in Batchelor (1967); Pope (2000). To this date, these equations are the gold standard in the mathematical description of fluid flows and, unfortunately, no general analytical solution is available.

The Navier-Stokes equations are coupled system of non-linear second order equations in four independent variables and their numerical solution present formidable challenges. From mathematical point of view, for the problem to be well posed, a set of initial and boundary conditions (b.c.) need to be prescribed. A various number of different boundary conditions could be specified. In the following a short description of the relevant boundary conditions will be presented.

For impermeable surfaces, as in the present study, the so called *kinematic boundary condition* is enforced for all positions at the fluid-solid interface:

$$\boldsymbol{n}\cdot\boldsymbol{u}_f(\boldsymbol{x},t) = \boldsymbol{n}\cdot\boldsymbol{u}(\boldsymbol{x}_p,t) \qquad \text{at} \qquad \boldsymbol{x}\in\partial\Omega_f. \tag{2.21}$$

The kinematic b.c. ensures that there is no mass transfer between the phases. Usually in viscous flows at the solid-fluid interface the so called *no-slip* boundary condition is imposed. The no-slip b.c. requires that the tangential velocity matches that of the boundary, i.e. the particle:

$$\boldsymbol{n}\times\left(\boldsymbol{u}_f(\boldsymbol{x},t) - \boldsymbol{u}(\boldsymbol{x}_p,t)\right) = 0 \qquad \text{at} \qquad \boldsymbol{x}\in\partial\Omega_f, \tag{2.22}$$

In combination with the kinematic boundary condition the no-slip b.c. reads then:

$$\boldsymbol{u}_f(\boldsymbol{x},t) = \boldsymbol{u}_p(\boldsymbol{x},t) \qquad \text{at} \qquad \boldsymbol{x}\in\partial\Omega_f. \tag{2.23}$$

The no-slip b.c. is a Dirichlet boundary condition, which specifies the value of a given flow variable at certain position, e.g. the fluid velocity at the particle surface.

In the context of homogeneous (isotropic) flows the domain is often assumed to be unbounded, i.e. a periodic boundary conditions in all three space directions are enforced. For a flow field variable $\phi(\boldsymbol{x},t)$ the periodic boundary condition is defined as:

$$\phi(\boldsymbol{x},t) = \phi(\boldsymbol{x}+\boldsymbol{m}\boldsymbol{L},t), \tag{2.24}$$

where $\boldsymbol{L} = (L_x, L_y, L_z)$ is the period of the domain and $\boldsymbol{m} = (m_x, m_y, m_z)$ are some integer numbers. The use of periodic boundary conditions allows a maximum symmetry. The unboundedness of the space does, however, lead to some mathematical difficulties (Frisch, 1995). In the textbook of Frisch (1995) the reader can find also a more detailed discussion on periodic b.c.. The periodicity in all three directions is particularly beneficial, since it reduces the complexity of the calculation of statistical quantities and number of samples is increased at the same time. However, when the length $\boldsymbol{L}$ of the domain is insufficient, the resulting flow field could be substantially influenced, since the largest scales present in the flow field are limited to the length of the periodicity.

For time dependent problems, one needs additionally to specify the initial conditions of the system. In most of the applications, however, one is interested in statistically stationary solutions of the Navier-Stokes equations. Statistically stationary solutions have the property that the statistics of the flow become independent of time and of the initial state of the system. When statistically stationary solution is found, then the flow is called to be in statistically stationary state. Thus, the initial state is often not specified explicitly and often the solution procedure is initialized from a random initial state. Although the statistically stationary state is expected to be for all initial configurations the same, the transition time to steady state may strongly vary depending on the initial state. Thus, a careful choice of the initial configuration needs to be made and different approaches to shorten the transition period exist.

2.2.2 Equations of the dispersed phase

As introduced in section §2.1.1, the scope of the present study is limited to monodispersed flows consisting of rigid heavy particles of spherical shape. The particles are further assumed to be of finite-size, thus for each particle the continuum assumption holds. This allows, among others, the meaningful definition of boundary conditions, e.g. no-slip boundary condition. These assumptions simplify the equations of motion for the particles and the resulting analysis by reducing the parameters that describe the configuration of the system (e.g. no surface deformation and surface tension), to the translation and rotation reference frames attached to the particles.

The dynamics of a system of rigid bodies is determined by its equations of motion, which can be derived from the Newton's second law of motion for linear and angular momentum:

$$\frac{\mathrm{d}(m_p \boldsymbol{u}_p)}{\mathrm{d}t} = \mathbf{F}_V + \mathbf{F}_\sigma + \mathbf{F}_{coll} \tag{2.25}$$

$$\frac{\mathrm{d}(\mathbf{J}_p \cdot \boldsymbol{\omega}_p)}{\mathrm{d}t} = \mathbf{T} \tag{2.26}$$

In the above equations m_p is the mass of the particle, $\mathbf{F}_V$, $\mathbf{F}_\sigma$, $\mathbf{F}_{coll}$ are forces exerted by the fluid upon the particle and will be discussed in more detail below, $\mathbf{J}_p$ is the moment of inertia tensor, $\boldsymbol{\omega}_p$ is the angular velocity of the particle and $\mathbf{T}$ is the torque on the particle. For mono-dispersed suspension of rigid and homogeneous spherical particles the equations can be simplified by taking advantage of the fact that particles have constant mass (i.e. $m_p = \rho_p V_p = \rho_p 1/6\pi d_p^3$), then the moment of inertia tensor reduces to a scalar quantity, viz. $J_p = \frac{2}{5} m_p (d_p/2)^2$. The above quantities are defined with respect to the center of mass of the particles. The position of the particles $\boldsymbol{x}_p$ and the orientation of the particle with respect to the Coordinate axis $\boldsymbol{\Omega}_p$, are functions of time and can be calculated according to the kinematic conditions:

$$\frac{\mathrm{d}\boldsymbol{x}_p}{\mathrm{d}t} = \boldsymbol{u}_p, \tag{2.27}$$

$$\frac{\mathrm{d}\boldsymbol{\Omega}_p}{\mathrm{d}t} = \boldsymbol{\omega}_p, \tag{2.28}$$

The solution of the set of equations 2.25 - 2.28 describes the temporal evolution of the particles and a unique solution can be found when the system of equations is completed by a set of initial conditions for the particle linear and angular position and velocity. The force and torque on the particle supplement the system as homogeneous forcing terms which are provided at each time instance.

Forces exerted upon a particle

In the context of the present work the total force exerted on the particle is linear superposition of the volume forces $\mathbf{F}_V$, such as the gravitational force, the hydrodynamic forces $\mathbf{F}_\sigma$, due to pressure and hydrodynamic stress $\boldsymbol{\sigma}$ and the forces due to particle-particle collisions $\mathbf{F}_{coll}$.

Volume or body forces are forces which are proportional to the particle mass and is defined as integral of a body force $\mathbf{f}_V$ over the volume of the particle:

$$\mathbf{F}_V = \int_V (\rho_p - \rho_f)\mathbf{f}_V \mathrm{d}V. \tag{2.29}$$

In this text the particle and fluid densities are assumed constant and the volume forces are assumed to be represented solely by the gravitational force. Other forces such as electromagnetic forces are neglected. Thereby, the body forces simplify to:

$$\mathbf{F}_V = (\rho_p - \rho_f)V_p\boldsymbol{g}, \tag{2.30}$$

with V_p being the volume of the particle.

The hydrodynamic forces $\mathbf{F}_\sigma$ are probably the most important and challenging forces in the present work. The hydrodynamic forces are forces which act on a surface element of the fluid-particle interface. Thus, only surface forces contribute to $\mathbf{F}_\sigma$. The fluid exerts a net force on the particle consisting of two sources, the pressure and the shear stress that results from the friction between the particle and the flow. Then the hydrodynamic force can be written in terms of the pressure p and the viscous stresses $\boldsymbol{\tau}$:

$$\mathbf{F}_\sigma = \oint_{\partial V} \boldsymbol{\sigma} \mathrm{d}s, \tag{2.31}$$

where ∂V is the particle-fluid interface and $\mathrm{d}s$ is a infinitesimal surface element. Supplementing equation 2.1 into the above equation results:

$$\mathbf{F}_\sigma = \oint_{\partial V} \boldsymbol{\tau} \cdot \boldsymbol{n} \mathrm{d}s - \oint_{\partial V} p\boldsymbol{n} \mathrm{d}s. \tag{2.32}$$

To determine the force $\mathbf{F}_\sigma$, one needs to calculate or estimate the pressure and the shear stress distribution on the fluid-particle interface. From historical point of view, the pressure part of the hydrodynamic force was less problematic than the shear stress, since for the evaluation of the hydrodynamic forces the flow field in vicinity of the particles is required, whereas the pressure was more accessible for measurements. The calculation of the shear stress distribution requires the inclusion of friction and viscous flow. That is one needs to solve the Navier-Stokes equations in order to obtain the shear stress distribution on the particle surface. Now days this is still a significant challenge for the numerical as well as experimental techniques.

Torque on a particle

Due to the the no-slip boundary condition on the particle surface, the fluid can formally grasp the particles and similar to the force on the particle the hydrodynamic stresses at the particle surface can generate particle torque. Then the particle torque can be written as:

$$\mathbf{T} = \int_{\partial V} \boldsymbol{l}_p \times (\boldsymbol{\tau} \cdot \boldsymbol{n}) \mathrm{d}s, \tag{2.33}$$

where $\boldsymbol{l}_p$ is the distance from the center of the particle to the surface element $\mathrm{d}s$ on the surface of the particle. The cross product between the two vectors in the above equation is marked by $\times$.

The particles considered in the present study have spherical shape (i.e. $|\boldsymbol{l}_p| = r_p$) with constant density across the particle cross section. Thus, due to the symmetry of the particle shape, the pressure acting on the particle surface does not contribute to the particle torque, since the pressure force acts always perpendicular to the surface element of the particle surface. Additionally the here employed collision force $\mathbf{F}_{coll}$ is also not contributing to the particle torque.

2.2.3 Coupling of fluid and particle equations

The Newton's second law[1] incorporated into the equations of particle motion showed that the particle motion is controlled by the forces acting on the particles at the particle-fluid interface. Further, the no-slip boundary condition imposed constraints on the fluid velocity field. From the above definitions it is evident that the interaction i.e. the coupling between the solid and fluid phase takes place at their boundary and is established over the boundary conditions. Then the set of equations for the continuous-phase (equations 2.18, 2.19) and for the dispersed-phase (equations 2.25 - 2.28) completely defines the problem of the motion of solid particles suspended in (turbulent) fluid flow.

2.3 Numerical model

In section §2.1 the physical scope of the here considered problem of solid heavy finite-size particles immersed in a fluid flow was defined. Then in section §2.2 the mathematical description of the physical problem was introduced in form of governing equations. In this section we will present the numerical methods applied in the present work in order to solve the governing equations for the continuous- and dispersed-phase. This is done in three steps. First we introduce the concept of direct numerical simulation and the numerical methods to solve the equations for the flow field. Then we present the numerical tools to solve for the particle motion and finally in a third step we introduce the numerical methods employed in this study to simulate the dispersed flow.

2.3.1 Direct numerical simulation of the carrier phase

In section §2.2.1 the Navier-Stokes equations were introduced and showed to describe the motion of the fluid. Their complexity restricted their applicability for the turbulence research for long time. Analytical solutions of the Navier-Stokes equations were only accessible for simplified configurations. First with the birth of computers in the 1950s researchers were able to provide solutions of the Navier-Stokes equations for more generic applications. Since then, the interest in numerical techniques and the development of numerical methods has increased at a spectacular rate. Although many of the ideas for the numerical methods used for the solution of the equations of fluid dynamics were established decades even centuries before computers appeared, their full potential was first harnessed with the introduction of computers. In this section we will discuss the numerical approaches to solve the system of equations for the flow field.

[1] *"The change of motion of an object is proportional to the force impressed and is made in the direction of the straight line in which the force is impressed"*, Newton (1687)

The numerical approaches to solve the Navier-Stokes equations are commonly classified based either on their accuracy or their simplicity. In most of the configurations the accuracy is directly proportional to the simplicity of the numerical approach, viz. the more exact the numerical approach, the more simple is the numerical ansatz. Here simple is used in the context of conceptual simplicity, i.e. the degree of numerical modeling required to obtain a solution.

Based on the degree of modeling, the numerical approaches can be divided in three groups. The first group is based on some type of averaging of the Navier-Stokes equations, either temporal or spatial. This approach leads to the so called *Reynolds averaged Navier-Stokes* equations (RANS). The second group solves the Navier-Stokes equations for the large scale motion of the fluid, and models the effect of the small flow scales. This group of simulations are known as *Large Eddy Simulations* (LES). The third group solves the governing equations for all scales of the fluid motion and avoids any use of modeling. The simulations of this group are referred as *Direct Numerical Simulations* (DNS). The simulations in the present thesis employ the method of DNS. In the following we will provide a short introduction and overview on DNS.

The most accurate of the above numerical approaches is the direct numerical simulation, whereby all scales of fluid motion, even the smallest length and time scales, are fully resolved. The Navier-Stokes equations are solved without any averaging or modeling other than the numerical discretization. In the past couple of decades, DNS proved to be valuable research tool and have provided a significant impact in our understanding of turbulence. This was acknowledged in a large number of works (see Moin and Mahesh (1998) for review).

As aforementioned, turbulent flows contain a large range of length, velocity and time scales. The use of DNS to solve the Navier-Stokes equations assures that all scales from the smallest Kolmogorov scales to the largest integral scales are captured and resolved. This places some numerical difficulties and constraints on the numerical methods and especially on the discrete representation of the Navier-Stokes equations, i.e. the spatial and temporal discretization. As result the computational domain should be large enough to accommodate the large scales of the fluid motion, while the numerical resolution should be as small as at least the half of the size of the smallest flow structures presented in the flow (e.g. for Fourier methods *Nyquist-Shannon sampling theorem*[2], Shannon (1998); Nyquist (2002)). The fine resolution of the flow field assures that the dissipation processes at which the effect of viscosity is dominant, are captured. Thus, the bigger the ratio Λ/η is, the higher is the number of discrete points needed for the correct representation of the flow. Dimensional considerations shows that this ratio is proportional to $\mathrm{Re}^{3/4}$ (Tennekes and Lumley, 1972). This considerations apply for each spatial direction increasing the resolution requirements by making the number of discrete points proportional to $\mathrm{Re}^{9/4}$. Considering the temporal resolution, the time step is related to the spatial resolution by the so called *Courant-Friedrichs-Lewy* (CFL) condition (Courant et al., 1928). When one is interested in the statistically stationary solution, the flow field needs to evolve long enough in time to assure the time independence of the flow statistics. Commonly direct numerical simulations require run-times of several months up to several years on high performance supercomputers. This in combination with the large number of discrete points makes the use of DNS extremely costly and this is one of the main limitation of DNS. Thus, direct numerical simulations are employed only for relatively small Reynolds numbers and

[2] *"If a function $x(t)$ contains no frequencies higher than B hertz, it is completely determined by giving its ordinates at a series of points spaced $1/(2B)$ seconds apart"*, Shannon (1998); Nyquist (2002)

for geometrically simple configurations. As computers become faster in processing speed and larger in memory, more complex and high Reynolds number flows have started to come in reach of DNS.

DNS provides very detailed information of the flow by capturing both the spatial and temporal characteristics of the flow. As a result, the amount of data provided by the simulation is significant and makes the statistical analysis and the tools needed a challenge. Additionally the space required for the storage of the simulation data is also a substantial constraint.

From the above consideration, we may conclude that DNS can be used primarily as research tool, providing detailed information about the physical processes involved in fluid dynamics. The results then can be used for the development of quantitative models, e.g. RANS, which will allow a more cost effective simulation of other similar flow configurations. A recent review on DNS for homogeneous isotropic turbulence can be found in Ishihara et al. (2009). A comprehensive introduction and review of the available numerical methods can be found in the excellent textbook of Ferziger and Perić (1996).

2.3.2 Immersed boundary method

As introduced in section §2.2.3, the interaction between the particles *immersed* in a viscous flow and the flow happens through out the boundary conditions at the fluid-solid interface. The immersed boundary method refers to a method which imposes the effects of an *immersed boundary* (IB) on the flow (Peskin, 1972). The term immersed boundary method (IBM) encompass all methods used for simulation of viscous flows with immersed objects on computational grids that do not conform to the shape of the immersed bodies. The aim of this section is to provide an overview on the immersed boundary method.

Before we introduce the IBM we would like to briefly discuss the methods which use body conformed grids. This will allow us a better assessment of the advantages and disadvantages of using the IBM. The methods employing body conformed grids precisely define the phase boundaries. Such grids are also known as *body-fitted* grids. Using a body-fitted grids greatly simplifies the discretization of the governing equation in the computational domain and the boundary conditions on the immersed boundary can be enforced directly. The main difficulties of this approach is the generation of the computational grid. The generation of body-fitted grids can become quite cumbersome. The problem becomes even more difficult when the fluid-solid interface moves throughout the enclosing fluid domain, requiring the generation of a new grid at each time step.

The IBM takes the opposite approach, the flow field is solved on a time independent, (usually) Cartesian grid which, in general, does not conform to the geometry of the fluid-solid interface. The IB is still represented by some means of surface grid, which in the present work is done by discretizing the particle surface by Lagrangian marker points, which are advected with the particle velocity. The enforcing of the boundary conditions on the IB is probably the most important part in the development of an IBM and a number of different approaches exist, whereas in the case of body-fitted methods the b.c. are clearly defined and enforced. The quality of the IBM method in use may depend strongly on the employed forcing approach. The boundary conditions representing the presence of the solid object are incorporated into the Navier-Stokes equations by modifying the equations in the vicinity of the boundary. The basic idea is to add an additional forcing term in the momentum equations for the fluid that represents the effect of the boundary. Ideally the forcing must be localized along the interface and probably extending into the solid.

The IBMs are commonly differentiated based on the calculation of the forcing term. There are two general variants. In the first the forcing term is introduced in the equations of fluid motions *a priori* to their discretization. This approach is known as the *continuous forcing approach* or *indirect forcing*. Usually in the process the interface is diffused over few mesh cells. In the second approach the forcing is introduced after the equations of motion are first discretized. This approach is often referred to as *discrete forcing approach* or *direct forcing*. The variant of the IBM used in the present work employs the continuous forcing approach.

The main advantage of IBMs is that simple time independent Cartesian grids can be used without resorting to grid transformations. The simulation of mobile highly complex interfaces can be performed relatively efficiently. For the representation of the interface no additional indicator functions for the interface are required. Additionally the Navier-Stokes equations are solved on the entire computational domain. This allows the use of highly adapted and efficient numerical solvers which were developed in the last decades in the context of turbulence research. Additional advantage is the simplicity of implementation and parallelization of the numerical codes allowing the use of high performance supercomputers. The IBMs are also highly flexible and allows the implementation of complex geometries in a short time compared to body-fitted approaches.

The IBM has been widely used in the past for the simulation of multiphase flows with various configurations (Uhlmann, 2008; Lucci et al., 2010; Lee and Balachandar, 2010; Tenneti et al., 2011; Chan-Braun et al., 2011) and a number of variants exist in the literature. For more detailed discussion of the different forcing approaches and variants we refer the reader to the comprehensive review of Mittal and Iaccarino (2005) and the textbooks of Prosperetti and Tryggvason (2007) and Tryggvason et al. (2011).

2.3.3 Details of the implementation

The present work investigates numerically the interaction and the resulting two-way coupled mechanisms of solid heavy particles settling under the influence of gravity in initially ambient or turbulent flow fields. The numerical experiments are performed by means of series of direct numerical simulations. The numerical code used in this thesis employs a variant of the immersed boundary method developed by Markus Uhlmann (Uhlmann, 2005). In the following a brief overview over the used numerical methods will be presented. For more detailed discussion of the various numerical aspects as parallelization techniques, linear system solver, scaling performance and etc. we refer the interested reader to a series of technical reports Uhlmann (2003a,b, 2004, 2010); Uhlmann and Doychev (2012).

Our methods employ a continuous forcing approach, where a localized volume force term is added to the momentum equations *a priori* to their discretization. Two separate discretizations for the Lagrangian and Eulerian quantities are employed. For the Eulerian flow field a uniform isotropic Cartesian grid is used, while the particle quantities are discretized in a Lagrangian manner and the fluid-solid interface is represented by a set of Lagrangian marker points, which are distributed evenly on the surface of the particles. The necessary interpolation of variable values from Eulerian grid positions to particle-related Lagrangian positions and vice-versa are performed by means of the regularized delta function approach of Roma et al. (1999). This procedure yields a smooth temporal variation of the hydrodynamic forces acting on individual particles while these are in arbitrary motion with respect to the fixed grid. The particles in the present work are free to visit any point in the computational domain and the use of

Cartesian grid with uniform isotropic mesh width ensures that the regularized delta function verifies important identities, such as the conservation of the total force and torque during interpolation and spreading. This allows the use of efficient numerical methods for the solution of the Navier-Stokes equations and especially the Poisson equation for the pseudo-pressure, which needs to be solved in each Runge-Kutta sub-step and is one of the most computational demanding parts of the solution procedure. The Poisson problem is solved by means of a parallel multigrid technique.

The Navier-Stokes equations are integrated in time by using a fractional step method. This implies a semi-implicit time discretization, the non-linear advection terms are treated explicitly by a low storage three-step Runge-Kutta scheme and the viscous terms are treated implicitly by a Crank-Nicholson scheme (Verzicco and Orlandi, 1996). This ensures that the total accuracy of the time discretization is of second order. This type of two-step time splitting scheme is not unique and another variants exist. For more detailed discussion the reader is referred to the textbook of Ferziger and Perić (1996). The particle motion is determined by the Runge-Kutta-discretized Newton equations for translational and rotational rigid-body motion, which are explicitly coupled to the fluid equations. The flow field is obtained with the particle position and velocities from the previous Runge-Kutta sub-step and then the particle equations are solved using the most recent flow field. This coupling between the two phases is also known as *weak-coupling*.

During the course of a simulation, particles can approach each other closely. However, very thin inter-particle films cannot be resolved by a typical grid and therefore the correct build-up of repulsive pressure is not captured which in turn can lead to possible partial 'overlap' of the particle positions in the numerical computation. In practice, we use the artificial repulsion potential of Glowinski et al. (1999), relying upon a short-range repulsion force in order to prevent such non-physical situations (s. section §2.2.2). This force is employed in order to avoid overlapping of particles or penetration into each other. This force is calculated as the sum of short-range repulsive forces between pairs of particles with distance between their centers falling below a threshold value. The model of Glowinski et al. (1999) is a rather simple model, however for dilute two-way coupling configurations where the particle-particle dynamics are negligible the model provides reasonable results (Uhlmann, 2008; Lucci et al., 2010; Chan-Braun et al., 2011). An overview and discussion of more elaborated models can be found in the textbook of (Crowe et al., 2011, section §5).

The spatial operators are discretized by means of central finite-differences on a staggered grid. The spatial discretization was carefully chosen in order to resolve all relevant scales of the flow, including the fluid-solid interface. The temporal and spatial accuracy of this scheme are of second order. The numerical code has been parallelized by means of the MPI standard and extensively testing and benchmarking of the code was performed. The numerical implementation of the above methods indeed achieved very good scaling on high performance supercomputers, e.g. JUGENE and JUQUEEN at the Jülich Supercomputer Center (JSC) Uhlmann (2010); Uhlmann and Doychev (2012). In a recent publication Uhlmann and Dušek (2014) the numerical model and method was benchmarked against a Fourier/spectral-element code (Ghidersa and Dušek, 2000; Jenny et al., 2004) showing good agreement with the spectral results. Another recent applications of the numerical method can be found in (Uhlmann, 2008; Lucci et al., 2010; Lee and Balachandar, 2010; Chan-Braun et al., 2011; Kidanemariam et al., 2013).

Chapter 3

Settling of a single particle in an ambient fluid

3.1 Introduction

The motion of particles under the simultaneous action of external and hydrodynamical forces is one of the most important topics with regard to dispersed flows. The understanding of the coupling mechanisms between the motion of the particle and the forces acting on the particle have been the subject of many studies in the past. Due to the perfect symmetry of spherical particles, the study of a single spherical particle settling under the influence of gravity in an ambient viscous fluid has been identified as one of the most essential configurations for studying the effect of the hydrodynamic forces on the particle and their implications to its motion. This configuration is particularly interesting, since all deviations from a rectilinear motion of the particle as well unsteadiness of the particle velocity are results from the interaction of the particle with its own wake. It is widely accepted that when a single particle settles in an a priori ambient fluid, a variety of motion patterns exist, from straight vertical to fully chaotic paths, for which the fluid motion in the near field around the particles, e.g. particle wake, plays a dominant role. The various wake structures that may exist in the wake of the particle could lead to unsteadiness of the hydrodynamic force and also to the onset of lift forces and torque on the particle. Their implications may have far reaching consequences, for example Dušek et al. (1994) showed that instabilities in the particle wake enhance the viscous effects, which in turn reduces the vertical velocity of the particle. Later Bouchet et al. (2006) showed that each change of the wake structure is accompanied with change of the drag with overall qualitative effect of the wake instabilities being an increase of the drag, which also would lead to a decrease of the particle vertical velocity. In the context of particle-induced turbulence generation and modulation, the answer to the question of whether the presence of particles will enhance or attenuate turbulence levels has been to our understanding inconclusive. Thus, elucidating the physical mechanism behind the wake shedding will contribute to the better understanding of the particle-induced turbulence modulation. A recent review on the wake structures and therewith introduced changes in the particle trajectories and drag coefficient can be found in Ern et al. (2012).

The study of a single settling particle could be seen as a necessary prequisite for the study of the settling of many particles in order to identify effects such as the presence of multiple particles and their influence upon the structure of the particle wake, the settling velocity or the hydrodynamic force. Additionally, it is of great interest to know if the motion of a single particle is relevant to the motion of particle suspensions.

In this chapter we report the results of two direct numerical simulations of a single heavy spherical particle settling under the influence of gravity in an initially ambient flow. The results of the simulations will serve as a reference data set for the following chapters. The comparison with simulations of many settling particles will allows us to isolate the different effects of the force on the particle such as the presence of multiple particles. The results here are not meant to provide a full qualitative study of the mechanisms involved in the settling of spherical particles. For such studies we refer the interested reader to the excellent works of Natarajan and Acrivos (1993); Johnson and Patel (1999); Ghidersa and Dušek (2000); Jenny et al. (2003, 2004); Bouchet et al. (2006); Horowitz and Williamson (2010), which are devoted specifically to the physics of a single particle.

The chapter is organized as follows, before we present the numerical set-up for the present simulations in section §3.3, we provide a short overview of the regimes of particle settling and wakes of a single heavy particle of spherical shape settling under the influence of gravity in section §3.2. Then we present the results for the two simulations in section §3.4, where the focus will be mainly on the hydrodynamic force on the particle, the particle settling velocity and the characteristics of the particle wake. The chapter closes with short summary and conclusions in section §3.5. Statistical quantities such as particle settling velocity, drag coefficient and various wake statistics will be compared with the literature, especially with data from Uhlmann and Dušek (2014).

3.2 Overview of the regimes of particle motion

In spite of the simple configuration of spherical particle settling in an uniform flow, depending on the particle relative velocity, the wake of the particle can be fully three dimensional and can contain highly complex vortical structures. As the Reynolds number increases, the flow around the particles undergoes a series of transitions which affect the motion of the particles. For very small Reynolds number, the flow exhibits a fore-and-aft symmetry around the particle and the flow is fully attached. With increasing Reynolds number the flow becomes progressively more asymmetric and vortex form at the rear stagnation point of the sphere. This is a sign of the beginning of a separation of the flow in vicinity of the particle surface. The flow separation is indicated by a change in the sign of the streamwise vorticity and first occurs at the rear stagnation point. It is widely accepted that the flow separation occurs at approximately $Re_p = 20$ (Clift et al., 1978). For Reynolds numbers in the region of $20 < Re_p < 130$ a steady wake region and a separation ring around the particle are formed. Visualizations evidenced that the wake in this regime consists of a rectilinear axi-symmetric thread downstream of the particle. Above some threshold value ($Re_p \gtrsim 210$), the wake of the particle becomes steady and oblique, symmetric about a plane inclined at some angle with respect to the vertical axis. As the Reynolds number further increases ($Re_p \gtrsim 270$), diffusion and convection of vorticity no longer keep pace with vorticity generation and a single-sided chain of vortex loops is shed from the particle wake. Instabilities originated in the free surface layer are convected downstream causing a periodic oscillation of the particle wake. Studies indicate that this periodic vortex shedding continuous until Reynolds number of around $Re_p = 420 - 500$ (Tomboulides and Orszag, 2000). When the particle Reynolds number increases beyond a value of about 600 the vortices are shed with varying orientation. In a recent study, Bouchet et al. (2006) performed detailed study of the demarcation points separating the different wake regimes for a fixed single particle subjected to uniform flow.

The different wake configurations and the transition of the particle wake for freely moving particles, was extensively studied in the work of Jenny et al. (2004). The authors compiled a large number of simulations sweeping the two-dimensional parameter space (Ga, ρ_p/ρ_f) of single particle settling or rising in an otherwise stagnant fluid. Please note that for such flow configurations the flow is fully characterized by the pair (Ga, ρ_p/ρ_f) (s. section 2.1.3). In order of increasing Galileo numbers, the authors identified the following wake regimes. For Galileo numbers below a value of approximately 155 ($Re_p < 212$) the wake of the particles was observed to be steady and axi-symmetric. When the Galileo number increases above this threshold, the particle wake loses its axi-symmetry via a regular bifurcation and axi-symmetry is replaced by planar symmetry. Ghidersa and Dušek (2000) showed that the symmetry plane is selected randomly by initial perturbations and is conserved until the Galileo number reaches a second threshold of Ga $= 173$ ($Re_p = 275$). In the interval of Ga $= 156 - 173$ the wake consists of separation bubble followed by two trailing counter-rotating vortices. The value of Ga $= 173$ marks a second bifurcation, of Hopf type, at which the wake becomes unsteady, but still planar symmetric, and the double threaded wake is replaced by the above mentioned chain of vortex loops. When the Galileo number is further increased, the amplitude of the oscillations is enhanced and the planar symmetry is replaced by chaotic shedding of hairpin vortices in random directions, which leads to the loss of periodicity. Here we would like to note that the above threshold values could vary depending on the particle-to-fluid density ratio. More precise values can be extract from the map provided by Jenny et al. (2004), which is reproduced in figure 3.1. More recently Horowitz and Williamson (2010) compiled similar diagram, but using the particle Reynolds number instead the Galileo number. The diagram is depicted in figure 3.2.

The particle motion, especially the trajectory of the settling sphere is a result of the mutual interaction between the particle's degree of freedom and the fluid in vicinity of the particle, i.e. the vortex formation in the particle wake. Depending on the particle Reynolds number or the Galileo number the particle trajectories undergo a series of transitions corresponding to the above introduced wake configurations. In the case of an axi-symmetric wake behind the particle, the wake is unable to generate any side force on the sphere and the trajectory of the particle does not deviate from the vertical path. As the wake changes to a steady oblique trailing vortex pair, the counter-rotating vortices will produce a constant lift force in direction lateral to the flow, which will cause the particles to settle on an oblique trajectory. When the steady oblique wake undergoes into one-sided vortex shedding, the orientation of the particle trajectory does not change, since the orientation of the vortex structures remains constant. In the chaotic regime the mean particle trajectory is vertical.

For the particles settling on steady rectilinear paths, the particle wakes tend to be similar to the wakes of a fixed sphere. Indeed early studies of Magarvey and Bishop (1961) showed that the wakes of freely moving spheres with rectilinear trajectories are closely resembled by the wakes of fixed particles. However, when particles are allowed to move freely, the additional degrees of freedom of the mobile particles affects the breaking of axi-symmetry (Jenny et al., 2003), primarily in changing the threshold values for the breaking of axi-symmetry.

For more detailed discussion and exhaustive bibliographical review on the topic of particle regimes of motion and transitions of the particle wake we refer the reader to the works of Jenny et al. (2004); Horowitz and Williamson (2010) and the references therein.

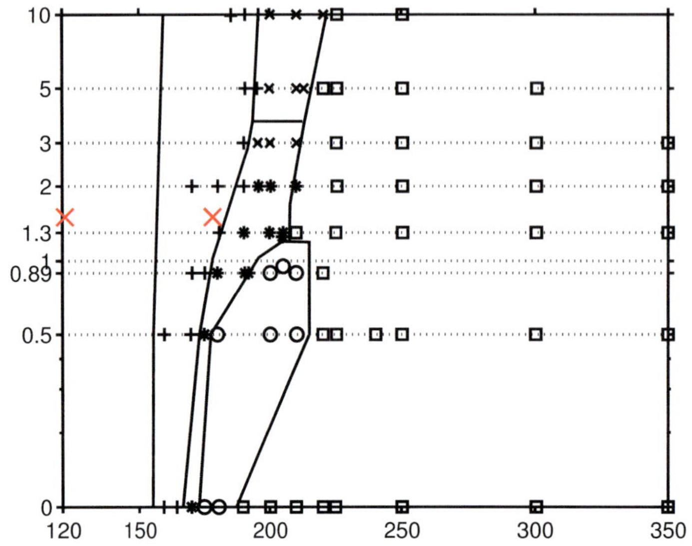

Figure 3.1: Diagram of flow regimes in the parameter space spanned by the Galileo number and particle-to-fluid density ratio (Ga, ρ_p/ρ_f). The diagram is digitalized from Jenny et al. (2004). The leftmost vertical line is the demarcation line between the the steady axi-symmetric regime on the left side and the steady oblique regime on the right side of the line. Different symbols corresponds to simulations classified as: $+$, steady oblique path; $*$, oscillating oblique path with low frequency; $\times$,oscillating oblique path with high frequency;, $\circ$, periodic zigzagging trajectory; $\square$ three-dimensional chaotic trajectory. The present simulations lie on the horizontal line at $\rho_p/\rho_f = 1.5$ and are marked by the two red crosses ($\times$).

3.3 Numerical set-up

In the present work we consider two of the above described regimes, the steady axi-symmetric and the steady oblique regime. For each regime the settling of a single heavy spherical particle under the influence of gravity in an initially stagnant fluid was performed. As aforementioned, this flow configuration is unambiguously characterized by two non-dimensional parameters, the fluid-to-particle density ratio (ρ_p/ρ_f) and the Galileo number (Ga as in equation 2.12). In the present work we consider only a single value for the fluid-to-particle density ratio, which was chosen as $\rho_p/\rho_f = 1.5$. This value corresponds to particles with slightly higher density than the fluid (e.g. polyester particles in water). Concerning the Galileo number, two values are considered corresponding to two of the regimes described above. For the steady axi-symmetric wake regime, the simulation of settling particle with Ga $= 121$ was performed, while the steady oblique regime was represented by a simulation of settling particle with Ga $= 178$. Henceforth, we refer to the simulation with Ga $= 121$ as $S121$ and with Ga $= 178$ as $S178$ (S for single particle). The physical parameters of the dispersed flow were kept in both cases constant (i.e. particle size d_p and kinematic viscosity of the fluid ν), except for the gravitational acceleration, which was adapted in order to achieve the desired Galileo numbers.

The parametrization of the problem by the two parameters, Ga and ρ_p/ρ_f, is particularly useful, since the calculation of the Galileo number requires only the properties of the particles and the fluid. Thus, the regime of settling can be determined *a priori*. However, another important dimensionless parameter widely used to characterize the motion of particles is the

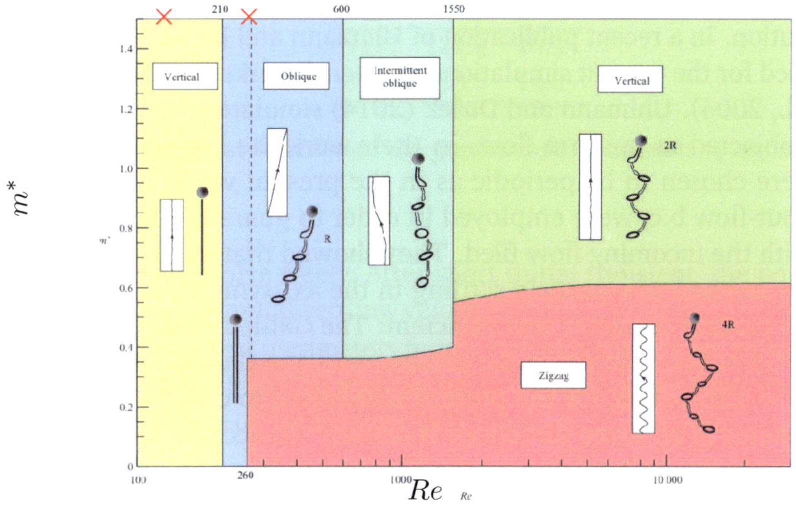

Figure 3.2: Map of regimes in the parameter space spanned by the mass ratio (defined as the relative density of the particles compared to the fluid) and the particle Reynolds number (m^*, Re). The diagram is digitalized from Horowitz and Williamson (2010). The diagram depicts the corresponding motion of the sphere and the associated wake patterns. The present simulations are marked by the two red crosses (×).

particle Reynolds number $Re_\infty = w_\infty D/\nu$, where w_∞ denotes the terminal relative velocity of the particles. In doing so, a good approximation of the settling velocity w_∞ is required, which is a product of the simulation and depends on the particle dynamics, e.g. forces acting upon the particle. An estimation of the particle Reynolds number prior the simulation can be made by using the terminal settling velocity w_∞^{clift} of a single particle in still fluid based on the balance between the drag on a fixed sphere (as given by Clift et al. (1978)) and the buoyancy force. The particle Reynolds number based on w_∞^{clift} and the particle diameter d_p corresponding to case $S121$ and case $S178$ are calculated to $Re_\infty^{clift} = 141$ and $Re_\infty^{clift} = 245$ respectively. The two data points of our simulations represented by the parameter duo (Ga, ρ_p/ρ_f) or ($Re_\infty^{clift}, \rho_p/\rho_f$) are depicted on the flow regime diagrams in figure 3.1 and 3.2.

The computational domain V_Ω for both cases $S121$ and $S178$ is a rectangular box elongated in the vertical direction in which also the gravitational acceleration acts. The elongated computational box takes into account the anisotropic character of the flow, which is dominated by the wake generated by the particle. The domain extends in the vertical direction to $L_z = 273d_p$ in case $S121$ and to $L_z = 341d_p$ in case $S178$. In the horizontal direction the domain boundaries lie $L_{x,y} = 17d_p$ apart from each other in case $S121$ and $L_{x,y} = 10d_p$ in case $S178$. Periodic boundary conditions in all three space directions were applied. The large extents of the computational domain in the vertical direction are owned to the periodic boundary conditions and were chosen in order to give the particle enough time to accelerate to their terminal settling velocity without periodic interference with its own wake. Since such a periodic system does not attain a steady state in the presence of settling particles, a constant vertical pressure gradient (i.e. a homogeneous body force) was imposed whose integral over the domain equals the total submerged weight of the suspended bodies.

As aforementioned the simulations in the present work are performed by means of DNS, i.e. all flow and particle scales of motion are resolved. In order to ensure that our simulations

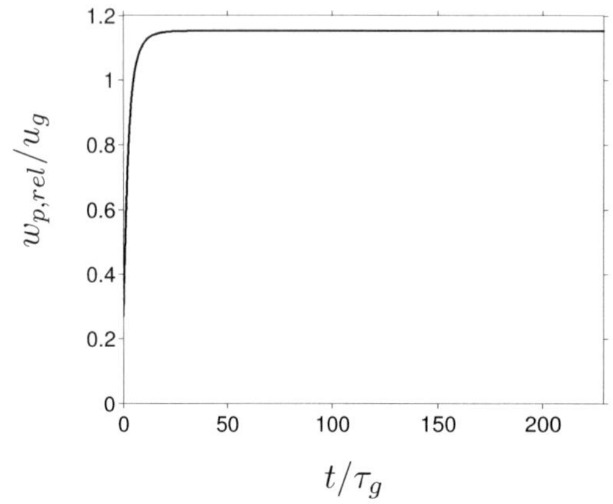

Figure 3.3: Temporal evolution of the vertical component of the particle relative velocity, $w_{p,rel} = w_p - w_f$ in case $S121$. Time and velocity scales are normalized by the gravitational scales τ_g (equation 2.15) and u_g (equation 2.11).

identically zero. The velocity $u_{h,p}$ is thereby defined as:

$$u_{h,p} = \sqrt{u_p^2 + v_p^2}, \tag{3.2}$$

where u_p and v_p are the horizontal components of the particle velocity $\boldsymbol{u}_p = (u_p, v_p, w_p)$. The vertical velocity, however, is not zero and results from the balance of the particle acceleration and the forces acting on the particle (cf. equation 2.25). The vertical component of the relative particle linear velocity is depicted in figure 3.3 as function of time. The linear velocity has been calculated as the difference between the particle and fluid velocity:

$$\boldsymbol{u}_{p,rel}(t) = \boldsymbol{u}_p(t) - \boldsymbol{u}_f(t). \tag{3.3}$$

The time and the particle relative velocity have been normalized by the gravitational time scale τ_g and the corresponding gravitational velocity scale u_g (cf. table 2.1). Here and in the following the gravitational scales are used for the normalization of all time and velocity plots, except where it is differently stated. For sake of completeness the various particle time scales defined in section §2.1.3 are depicted in table 3.3 as function of the gravitational time scale. As can be seen the particle accelerates in vertical direction for approximately $30\tau_g$ where the vertical component of the particle relative velocity reaches its maximum value, which remains approximately constant for the entire duration of the simulation.

The simulation in case $S121$ was terminated after approximately $270\tau_g$. This time is the approximate particle return time, which is the time the particle needs to travel the distance corresponding to the vertical extension of the computational domain L_z. The return time is defined as the ratio of the vertical extension of the computational domain L_z and the particle relative velocity $w_{p,rel}$. This is required in order to prevent the particle wake to interfere with the fluid velocity seen by the particle. Due to the periodic b.c. in the vertical direction, the particle will settle through the computational domain periodically and may interact with the fluctuations in the flow field which were generated by the particle themselves.

As can be seen in figure 3.3 the vertical component of the particle relative velocity does not experience any significant fluctuations and remains constant for the time period $t/\tau_g = [30, 270]$. The particle Reynolds number based on the terminal settling velocity in case $S121$ was calculated to $Re_p = 139.8$, which is slightly smaller than the terminal settling velocity estimation of Clift et al. (1978), viz. $Re_{\infty}^{clift} = 141$.

Regarding case $S178$, the particle with the parameter pair (Ga $= 178, \rho_p/\rho_f = 1.5$) is expected to settle in the steady oblique regime. The oblique motion of the particle is characterized by an oblique angle α of the particle velocity to the vertical axis and the axi-symmetry of the particle trajectory and wake is replaced by planar symmetry, where the plane is inclined at the oblique angle α. The angle α can be defined as:

$$\alpha = \tan^{-1}\left(\frac{u_{h,p}}{w_{p,rel}}\right) \tag{3.4}$$

For the present case the angle was calculated to $5.21°$ degree which is in very good agreement with the results of Uhlmann and Dušek (2014), where they observed an oblique angle of $5.22°$. The oblique motion of the particle implies that the particle motion is not entirely in the vertical direction and that the particle drifts also in the horizontal direction with a non-zero horizontal component of the particle velocity. The horizontal ($u_{h,p}$) and vertical ($w_{p,rel}$) components of the particle velocity are depicted in figure 3.4a-b as function of time. Similar as in case $S121$, the simulation was terminated at a time point before the return time of the particle in order to avoid interferences with the wake of the particle. For case $S178$ we estimated this time to be $300\tau_g$. After it is released to move freely, the particle accelerates for approximately $30\tau_g$ gravitational time units, where it reaches its maximum value. During this time the horizontal particle velocity remains initially zero and begins slowly to increase as soon as the vertical component of the particle relative velocity reaches the first bifurcation point (Ga $= 155$, $Re_p = 212$) at which the loss of axi-symmetry and the oblique motion sets in. A small instability in the particle motion sets in as soon as both a sufficient vertical velocity and a sufficiently long wake has developed, which level off at a steady value after a transient period of approximately $100\tau_g$. As reported in Dušek et al. (1994) the instabilities enhance the viscous effects and this results in decrease of the vertical particle velocity. As can be seen after reaching its maximum value, the particle vertical velocity decreases slightly. The time at which $w_{p,rel}$ decreases coincides with the time at which the instabilities in the horizontal particle velocity sets in. This observations are in line with the observations made in Jenny et al. (2004, figure 4) for Galileo number Ga $= 170$. After the initial fluctuations in the particle velocity decayed, the particle velocity reaches its asymptotic steady value in the time interval $t/\tau_g = [130, 300]$. It can be observed that the gravitational velocity u_g in case $S178$ is lower than the particle terminal settling velocity by as much as 30%. The particle Reynolds number based on the particle terminal velocity calculates then to a value of $Re_p = 233.06$.

The oblique motion of the particle will cause a rotation of the particle due to the torque exerted on the particle. The rotation of the particle is expected to be around an axis which is orthogonal to the plane of the particle motion. Indeed the horizontal angular velocity is non-zero and its time evolution is depicted in figure 3.4c. The horizontal component of the particle angular velocity $\omega_{h,p}$ was calculated in the same manner as $x_{h,p}$ and $u_{h,p}$, viz. $\omega_{h,p} = (\omega_x^2 + \omega_y^2)^{1/2}$. Similar to the velocity $u_{h,p}$ the particle first begins to rotate as the oblique motion sets in ($t/\tau_g \approx 30$) and experience similar instabilities over the time period $t/\tau_g = [30, 130]$, which level off at a mean value of 0.027. Slight oscillations of the angular velocity in this interval are of numerical origin, with amplitude of $\mathcal{O}(10^{-4})$.

The results obtained in this sections for case $S178$ compares well with the results of Uhlmann and Dušek (2014). The relative difference to the results of Uhlmann and Dušek

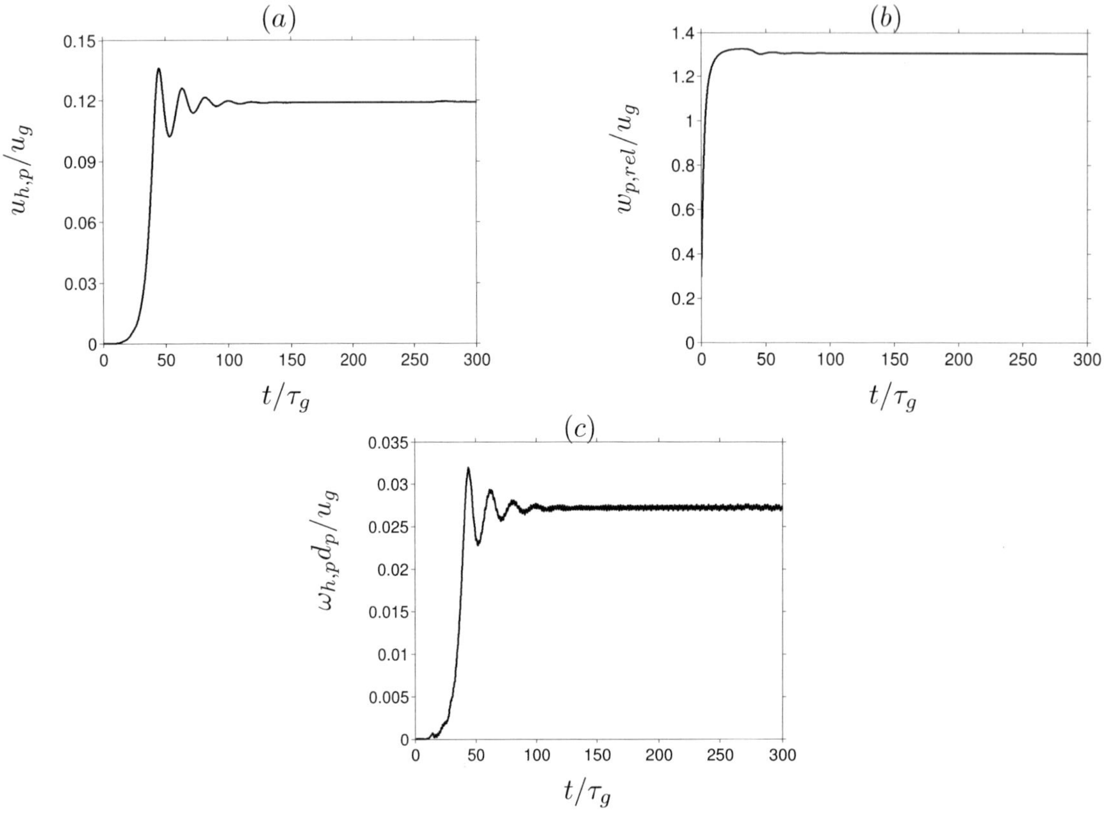

Figure 3.4: Results for case $S178$. Time evolution of the (a) particle horizontal velocity, $u_{h,rel} = (u_{p,rel}^2 + v_{p,rel}^2)^{1/2}$. (b) vertical component of the particle relative velocity, $w_{p,rel}$. (c) the particle angular velocity. Here the component orthogonal to the incoming flow is showed, $\omega_{h,p} = (\omega_x^2 + \omega_y^2)^{1/2}$. For the normalization of all velocities the gravitational velocity scale u_g was used. Time was normalized by the gravitational time scale τ_g.

(2014, table 7) was evaluated for the particle velocity components to $\Delta_{u_{h,p}} = 7.8\%$ for the horizontal component and to $\Delta_{w_{p,rel}} = 1.7\%$ for the vertical component. For the angular velocity, however, the difference was calculated to $\Delta_{\omega_{h,p}} = 30\%$. The difference Δ_ϕ for some relevant quantity ϕ to a reference quantity ϕ_{ref} is defined as:

$$\Delta_\phi = \frac{|\phi - \phi_{ref}|}{|\phi_{ref}|}. \tag{3.5}$$

One reason for the observed deviation from the reference data could be the temporal discretization, i.e. the size of the time step. For the simulations performed in Uhlmann and Dušek (2014) a time step corresponding to a CFL number of 0.3 was used, whereas for the present simulations a CFL number of 0.5 was employed. Uhlmann and Dušek (2014) identified two reasons for the time dependence of the results. The first is the use of the fractional step method, which introduces a “slip-error” on the fluid-solid interface of the order of Δt (Δt being the time step). The second reason is that in the statically steady state the flow is non-trivially unsteady in a fixed frame of reference, due to the particle motion relative to the computational grid.

In the steady state where no changes in the particle velocity are present the force on the particle cancels and the total sum becomes zero. This implies that the drag and the buoyancy force are in equilibrium. Then we can easily estimate the drag force f_z on the particle through the following relation:

$$f_z = (\rho_p - \rho_f)|\boldsymbol{g}|\frac{d_p^3\pi}{6}. \quad (3.6)$$

It is common to present the drag force in a form of the drag coefficient, which can be calculated form the force f_z by normalizing it with the reference force f_{ref} which is defined as the far-field dynamic pressure acting on the projected cross-sectional area of the particle, i.e.:

$$C_d = \frac{f_z}{\frac{1}{2}\rho_f w_\infty^2 A_p}, \quad (3.7)$$

where A_p is the projected cross-sectional area of the particle and w_∞ is the terminal settling velocity of the particle. We calculated the drag coefficient of the particle in case $S121$ to $C_d = 1.002$ and in case $S178$ to $C_d = 0.78$. The total deviation from the drag coefficient of a single fixed particle subjected to uniform flow (Clift et al., 1978) was evaluated in both cases to be less than 9%. This is most probably to the time step used in the current simulations.

Before we turn to the wakes of the particle, we can conclude in the meantime that the spatial and temporal resolutions used in case $S121$ and $S178$ were sufficient to correctly reproduce the motion of the particles.

3.4.2 Particle wakes

In this section we report results on the flow field conditioned on the particle position, e.g. the particle wake. Before we proceed with the results, let us first fix some geometrical definitions and the appropriate notation. First we define a particle centered coordinate system with fixed spatial orientation as the Eulerian fixed coordinate system used to describe the flow field, i.e. $\tilde{\boldsymbol{x}} = \boldsymbol{x} - \boldsymbol{x}_p$. The origin of the coordinate system is attached to the center of the particle and translates with the particle motion without rotation. Any position vector in this coordinate system will be denoted by $\tilde{\boldsymbol{x}} = (\tilde{x}, \tilde{y}, \tilde{z})$. For the presentation of the results in the present section, one additional coordinate system is required, $\hat{\boldsymbol{x}} = (\hat{x}, \hat{y}, \hat{z})$. This coordinate system is also particle centered, but with the following orientation of the coordinate axes (unit normal vectors):

$$\hat{\mathbf{e}}_z = -\boldsymbol{u}_{p,rel}/||\boldsymbol{u}_{p,rel}||, \quad (3.8)$$

$$\hat{\mathbf{e}}_y = \mathbf{e}_z \times \boldsymbol{u}_h/||\boldsymbol{u}_h||, \quad (3.9)$$

$$\hat{\mathbf{e}}_x = \hat{\mathbf{e}}_y \times \hat{\mathbf{e}}_z, \quad (3.10)$$

where $\boldsymbol{e}_z$ denotes the unit normal vector of the Eulerian z axis. The vector $\hat{\mathbf{e}}_z$ is aligned with the particle relative velocity vector in the opposite direction, $\hat{\mathbf{e}}_y$ lies in the horizontal plane and is perpendicular to the vertical axis $\mathbf{e}_z$ and the vector pointing in the direction of the horizontal velocity $\boldsymbol{u}_h$. The vector $\hat{\mathbf{e}}_x$ is perpendicular to the vectors $\hat{\mathbf{e}}_y$ and $\hat{\mathbf{e}}_z$. A sketch of the used coordinate systems can be found in figure 3.5. The evaluation of any particle or fluid quantity ϕ at some position $\tilde{\boldsymbol{x}}$ or $\hat{\boldsymbol{x}}$ in the above defined coordinate systems will be denoted as $\tilde{\phi}$ or $\hat{\phi}$, except when it is specifically stated otherwise.

In the steady axi-symmetric regime the two coordinate systems $(\tilde{x}, \tilde{y}, \tilde{z})$ and $(\hat{x}, \hat{y}, \hat{z})$ coincide, due to alignment of the particle velocity with the vertical z axis. On the other hand, in the steady oblique regime the two coordinate systems deviate from each other, due to differences in the orientation of the particle velocity and the vertical axis.

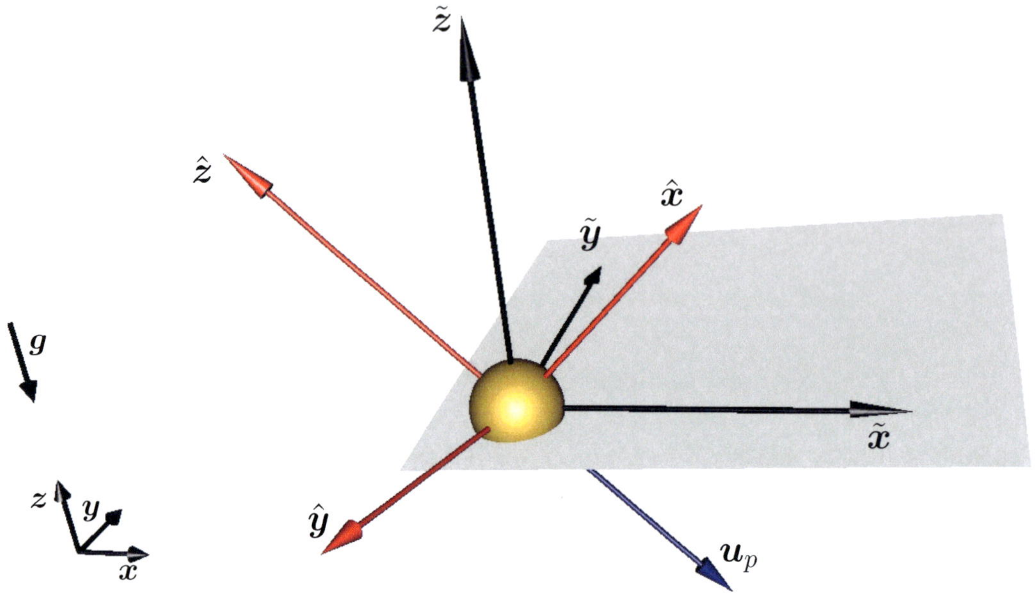

Figure 3.5: Sketch of the coordinate systems used for the evaluation of the flow field conditioned on the particle position, i.e. particle wakes. The coordinate system $(\tilde{x}, \tilde{y}, \tilde{z})$ is attached to the particle center and has the same orientation as the Eulerian coordinate system used for the representation of the flow field (x, y, z). The definition of the coordinate system $(\hat{x}, \hat{y}, \hat{z})$ is given by the set of equations 3.8-3.10.

The wake structures reported in the literature corresponding to the two regimes considered in the present work are identified as a vortex bubble around the particle in case $S121$ and a vortex bubble followed by two counter-rotating vortices in case $S178$. The structure for the wake in case $S121$ at steady state is depicted in figure 3.6a-c. The graph in figure 3.6a shows an iso-surface of the vertical component of the relative velocity $\tilde{w}_{p,rel}/u_g$. The vortical structure of the wake is also depicted in figure 3.6b by means of the λ_2 criterion of Jeong and Hussain (1995). The λ_2 criterion consist of plotting an iso-contour of the second largest eigenvalue of the tensor $\boldsymbol{S}^2 + \boldsymbol{\Omega}^2$, where $\boldsymbol{S}$ and $\boldsymbol{\Omega}$ are the symmetrical and anti-symmetrical parts of the velocity gradient tensor. It can be seen that the structure and orientation of the particle wake are in line with the observations in the literature. Contour lines of the velocity $\tilde{w}_{p,rel}/u_g$ through the particle's center are shown in figure 3.6c. The graph reveals the symmetrical structure of the particle wake around z axis. The region denoted by the red contour line marks the recirculation region, where $\tilde{w}_{p,rel}/u_g \leq 0$. The maximal extent of the recirculation region from the particle downstream faced surface defines the so called *recirculation length* l_e of the particle wake. The position on the zero iso-conour definig the recirculation length is marked by the blue cross. Due to the symmetry of the wake around the vertical axis in case $S121$ the point on the recirculation region defining the recirculation length lies on the vertical axes $\tilde{z}$ and $\hat{z}$.

The pendant plot for the wake in case $S178$ is shown in figure 3.7a-c. It can be seen that the wake is indeed inclined to the vertical axis and that it consist of a vortex bubble followed by two counter-rotating vortices. This is also confirmed in the contour plot of the velocity $\tilde{w}_{p,rel}/u_g$, one can clearly see the oblique orientation of the wake with its recirculation region. In contrast to case $S121$ the maximal extent of the recirculation region is found to be positioned off-center, i.e. not on the vertical axis $\tilde{z}$. It is interesting to note that its position is not on the vertical axis $\hat{z}$ either.

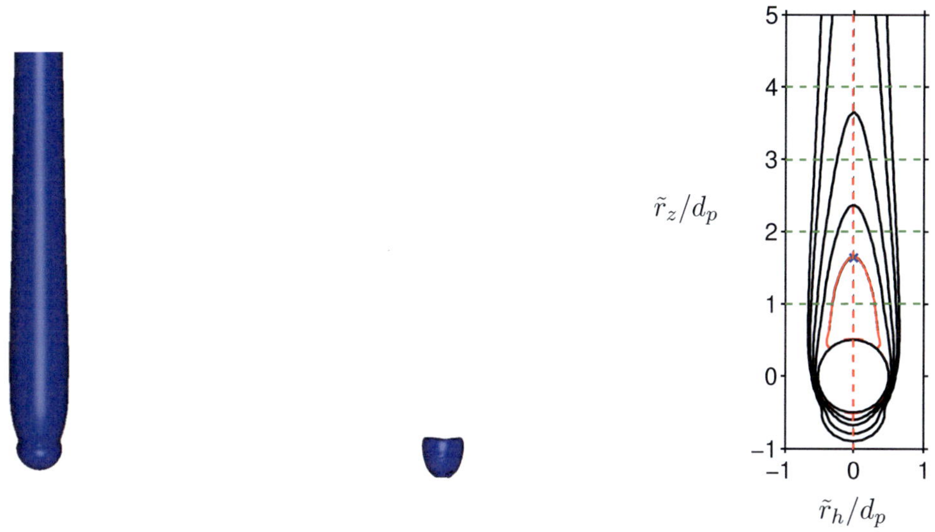

(a) (b) (c)

Figure 3.6: Results for the wake in case $S121$. (a) Iso-contour of the relative relative velocity $\tilde{w}_{p,rel}/u_g$. The iso-contour value was chosen as $\tilde{w}_{p,rel}/u_g = 1$. (b) Iso-contour of the vortical structure around the particle represented by λ_2 criterion. (c) Contour plot of the projected relative velocity $\tilde{w}_{p,rel}/u_g$ in the plane parallel to the vertical axis passing trough the particle center $\boldsymbol{x}_p$. Contours are shown for values $(-0.2{:}0.2{:}0.8)$. The red line (——) marks the extent of the recirculation region, i.e. $\tilde{w}_{p,rel}/u_g = 0$. The blue cross x marks the maximal distance of the recirculation region to the surface of the particle and blue dashed line (- - - -) represents the plane of symmetry at which the wake is symmetric. The red dashed line (- - - -) marks the vertical coordinate $\hat{z}$ (the vector aligned with the particle relative velocity $w_{p,rel}$). Green dashed line is the axis perpendicular to $\hat{z}$. The position of the z-axis is denoted by the black dashed line (- - - -). Due to axi-symmetry of the particle wake the lines (- - - -), (- - - -) and (- - - -) coincide.

Figure 3.8 shows profiles of the velocity $\tilde{w}_{p\ rel}$ on the axis $\hat{z}$ passing through the particle's center. In both cases, the flow upstream of the particle starts decelerating with approaching the particle at approximately $\tilde{x}/d_p = -3$ and vanishes at the surface of the sphere at $\tilde{x}/d_p = -0.5$. Immediately downstream of the sphere the particle relative velocity becomes negative after which starts to increase slowly towards the value of the undisturbed upstream relative velocity. The negative values of $\tilde{w}_{p,rel}$ indicate the presence of the recirculating region. Due to the axi-symmetry of the particle wake in case $S121$, the recirculation length can be easily estimated by locating the position at which the particle relative velocity $\tilde{w}_{p,rel}$ cancels again. We calculated the value of the recirculation length in case $S121$ to $l_e/d_p = 1.19$ ($l_e/r_p = 2.38$). For the calculation of l_e in case $S178$, one needs to consider the oblique orientation of the recirculation region in the particle wake. Thus, the recirculation length for case $S178$ cannot be directly evaluated from the profile of $\tilde{w}_{p,rel}$ on the $\hat{z}$ axis. In order to calculate l_e for case $S178$, we employ the following definition. The recirculation length l_e is calculated as the maximal distance from any point on the line encapsulating the recirculation region to a surface element on the downstream faced side of the particle. The line connecting the surface element with the the point on the recirculation region boundary is aligned with the unit normal vector of the surface element. The value of the recirculation length for case $S178$ calculates then to $l_e/d_p = 1.57$ ($l_e/r_p = 3.14$). This value is in line with the reference value reported in Uhlmann and Dušek (2014) with deviation of $\Delta_{l_e} = 1.5\%$.

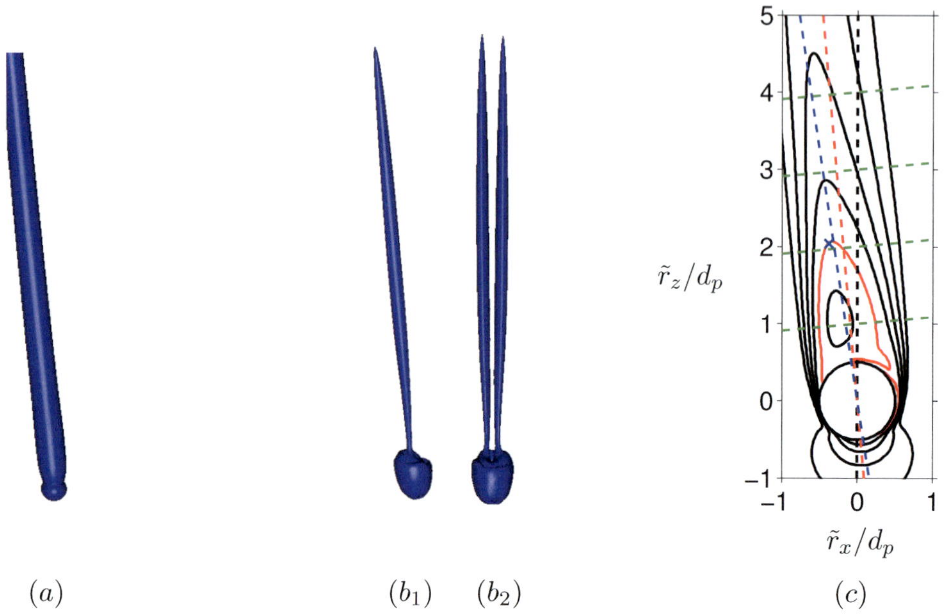

Figure 3.7: Results for case $S178$. (a) Same as in figure 3.6, but for case $S178$. The iso-contour value was chosen as $\tilde{w}_{p,rel}/u_g = 1$. (b) Front and side view of the vortical structure in the particle wake. The oblique orientation and the double threaded structure of the wake are clearly visible. The colors indicate the orientation of rotation of the vortices as in figure 3.6. (c) Same as in figure 3.6, but for case $S178$. The contour values are shown for $(-0.2{:}0.2{:}0.8)$. Color coding of the dashed lines as in figure 3.6.

The recovery of the fluid velocity in the particle wake can be further investigated by considering the velocity defect $\hat{u}_{d0}$ on the axis $\hat{z}$. The velocity $\hat{u}_{d0}$ is defined as the velocity deficit of the relative velocity $w_{p,rel}$:

$$\hat{u}_{d0}(0,0,\hat{z}) = \frac{\hat{w}_\infty(0,0,\hat{z}) - \hat{w}_{p,rel}(0,0,\hat{z})}{\hat{w}_\infty(0,0,\hat{z})}, \tag{3.11}$$

where $\hat{w}_\infty(0,0,\hat{z})$ is defined as the maximum of the incoming relative velocity on the $\hat{z}$ axis, viz. $\hat{w}_\infty(0,0,\hat{z}) = \max_{\hat{z}} \hat{w}_{p,rel}(0,0,\hat{z})$. The downstream evolution of $\hat{u}_{d0}$ with increasing particle distance for both cases is depicted in figure 3.10. As expected form its definition, the velocity defect $\hat{u}_{d0}$ in the particle wake becomes unity at the particle surface, due to the no-slip boundary condition on the particle surface. The velocity defect is evaluated for distances up to $100d_p$ downstream of the particle. The first feature evident form the graph in figure 3.10 is that with increasing distance from the particle surface the velocity deficit increases initially (due to recirculation in the wake of the particle) and reaches its maximum at $0.9d_p$ in case $S121$ and at $1.085d_p$ in case $S178$. Similar observation was made in the work of Wu and Faeth (1993), where the wake of a mobile sphere in a still surrounding was experimentally investigated. The authors noticed that the position at which the maximum of $\hat{u}_{d0}$ is observed varied with the particle Reynolds number with tendency to be closer to the particle surface for increasing Reynolds numbers. It is well accepted that the velocity deficit in the particle wake may exhibit regions with different decay rates of the form $\hat{u}_{d0} \sim \hat{r}_z^{-a}$, where a can have different values depending on the particle Reynolds number and the flow conditions in the particle wake. The most commonly reported values for a in the literature are $a = 1$, $a = 2/3$ and $a = 2$. Power

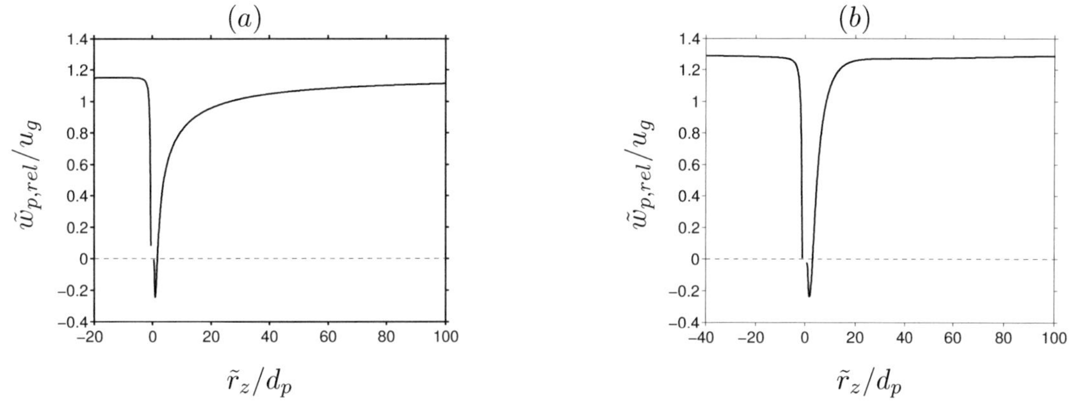

Figure 3.8: Vertical component of the relative velocity on the center line passing trough the particle center $\boldsymbol{x}_p$ and parallel to the vertical axis. (a) Case $S121$. (b) Case $S178$. The coordinate system is particle centered, i.e. origin at $\boldsymbol{x}_p$. Negative values of the x axis represent the upstream velocity in front of the particle and positive values represent the velocity downstream. The distance from the particle was normalized by the particle diameter d_p. The velocity is made non-dimensional with the gravitational velocity u_g.

law decay with $a = 1$ is known to be present in laminar wakes (Schlichting, 1979; Wu and Faeth, 1993; Bagchi and Balachandar, 2004), while a power law with $a = 2/3$ is characteristic for turbulent wakes (Tennekes and Lumley, 1972; Batchelor, 1972; Parthasarathy and Faeth, 1990a; Wu and Faeth, 1993; Pope, 2000). Recent investigations of the particle wake subjected to turbulent environment observed decay rates in the far wake with the power law of $a = 2$ (Wu and Faeth, 1994; Legendre et al., 2006; Amoura et al., 2010). The corresponding power laws are depicted in figure 3.10 by the dashed lines. As can be seen the velocity $\hat{u}_{d0}$ in case $S121$ exhibits immediately after reaching its maximum value a power law decay characteristic for turbulent wakes ($a = 2/3$) which is followed by a decay rate custom to laminar wakes ($a = 1$). The change of slope from $\hat{r}_z^{-2/3}$ to $\hat{r}_z^{-1}$ takes place at downstream location in a distance of $15d_p$. The extension of the turbulent power law in the present work is inline with the observations made in previous studies, e.g. Wu and Faeth (1994); Bagchi and Balachandar (2004). It should be noted that the local Reynolds number decreases with the downstream distance in the wake of an axi-symmetric body (Pope, 2000). The velocity deficit in the particle wake in case $S178$ experiences similar decay rate in the near-wake zone for $1 < \hat{r}_z/d_p < 5$ as in case $S121$, after which the slope changes and the velocity $\hat{u}_{d0}$ decays roughly with the power law of $\hat{r}_z^{-2}$.

Let us now turn to the profiles of the velocity $\tilde{w}_{p,rel}$ in cross-stream direction taken along the axis in direction of $\hat{\boldsymbol{e}}_y$ at different distances from the particle surface. The profiles are depicted in figure 3.11. The radial axis at which the profiles are shown are visualized by the green dashed lines in figure 3.6c and figure 3.7c. The profiles are shown at four different distances form the particle surface, i.e. $(\boldsymbol{x} \cdot \hat{\boldsymbol{e}}_z)/d_p = [0.5,\ 1.5,\ 2.5,\ 3.]$.

It can be observed that the radial profiles in case $S121$ (figure 3.11a) are symmetric with respect to the particle position, resembling the symmetry of the particle wake. With increasing distance the influence of the particle on the fluid velocity accordingly decreases and is reflected by flattening of the velocity profile. With increasing radial distance from the particle center, the effect of the particle diminishes quickly and the value of the undisturbed velocity is reached at approximately half particle diameter.

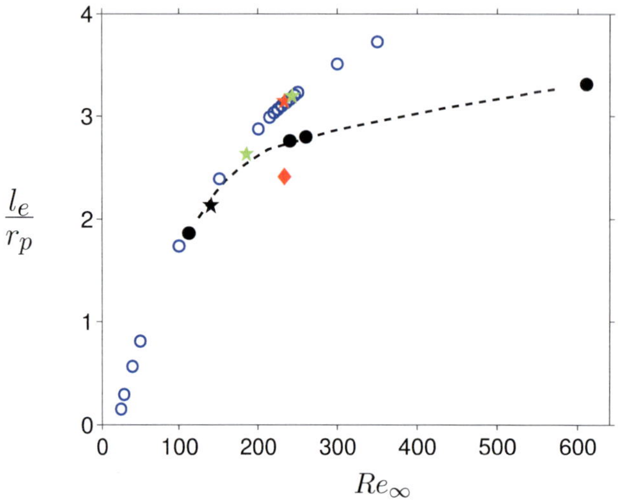

Figure 3.9: Recirculation length l_e in the wake of the particles for both cases $S121$ (★) and $S178$ (★, ♦) as function of the Reynolds number based on the terminal settling velocity Re_∞. The symbols marked by the pentagrams (★, ★) represent the length of the recirculation region by considering the obliqueness of the wake. The red diamond symbol (♦) marks the length of the recirculation length in case $S178$ evaluated on the vertical axis $\tilde{z}$. The difference between the two definitions is best illustrated in figure 3.7c. The results are plotted along the data from Uhlmann and Dušek (2014) (★) and the digitalized data for single fixed particle in uniform flow from Bouchet et al. (2006), (○) and Bagchi and Balachandar (2004), (●), (- - - -).

Considering case $S178$, the corresponding profiles of $\tilde{w}_{p,rel}$ in figure 3.11b reveals that the symmetry in the velocity with respect to the center of the particle is lost. However, when the fluid velocity profiles are evaluated in the coordinate system defined by $(\hat{x}, \hat{y}, \hat{z})$, the symmetry of the streamwise velocity component is restored.

It is well known that the cross-stream profiles of the velocity deficit $\hat{u}_{d0}$ in the wake of an isolated sphere become self-similar and follow a Gaussian function (Tennekes and Lumley, 1972; Wu and Faeth, 1994; Bagchi and Balachandar, 2004; Legendre et al., 2006; García-Villalba et al., 2012):

$$\frac{\hat{u}_d}{\hat{u}_{d0}} = \exp[-(\hat{z}/h_{wz})/2], \tag{3.12}$$

where $\hat{u}_d$ is the mean velocity deficit in the vertical plane perpendicular to $\hat{\mathbf{e}}_x$ and is defined as:

$$\hat{u}_d(0, \hat{y}, \hat{z}) = \frac{\hat{w}_\infty(0, \hat{y}, \hat{z}) - \hat{w}_{p,rel}(0, \hat{y}, \hat{z})}{\hat{w}_\infty(0, \hat{y}, \hat{z})}, \tag{3.13}$$

with $\hat{w}_\infty(0, \hat{y}, \hat{z})$ being the maximum of $\hat{w}_{p,rel}(0, \hat{y}, \hat{z})$ over $\hat{x}$. In the above equation h_{wz} denotes the cross-stream location where the velocity deficit becomes $\hat{u}_{d0} \exp(-1/2)$ and is known as the *half-width* of the particle wake.

Figure 3.12 shows the cross-stream profiles of the velocity deficit in the particle wake for both cases. The profiles are taken at the same distance from the particle surface as the velocity profiles in figure 3.11. It can be observed, that in case $S121$ the velocity deficit across the wake seem to be reasonably correlated with each other and with the Gaussian velocity distribution, whereas in case $S178$ deviation from the Gaussian function.

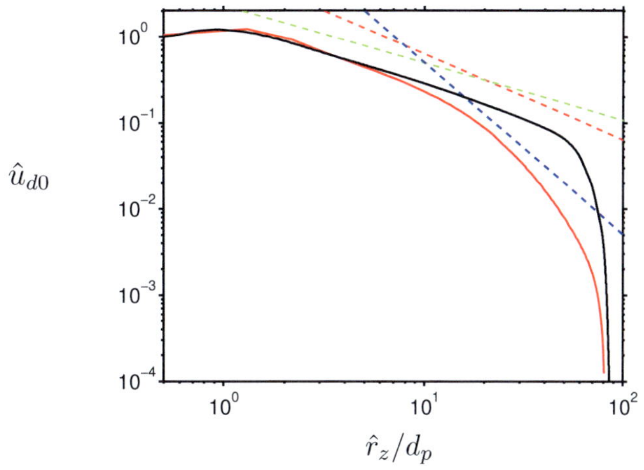

Figure 3.10: Streamwise velocity deficit $\hat{u}_{d0}$ defined as in equation 3.11 along the vertical axis $\hat{z}$ as function of the downstream distance $\hat{r}_z$ from the particle surface. Case $S121$ ——, case $S178$ ——. The downstream distance $\hat{r}_z$ from the particle surface is defined as $\hat{r}_z = \hat{z} - z_p$. The dashed straight lines indicate decay rates proportional to $\tilde{r}_z^{-1}$ (- - - -), $\tilde{r}_z^{-2}$ (- - - -) and $\tilde{r}_z^{-2/3}$ (- - - -).

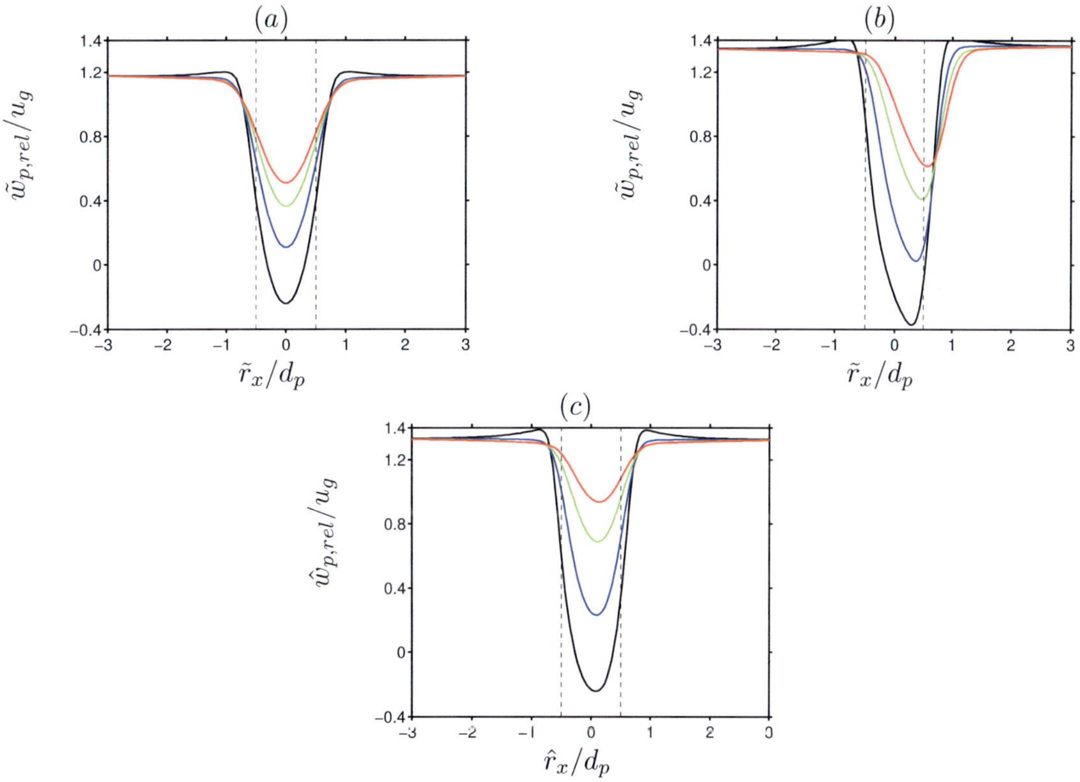

Figure 3.11: Cross-profiles of the vertical component of the relative velocity $\hat{w}_{p,rel}$ at different distances from the particle surface. Profiles are taken along the axis, which is aligned with the unit normal vector $\hat{\boldsymbol{e}}_x$ (cf. figure 3.6c and figure 3.7c), except in subfigure (b) where the profile is taken along horizontal axis $\boldsymbol{e}_h$. (a) Case $S121$. (b) - (c) Case $S178$. The black dashed lines (- - - -) mark the extension of the particle. The color indicates the distances downstream from the particle: (——), $\hat{\boldsymbol{e}}_z|\hat{\boldsymbol{x}}|/d_p = 0.5$; (——), $\hat{\boldsymbol{e}}_z|\hat{\boldsymbol{x}}|/d_p = 1.5$; (——), $\hat{\boldsymbol{e}}_z|\hat{\boldsymbol{x}}|/d_p = 2.5$; (——), $\hat{\boldsymbol{e}}_z|\hat{\boldsymbol{x}}|/d_p = 3.5$. The gravitational velocity u_g was used for non-dimensionalizing the velocity profile.

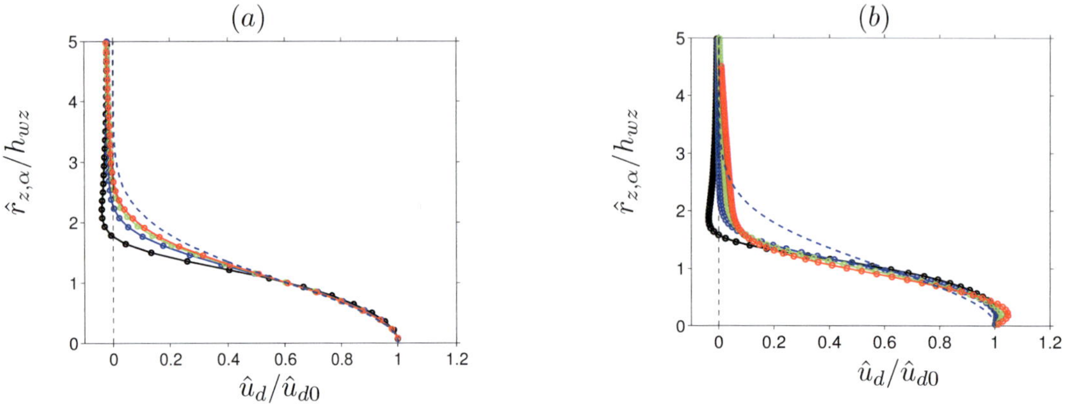

Figure 3.12: Cross-profiles of the velocity deficit $\hat{u}_d/\hat{u}_{d0}$ in the particle wake as function of the cross-stream coordinate $\hat{r}_{z,\alpha}$. The profiles are taken at 4 different downstream locations $\hat{z}/d_p = [1, 3, 5, 7]$. The blue dashed (----) line corresponds to a Gaussian function. The velocities $\hat{u}_d$ and $\hat{u}_{d0}$ are defined as in equations 3.11, 3.13. (a) Case $S121$. (b) Case $S178$. The distance from the particle was normalized by the wake half-width h_{wz}.

	Ga	$u_{h,p}/u_g$	$w_{p,rel}/u_g$	α°	C_d	l_e/d_p
$S121$	121.24	0	1.15	0°	1.002	1.19
$S178$	178.46	0.12	1.30	5.21°	0.78	1.57

Table 3.4: Various statistical results for the direct numerical simulations of case $S121$ and $S178$. The results are presented for the time period at which the simulations are at steady state, i.e. $t/\tau_g = [30, 270]$ for case $S121$ and $t/\tau_g = [130, 300]$ for case $S178$.

3.5 Conclusion

In this chapter we presented the direct numerical simulations of the sedimentation of single heavy finite-size particle under the influence of gravity in an initially ambient flow field. Here we considered the settling of the particles in two regimes. In the first one the particle settles in the steady axi-symmetric regime with Galileo number of Ga $= 121$, whereas in the second the particle settles in the steady oblique regime with Ga $= 178$. The particle density-to-fluid ratio in both cases was kept constant to $\rho_p/\rho_f = 1.5$. Special care was taken to determine the appropriate spatial and temporal resolution of the particle motion in order to capture the small scale flow structures in vicinity of the particle. For the case with Ga $= 121$ the resolution of 15 grid points per particle diameter was determined to be sufficient while for Ga $= 178$ the resolution of 24 grid points per diameter was used. The present simulations showed that this numerical method and the chosen resolution are able to capture the characteristic features of the particle motion and the flow field in vicinity of the particle. The motion of the particle was represented with sufficient accuracy and was found to be in both cases, as predicted from previous studies, in equilibrium with the surrounding flow field, i.e. steady. The particle in case $S121$ settled on a rectilinear trajectory, whereas the path of the particle in case $S178$ was steady rectilinear in an oblique angle to the vertical direction. The double threaded wake is, however, randomly oriented with respect to the horizontal direction. Further, the structure of the particle wake was also accurately reproduced. The wake in case $S121$ was found to be axi-symmetric and consists of a toroidal vortex around the particle, while in case $S178$ a double threaded wake with counter-rotating vortices was observed. The wake in case $S178$ was further found to be symmetric in a two-dimensional plane inclined at an oblique angle, which resulted from the oblique motion of the particle. The statistical description of the particle wake and particle motion, such as the wake structure and wake recirculation length as well as the particle terminal settling velocity, particle rotation and steadiness of the particle motion were compared with the recent study of Uhlmann and Dušek (2014). The present results appear to be in a good agreement with the reference data. The results of the simulations in this chapter are meant to be used as reference for the following chapters §4 - §8 where the settling of multiple particles is considered. The results for various statistical quantities obtained at steady state are summarized for convenience in table 3.4.

Chapter 4

Flow through a stationary array of randomly distributed spheres

4.1 Introduction

As described in the previous chapter, when the particle Reynolds number becomes finite, a wake behind the particles is formed, which can perturb the flow field downstream of the particle and for sufficiently large Reynolds numbers the flow may even become turbulent. The problem becomes even more complicated, when multiple particles are present in the flow. In the context of particle-induced turbulence, the wake generated by the particles can contribute significantly to the turbulence production in the flow. Thus, the characterization of the flow fluctuations in the particle wake is crucial for the understanding of the turbulence-production mechanism. As introduced in section §1.2.2, the flow field induced by the suspension of multiple particles could have statistical properties different from that of classical turbulence and is often referred to as pseudo-turbulence. For example, in large number of studies (Lance and Bataille, 1991; Martínez-Mercado et al., 2007, 2010; Risso et al., 2008; Riboux et al., 2013) the power spectral density of the vertical fluid velocity component was observed to exhibit regions where it evolves roughly as the power of -3 of the wave number. Despite the recent effort on the subject, to our knowledge there is still not a conclusive theoretical interpretation of the existence of the -3 spectral range. A possible reason for this was recently proposed by Risso et al. (2008). The authors argued that the fluid velocity fluctuations induced by the particles involve two different contributions. The first contribution is related to the spatial variation of the velocity and is called *spatial fluctuation*, since it results from the local particle distribution. The second contribution corresponds to *temporal fluctuations* in the velocity field that results from hydrodynamic instabilities in the flow. According to the authors, in a laboratory frame, where the particles are allowed to move freely, this two contributions cannot be easily distinguished and there is a lack of diagnostic tools for the decomposition of the velocity fluctuations in their spatial and temporal parts. However, the decomposition could be easily performed for the case of a flow through a random array of fixed particles. In a following work, Risso (2011) showed that a superposition of independent random bubble disturbances generates a power spectrum with -3 power law. He identified the spatial distribution of the bubbles as the main reason for the power law. More recently Riboux et al. (2013) performed a numerical study of the flow past a random array of stationary bubbles at low solid volume fractions. The bubbles were modeled by time depending momentum sources fixed in space at

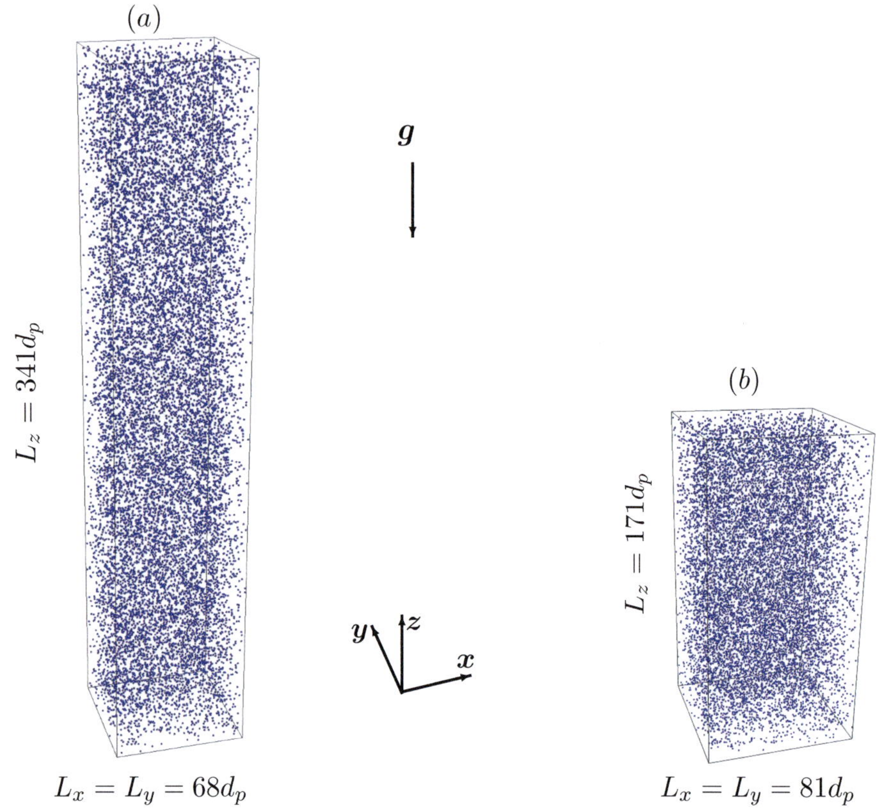

Figure 4.1: Dimensions of the computational domain with the initial particle distribution for (a) case $F121$ and (b) case $F178$. d_p denotes the particle diameter. Periodic boundary conditions are applied in all three directions in both cases.

those in case $F121$. Most probably the larger fluctuations in case $F178$ stem from the oblique character of the individual particle wake, as shown in chapter §3. On the other hand, the fluctuations of the vertical velocity component show the opposite behavior, i.e. the fluctuations in case $F121$ are larger than in case $F178$.

Figure 4.3 depicts the normalized probability distribution function (p.d.f.) of the horizontal and vertical components of the fluid velocity fluctuations. As can be seen both components do not resemble a Gaussian distribution. As expected the fluctuations of the horizontal component in both cases are symmetric about their mean value. However, the p.d.f. evidences that the fluctuations exhibit strong intermittency. It is interesting to note that the fluctuations of the horizontal velocity component in case $F121$ are more intermittent than in case $F178$. This is confirmed by evaluating the fourth moment of the distribution, which is known as the kurtosis (flatness). We measured the kurtosis in case $F121$ to $K(u) = 27.05$ and in case $F178$ to $K(u) = 14.95$. Regarding the vertical component, the velocity fluctuations in both cases are not symmetric with large downward fluctuations being more probable than upward ones. This is manifested in the significant bump observed in the p.d.f. around the negative value of six standard deviations. As a result, the p.d.f. becomes negatively skewed and indeed the skewness (i.e. the third moment of the distribution function) in both cases showed to be negative with very similar values, $S(w) = -2.40$ and $S(w) = -2.60$ in case $F121$ and case $F178$ respectively. This is a characteristic feature of wake flows where fluid is entrained in the wakes of the particles. Considering the intermittency of the vertical velocity fluctuations, it appears that the

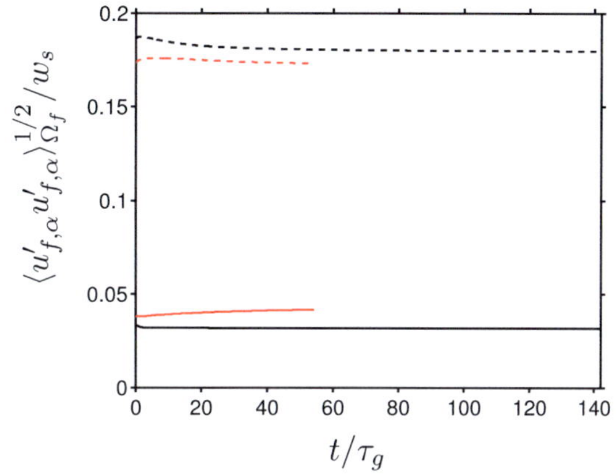

Figure 4.2: Intensity of the fluid velocity fluctuations for case $F121$ (——, - - - -) and for case $F178$ (——, - - - -) as functions of time. Solid lines correspond to horizontal velocity component, dashed lines show the vertical component. The quantities are normalized with the terminal settling velocity w_s of the corresponding single-particle cases ($S121$ and $S178$, respectively).

fluctuations in case $F178$ are slightly more intermittent than in case $F121$. We calculated the kurtosis in case $F121$ to $K(w) = 12.66$ and in case $F178$ to $K(w) = 14.07$.

These results are in line with the observations made in Risso and Ellingsen (2002); Riboux et al. (2010). In both studies the agitation in the fluid by the rise of bubbles was investigated. Nevertheless, the authors observed local increase of the probability density function of the fluid velocity at large velocity fluctuations. Risso and Ellingsen (2002) noted that this increase of the p.d.f. is strongly affected by events in the vicinity of the particles and that the velocity fluctuations are controlled by non-linear interactions between the wakes of all bubbles. Similar shape of the probability density function was observed in the study of Riboux et al. (2013), where they studied the flow field fluctuations induced by an assembly of stationary bubbles.

As introduced in section §4.1, the velocity fluctuations in the flow field induced by the particles could be decomposed in their spatial and temporal fluctuations (Risso et al., 2008). In order to investigate the relative influence of the two contributions for the present simulations, we decomposed the flow field fluctuations in their spatial and temporal parts. To this end, we first define the following decompositions of the flow field with respect to three different averaging operators:

$$\boldsymbol{u}_f(\boldsymbol{x},t) = \langle \boldsymbol{u}_f\rangle_{\Omega_f}(t) + \boldsymbol{u}'_f(\boldsymbol{x},t) \tag{4.1}$$
$$\boldsymbol{u}_f(\boldsymbol{x},t) = \langle \boldsymbol{u}_f\rangle_t(\boldsymbol{x}) + \boldsymbol{u}''_f(\boldsymbol{x},t) \tag{4.2}$$
$$\boldsymbol{u}_f(\boldsymbol{x},t) = \langle \boldsymbol{u}_f\rangle_{\Omega_f,t} + \boldsymbol{u}'''_f(\boldsymbol{x},t). \tag{4.3}$$

The above three equations define implicitly the fluctuations of the fluid velocity field with respect to (i) the instantaneous space phase averaging of the flow field $\boldsymbol{u}'_f(\boldsymbol{x},t)$, (ii) the three dimensional time averaging flow field $\boldsymbol{u}''_f(\boldsymbol{x},t)$ and (iii) the space and time averaging of the fluid velocity $\boldsymbol{u}'''_f(\boldsymbol{x},t)$. A detailed definitions of the averaging operators $\langle\cdot\rangle_{\Omega_f}$, $\langle\cdot\rangle_t$ and $\langle\cdot\rangle_{\Omega_f,t}$ in equations 4.1 - 4.3 can be found in appendix A. Based on these equations, the instantaneous fluctuating velocity field with respect to the space and time average can be expressed alternatively as follow:

$$\boldsymbol{u}'''_f(\boldsymbol{x},t) = \langle \boldsymbol{u}'_f\rangle_t(\boldsymbol{x}) + \boldsymbol{u}''_f(\boldsymbol{x},t). \tag{4.4}$$

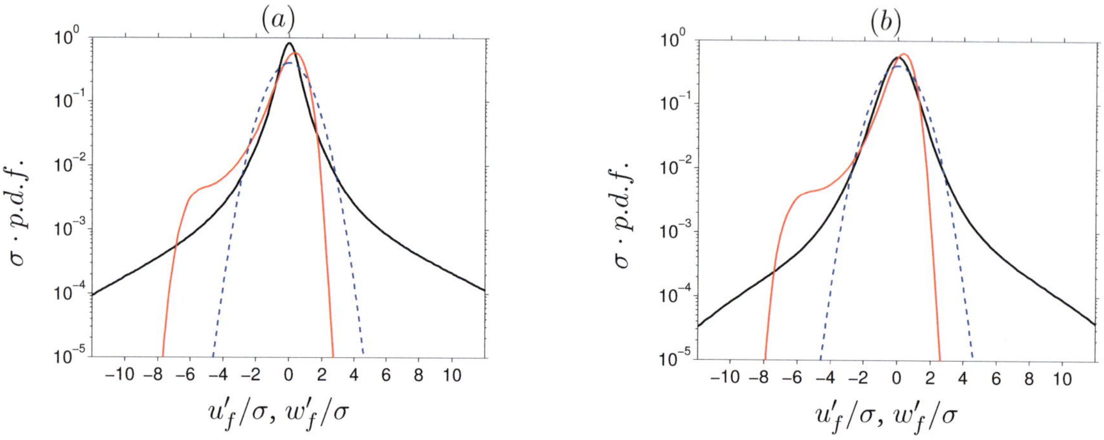

Figure 4.3: Probability density function of the flow velocity field for (a) case $F121$ and (b) case $F178$. Horizontal component (———), vertical component (———). The blue dashed line (- - - -) corresponds to a Gaussian distribution.

For detailed derivation of equation 4.4 the reader is referred to appendix B. In view of the work of Risso et al. (2008); Riboux et al. (2013) the expression of $\langle \boldsymbol{u}'_f \rangle_t(\boldsymbol{x})$ will be referred hereafter to as *spatial fluctuations* and the expression $\boldsymbol{u}''_f(\boldsymbol{x}, t)$ to as *temporal fluctuations*. As noted by Risso et al. (2008), the spatial fluctuations originate from the local inhomogeneity of the velocity field, which is a result of the particle spatial distribution. The temporal fluctuations on the other hand are generated by the collective instability of the flow around the randomly distributed particles, since the flow induced by a single particle at this Galileo number would be steady. Pleas note that while the spatial fluctuations are only a function of the position $\boldsymbol{x}$, the temporal fluctuations depend on both the position $\boldsymbol{x}$ and the time t.

The variance of the fluctuations with respect to the space and time average can be now expressed as the additive contributions of the variance of the spatial and the temporal fluctuations, viz.:

$$\langle \boldsymbol{u}'''_f \boldsymbol{u}'''_f \rangle_{\Omega_f,t} = \langle \langle \boldsymbol{u}'_f \rangle_t \langle \boldsymbol{u}'_f \rangle_t \rangle_{\Omega_f,t} + \langle \boldsymbol{u}''_f \boldsymbol{u}''_f \rangle_{\Omega_f,t} \tag{4.5}$$

A detailed derivation of the relation 4.5 can be found in appendix B. As noted by Risso et al. (2008) the spatial contribution $\langle \langle \boldsymbol{u}'_f \rangle_t \langle \boldsymbol{u}'_f \rangle_t \rangle_{\Omega_f,t}$ is related to the spatial variations of the time averaged velocity field and is not related to turbulence. On the other hand, the contribution $\langle \boldsymbol{u}''_f \boldsymbol{u}''_f \rangle_{\Omega_f,t}$ is related to the temporal fluctuations of the flow field and measures the turbulence intensity in the flow.

The decomposition in equation 4.4 is depicted in figure 4.4 in a two dimensional plane at some instant in each of our two simulations. The exemplary data shown is illustrative of the entire data set. It can be seen that in both cases $F121$ and $F178$ the dominant contribution to the fluctuating vertical fluid velocity $\boldsymbol{u}'''_f$ stems from the spatial fluctuations $\langle \boldsymbol{u}'_f \rangle_t$. The fluctuations with respect to the time averaged field, $\boldsymbol{u}''_f$, on the other hand, are much less significant in the present range of Reynolds numbers and at the chosen solid volume fraction. Compering the two flow cases, it is observed that the temporal fluctuations $\boldsymbol{u}''_f$ are significantly larger in case $F178$ than in case $F121$, which is consistent with the expectations (increase in Reynolds number). Moreover, the spatial fluctuations $\langle \boldsymbol{u}'_f \rangle_t$ in both cases resemble closely the particle wakes in the total fluctuations $\boldsymbol{u}'''_f$. On the other hand, the temporal fluctuations $\boldsymbol{u}''_f$ and the particle wakes are only weakly correlated with each other and they seem to be

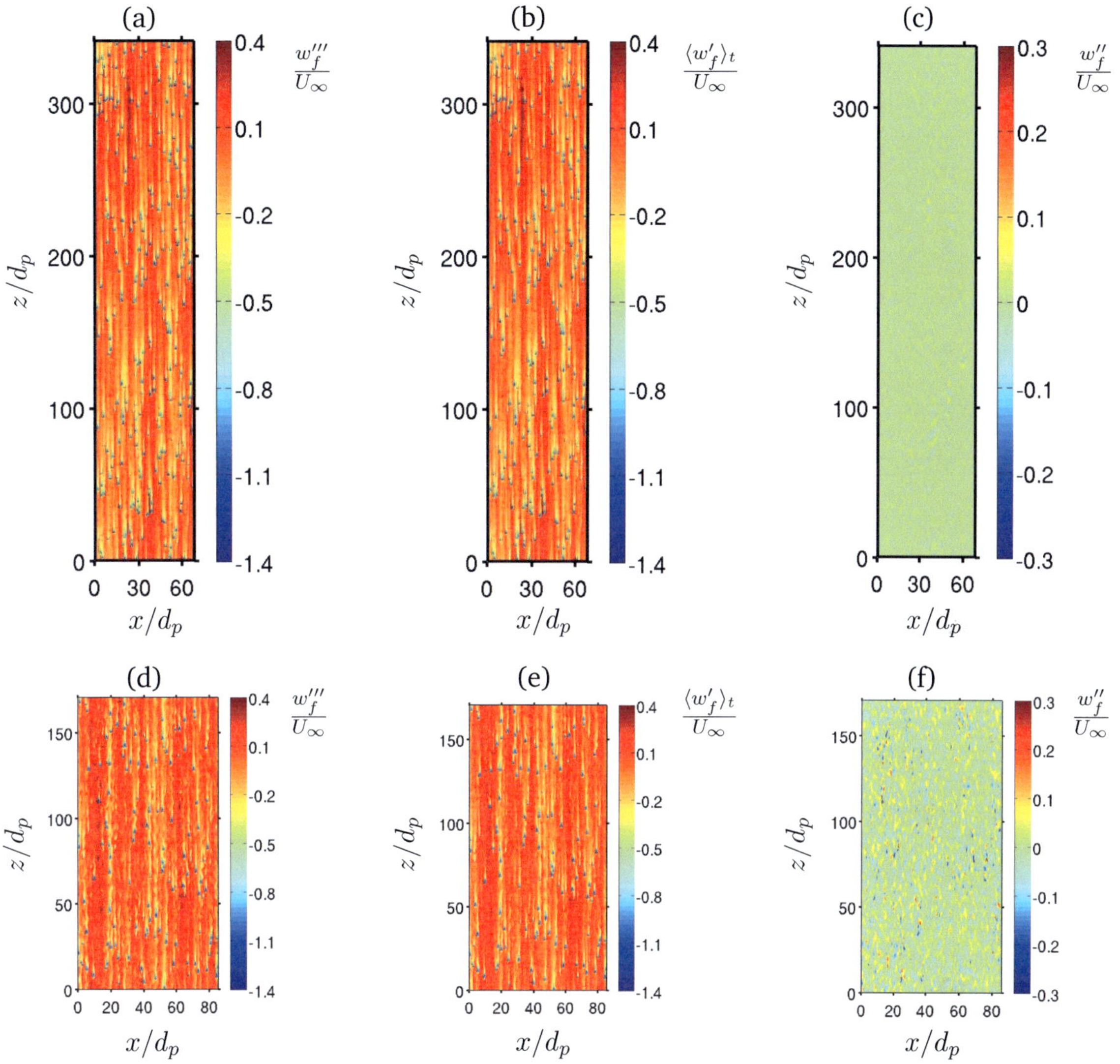

Figure 4.4: Fluid velocity fluctuations of the vertical velocity component normalized by the undisturbed incoming velocity U_∞ for (a,b,c) case $F121$ and for (d,e,f) case $F178$. Total instantaneous fluctuations $\boldsymbol{u}_f'''(\boldsymbol{x},t)$, (a,d). Spatial fluctuations $\langle \boldsymbol{u}_f' \rangle_t(\boldsymbol{x})$, (b,e). Temporal fluctuations $\boldsymbol{u}_f''(\boldsymbol{x},t)$, (c,f). The fluctuations are shown in a two dimensional vertical plane. Note the difference in the color range between (a,b,d,e) and (c,f).

homogeneously distributed over the particle assembly, while the spatial fluctuations show highly inhomogeneous distribution. It is important to note that, opposite to the temporal fluctuations, the spatial fluctuations are not statistically symmetric. Regarding the horizontal velocity component (figure omitted), the mean particle wake has only a minor contribution to the velocity fluctuations. Thus, the total fluctuations are represented almost entirely by their temporal part. As a result the spatial fluctuations are negligible. The temporal contribution, however, originates from the instabilities in the particle wakes caused by the presence of the particle assembly.

To estimate more quantitatively the contributions of the spatial and temporal fluctuations, we have computed the p.d.f. of the different contributions, which are depicted for the vertical

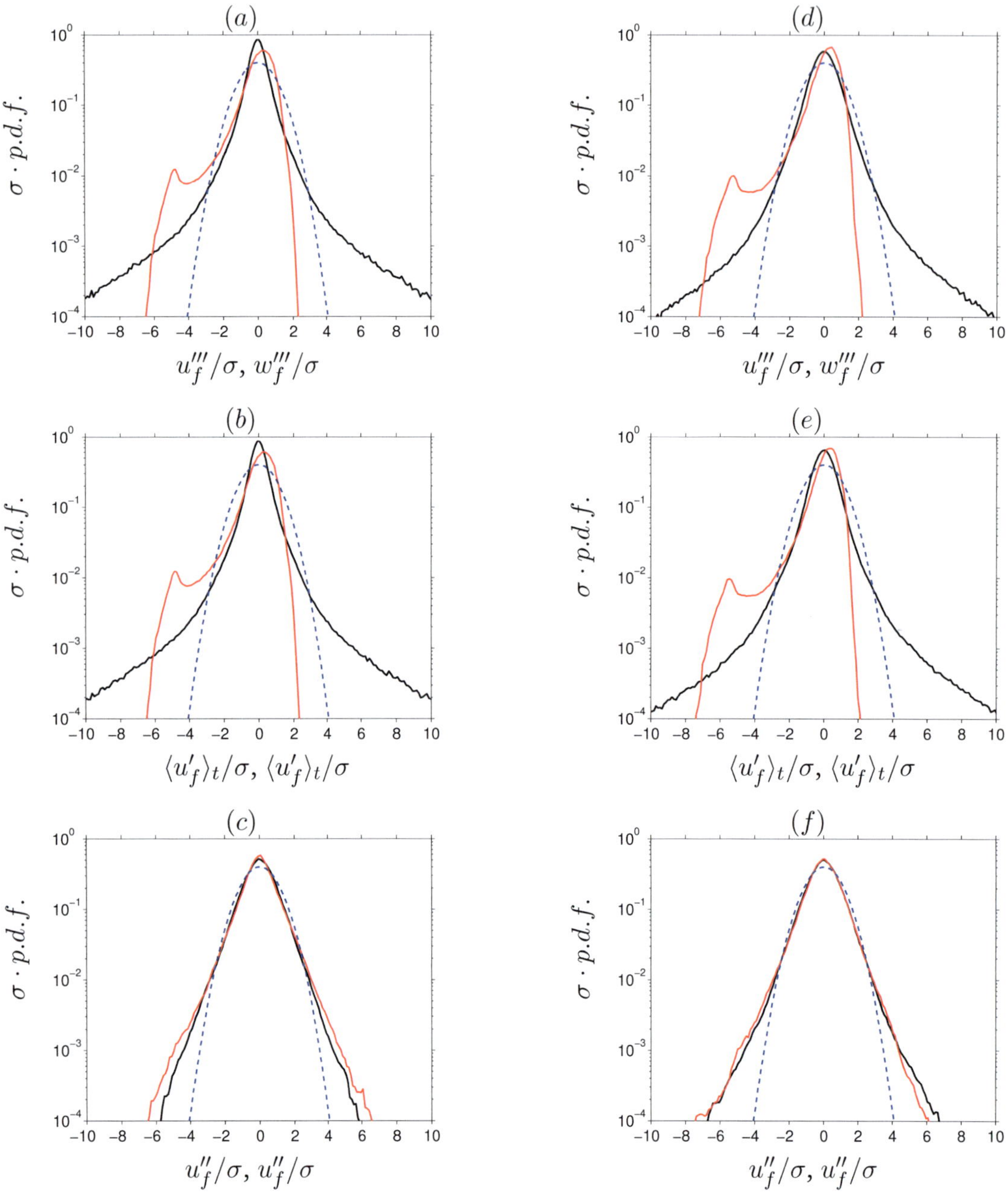

Figure 4.5: Probability density function of the fluid velocity field for (a,b,c) case $F121$ and (d,e,f) case $F178$. The p.d.f. is normalized by the standard deviation of the velocity. Total instantaneous fluctuations $\boldsymbol{u}'''_f(\boldsymbol{x},t)$ (a,d). Time average of the spatial fluctuations $\langle \boldsymbol{u}'_f\rangle_t(\boldsymbol{x})$ (b,e). Temporal fluctuations $\boldsymbol{u}''_f(\boldsymbol{x},t)$ (c,f). Horizontal velocity component (———). Vertical velocity component (———). Gaussian distribution (- - - -).

velocity component in figure 4.5. It can be observed, that in both cases the p.d.f. of the spatial fluctuations resembles very closely the p.d.f. of the total fluctuations (cf. figure 4.3) for both the vertical and the horizontal velocity components. This is best observed in the third and fourth moments of the p.d.f.. We have calculated the skewness and the kurtosis of the distributions, which can be found in table 4.3 for case $F121$ and in table 4.4 for case $F178$. Regarding the temporal fluctuations, figure 4.5 reveals that in both cases the distribution of the horizontal and the vertical velocity components is very similar. It is interesting to note that the distribution of the vertical component is as symmetric as of the horizontal component. This implies that the asymmetry in the total fluctuations of the vertical component is inherited from the distribution of the spatial fluctuations.

Overall, the results confirm the observations made recently by Riboux et al. (2013), that the mean wake of the particles is mainly responsible for the fluid velocity fluctuations.

	$S(u)$	$S(w)$	$K(u)$	$K(w)$
$\boldsymbol{u}_f'''(\boldsymbol{x},t)$	0.006	−2.23	25.86	10.72
$\langle\boldsymbol{u}_f'\rangle_t(\boldsymbol{x})$	0.002	−2.24	26.06	10.77
$\boldsymbol{u}_f''(\boldsymbol{x},t)$	−0.047	0.030	4.64	6.35

Table 4.3: Skewness and kurtosis for case $F121$ of the probability distribution functions shown in figure 4.5 (a,b,c). $S(u)$ and $K(u)$ correspond to the p.d.f. of the horizontal velocity components and $S(w)$ and $K(w)$ to the vertical velocity component.

	$S(x)$	$S(z)$	$K(x)$	$K(z)$
$\boldsymbol{u}_f'''(\boldsymbol{x},t)$	0.07	−2.70	14.64	13.99
$\langle\boldsymbol{u}_f'\rangle_t(\boldsymbol{x})$	0.076	−2.90	17.79	15.36
$\boldsymbol{u}_f''(\boldsymbol{x},t)$	−0.002	−0.18	5.95	6.00

Table 4.4: Same as in table 4.3 but for case $F178$. The corresponding p.d.f.s can be found in figure 4.5 (d,e,f).

4.3.2 Energy spectrum and two-point correlations of the flow field

Figure 4.6 depicts the one-dimensional energy spectrum of the flow field for case $F121$ and $F178$. In the present work the one-dimensional energy spectrum is computed as:

$$E_{ij}(\kappa,t) = \sum_{\boldsymbol{\kappa}} \langle \hat{u}^*_{f,i}(\boldsymbol{\kappa})\hat{u}_{f,j}(\boldsymbol{\kappa})\rangle, \tag{4.6}$$

where $\boldsymbol{\kappa}$ is the wave number vector, $\hat{\boldsymbol{u}}_f(\boldsymbol{\kappa})$ denotes the Fourier transform of the flow velocity $\boldsymbol{u}_f$ and $\hat{\boldsymbol{u}}^*_f(\boldsymbol{\kappa})$ the complex conjugate of $\hat{\boldsymbol{u}}(\boldsymbol{\kappa})$. Due to the presence of the dispersed phase in the flow, the volume occupied by the fluid Ω_f may be not continuous and encloses the volume which is occupied exclusively by the particles. Thus, difficulties may arise by the computation of the Fourier transform of the fluid velocity $\boldsymbol{u}_f$, since the velocity signal would exhibit discontinuities at the particle positions. In order to avoid this the velocity signal at the position of the particles is approximated by means of a tri-linear interpolation. This procedure leads to smoothing of the well known "*step-noise*" at wave numbers (wavelengths) around the particle scale (Parthasarathy and Faeth, 1990a; Lucci et al., 2010), e.g. $\kappa_x d_p \sim 2\pi$ ($\lambda_x/d_p \sim 1$). In order to better compare the energy contents between the different cases, the spectra are normalized by their respective maximum value. For more detailed discussion on the calculation of E_{ij} and the interpolation procedure the reader is referred to appendix C.

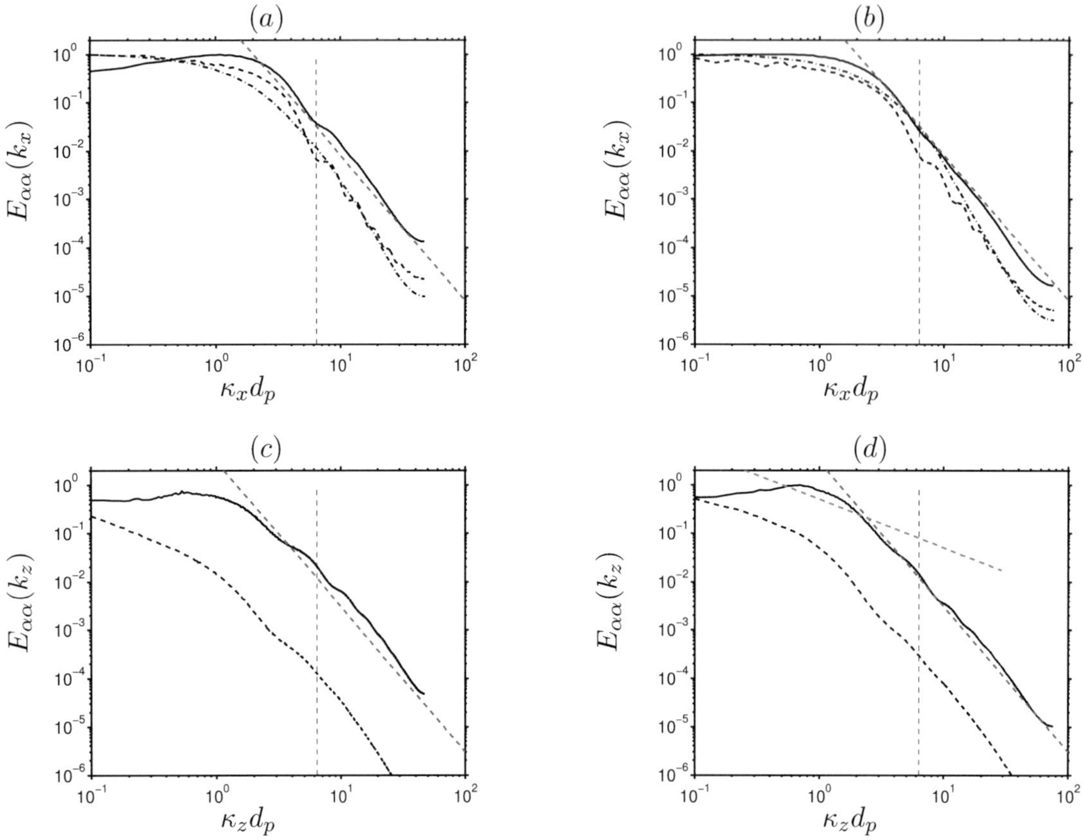

Figure 4.6: One dimensional energy spectrum $E_{\alpha\alpha}$ ($\alpha = 1, 2, 3$) of the flow field for case $F121$ (a,c) and case $F178$ (b,d). Horizontal direction (a,b) and in vertical direction (c,d). Lines show velocity components in horizontal directions (u: ----, v: —·—), and in vertical direction (w: ——). Note that the graphs in (c) and (d) show only the two statistically independent components. Vertical dashed line (----) represents the wave number corresponding to the particle diameter d_p. Magenta lines (----) represents the slope corresponding to κ_α^{-3}.

As introduced in section §4.1 the energy spectrum in dispersed bubble flows was observed to experience a decay power law as the power of -3 of the wave number $\boldsymbol{\kappa}$. The main question here is whether the energy spectrum of the flow field experience power low decay for the present flow configurations.

First let us consider the spectrum in the horizontal x-direction. Please note that in the y-direction the spectrum can be easily obtained by replacing the u velocity component with the v-component. As can be seen, the slope of the energy spectrum of the u velocity component exhibits in both cases a κ_x^{-3} sub-range, κ_x being the wave number in the horizontal direction. We would like to stress that this scaling behavior is maintained in a wide range of the wave number, ranging from approximately $\kappa_x d_p = 3.5$ to $\kappa_x d_p = 40$. The decay rate of the energy spectra of the remaining two velocity components also seems to be approximated by a constant decay law. We measured the decay rate to be approximately -3.5, indicating faster decay than the u-component. The influence of the particle interface on the spectra can be clearly depicted in the spectrum by the presence of the above mentioned "step noise", which is observed for wave numbers larger than the wave number corresponding to the particle diameter. In the vertical direction, similar decay rate of the spectra for both the horizontal and vertical

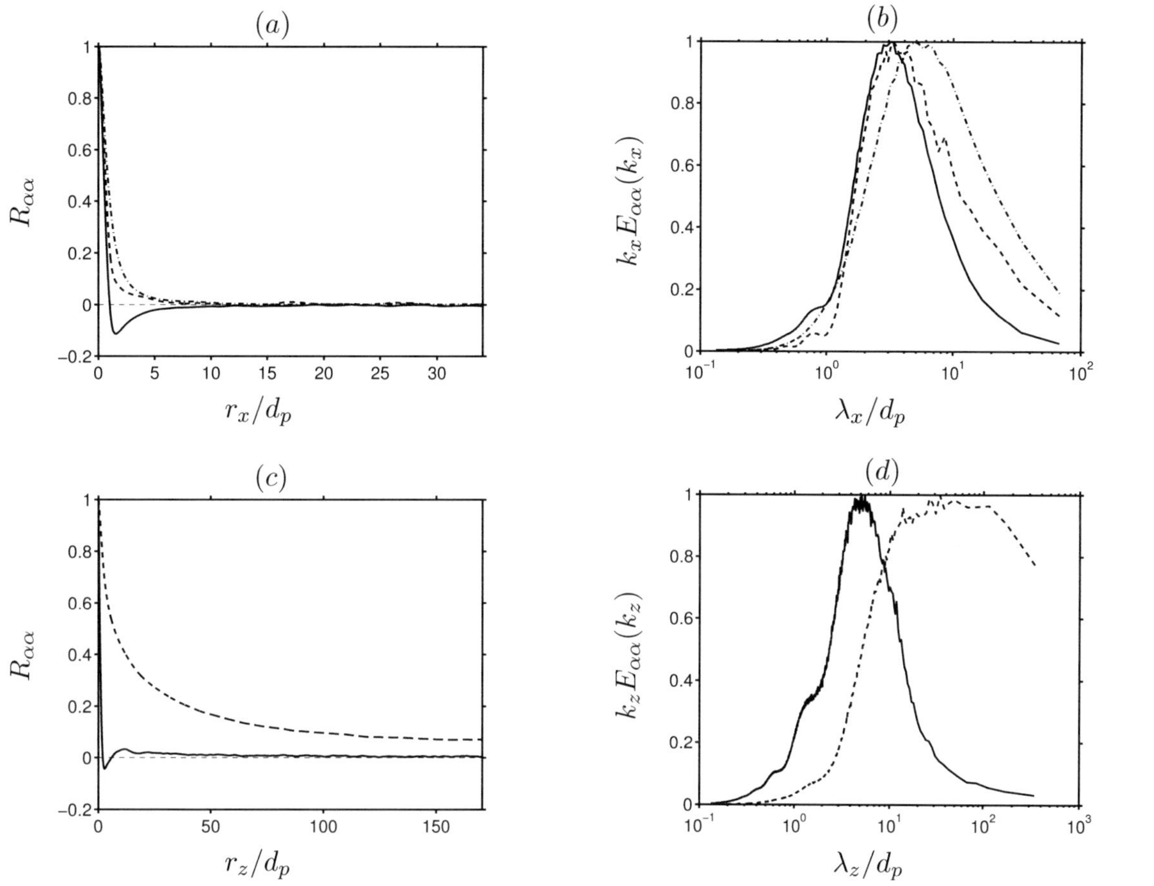

Figure 4.7: Two-point auto-correlation function of the fluid velocity fluctuations in case $F121$ for separations in x-direction (a) and z-direction (c) with corresponding pre-multiplied spectra in (b) and (d). Lines show velocity components in horizontal directions (u: ——, v: —·—), and in vertical direction (w: ----). r_x and r_z denote the separation in the horizontal and vertical direction. λ_x and λ_y denote the wave length. All spectra are normalized to a maximum value of unity.

velocity components can be observed. Please note that corresponding plots show only the two statistically independent velocity components. However, a difference between the two cases can be observed, with the power law being more distinctive in case $F178$ than in case $F121$. In case $F178$ the -3 sub-range in the spectrum for both velocity components is maintained for nearly two decades of the wave number, i.e. $\kappa_z d_p \in [3, 80]$. The energy spectra in both, the horizontal and the vertical direction clearly show a reverse cascade process, since energy is present at scales more than two orders of magnitude larger than the particle scale, where energy input takes place.

In order to better characterize the flow field induced by the particles, we have calculated the two-point correlation function of the flow field and the corresponding pre-multiplied spectrum. Figures 4.7 and 4.8 show the two-point correlation function of the fluid velocity fluctuations for both cases $F121$ and $F178$. It can be seen from the figures that in the horizontal directions all velocity components decorrelate for separations larger than ten particle diameters, which is smaller than the length of the computational domain in the horizontal direction. Concerning the correlation function in the vertical direction, the figure shows that the vertical velocity component is not fully decorrelated at separations of half the box vertical length. At the largest

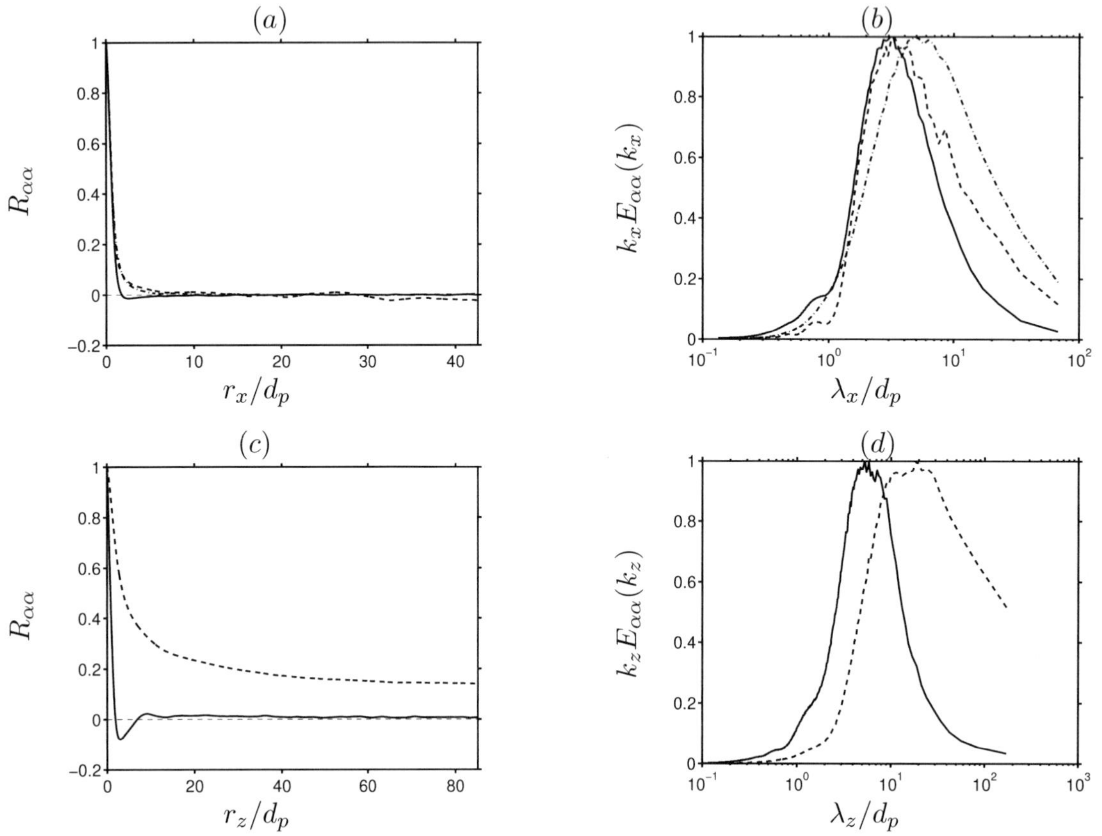

Figure 4.8: Two-point auto-correlation function of fluid velocity fluctuations in case $F178$ for separations in x-direction (a) and z-direction (c) with corresponding pre-multiplied spectra in (b) and (d). Lines show velocity components in horizontal directions (u: ——, v: —·—), and in vertical direction (w: ----). All spectra are normalized to a maximum value of unity.

separation r_z the correlation coefficient measures 0.07 in case $F121$ and 0.14 in case $F178$ (please note that in terms of the particle diameter, the computational domain in case $F121$ is two times larger than in case $F178$). A possible reason for the long distance correlation of the flow field is the strong anisotropy of the particle wakes.

The corresponding pre-multiplied spectra show that the smallest scales of the flow in both cases are well resolved. For wave numbers larger than the particle scale distinct maxima can be observed. Concerning the horizontal direction, the maximum in the pre-multiplied spectra for both cases is located at around $\lambda_x/d_p = 3$, where λ_x is the wave length in the horizontal direction. This value is most probably related to the average lateral extension of the particle wakes. After reaching its maximum the spectrum diminishes rapidly for all three velocity components. Although the velocity fluctuations completely decorrelate, the energy spectrum in the horizontal direction does not drop entirely to zero, indicating the presence of very large flow structures in the flow. Similar trend can be observed for the spectrum in the vertical direction. The spectrum of the horizontal velocity component experience a global maximum at slightly larger wave-lengths than in the horizontal direction, i.e. $\lambda_z/d_p = 4$. After reaching its maximum value the spectra drops in both cases quickly to a value of approximately 0.03. Regarding the vertical velocity component the spectra in both cases shows somehow different behavior. While in both cases the spectra exhibits a global maximum, the maximum in case

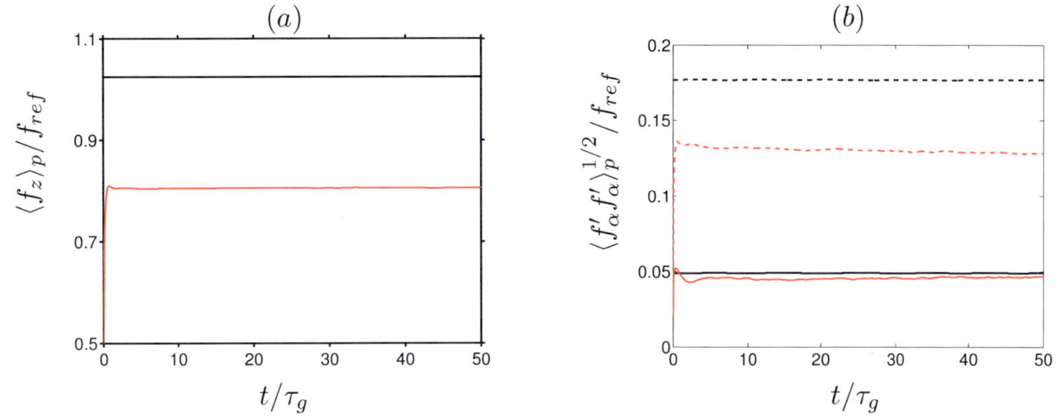

Figure 4.9: (a) Time evolution of the vertical component of the mean hydrodynamic force $\langle f_z \rangle_p$ for case $F121$ (——) and case $F178$ (——). The force is normalized by the reference force f_{ref} as in equation 3.7. (b) Variance of the horizontal and the vertical components of the particle hydrodynamic force as function of time ($\alpha = x, z$). Color coding as in (a). Solid lines correspond to the horizontal component and dashed lines to the vertical component.

$F121$ is observed over a wider range of wave lengths ($\lambda_z/d_p = [10, 200]$), than in case $F178$ ($\lambda_z/d_p = [7, 30]$). In case $F121$ the spectrum diminishes to 0.77, while in case $F178$ a value of 0.51 was observed. It is interesting to note that the wave number at which the spectra in figure 4.6 begins to decay as -3 power law corresponds roughly with the occurrence of the maximum value of the pre-multiplied spectra in figure 4.7 and 4.8.

4.3.3 Particle hydrodynamic force

Figure 4.9a shows the mean of the hydrodynamic force (vertical component) for both cases $F121$ and $F178$. The force has been normalized with the reference force from equation 3.7. Thus, the values in figure 4.9a can be interpreted as a mean drag coefficient of the dispersed phase. As can be seen the average hydrodynamic force in both cases does not experience significant fluctuations over the simulated time period. The average drag coefficient in case $F121$ is evaluated to 1.02 and in case $F178$ to 0.80. The difference between the value of the drag coefficient for the single settling particle from the simulation $S121$ ($S178$) and the mean value in case $F121$ ($F178$) is then calculated to 1.7% (2.5%). The relatively small difference between the drag coefficient for the single particle and the assembly of particles for the present configurations evidences that the hydrodynamic interactions between the particles are relatively small. As result the average force acting on the particles resembles very closely the force of the single particle.

Based on numerical simulations, recently Tenneti et al. (2011) proposed a correlation for the mean drag force exerted by the fluid on the particles. They considered a uniform flow past an assembly of stationary particles with particle Reynolds number in the range of $Re_p \in [0.01, 300]$. However, they considered relatively large solid volume fractions ($\Phi_s > 0.1$). By applying their correlation to the present parameter values for Re and Φ_s, the average drag coefficient was estimated to 0.95 and 0.776 for case $F121$ and case $F178$. When compared with the present data, the correlation for the drag force under-predicts the values by approximately 6.8% in case $F121$ and by 3.6% in case $F178$.

The intensities of the fluctuations of the force components with respect to the instantaneous mean force are shown in figure 4.9b. The intensity of the particle force are represented by its root mean square, which is calculates as:

$$\langle f_i' f_i' \rangle_p(t) = \langle (f_i(t) - \langle f_i(t) \rangle_p)^2 \rangle_p, \qquad i = 1, 2, 3. \tag{4.7}$$

The horizontal component of the hydrodynamic force variance in both cases is small, approximately 5% of the reference force. On the other hand, the fluctuations of the vertical force component experience different intensities, where fluctuations in case $F121$ are almost 30% larger than in case $F178$. Corresponding to the anisotropy of the flow field, the fluctuations of the hydrodynamic force experience similar anisotropy levels. The hydrodynamic force in case $F178$ appears to be more evenly distributed among the horizontal and the vertical components with anisotropy level of 2.77, whereas in case $F121$ the anisotropy level was evaluated to 3.6.

The statistics of the average hydrodynamic force in figure 4.9 are integral quantities describing the force exerted on average by the fluid on the particles, providing a kind of macroscopic perspective on the hydrodynamic force. In order to gain more insight into the effects of the hydrodynamic interaction between the particles on the force exerted on an individual particle in the particle assembly, we have performed particle-conditioned averaging of the hydrodynamic force. This allows us to provide a local (microscopic) view of the particle forces. First we define a discrete solid phase indicator function, which is unity for computational nodes occupied exclusively by the dispersed phase and zero otherwise. The indicator function is then applied to the computational domain and multiplied by the hydrodynamic force of the particle located at the respective spatial position. The averaging domain is then shifted to the location of each particle, taking into account the periodicity of the domain, and the values in the domain are added up. This procedure was repeated for $3 \cdot 10^6$ and 10^6 particle snapshots in case $F121$ and case $F178$ respectively. The so calculated two dimensional field was then divided by the number of samples at each discrete node, yielding an average force field. Here and in the following we will refer to this quantity as the particle conditioned hydrodynamic force and will be denoted as $\langle \tilde{\boldsymbol{f}}(\tilde{x}, \tilde{z}) \rangle_{p,t}$. Henceforth we will use the notation from section 3.4.2 for the coordinates $(\tilde{x}, \tilde{z})$ to indicate that the coordinates are defined relative to the position of the test particle. Any quantity $\psi(\tilde{\boldsymbol{x}})$ evaluated at position $(\tilde{\boldsymbol{x}})$ relative to the particle center will be denoted as $\tilde{\psi}(\tilde{\boldsymbol{x}})$. Detailed definition of the averaging procedure can be found in appendix A.

A map of the vertical component of the average force $\langle \tilde{\boldsymbol{f}}(\tilde{x}, \tilde{z}) \rangle_{p,t}$ is shown in figure 4.10. Note that the quantity $\langle \tilde{\boldsymbol{f}}(\tilde{x}, \tilde{z}) \rangle_{p,t}$ is not invariant with respect to the vertical direction $\tilde{z}$. As a consequence, the map of the particle-conditioned hydrodynamic force does not posses symmetry with respect to the horizontal axis $\tilde{x}$. The map in figure 4.10 shows that indeed in both cases the average force upstream and downstream of the reference particle differ. Downstream of the particle one can observe that the vertical component of the hydrodynamic force is on average smaller than the mean over all particles. At immediate proximity behind the reference particle the hydrodynamic force is on average reduced by as much as 60%, when compared to the mean force $\langle f_z \rangle_p$. On the other hand upstream of the reference particle, the hydrodynamic force experience values similar to that of $\langle f_z \rangle_p$. In both cases the domain of influence of the test particle is confined in a narrow region of the order of the particle diameter elongated in the vertical direction. As can be seen the particle influence on the force is visible for distances up to 80 particle diameters downstream of the particle.

In order to provide more qualitative insight into the spatial variation of the particle-conditioned drag force, we consider the horizontal and vertical profiles of the force through

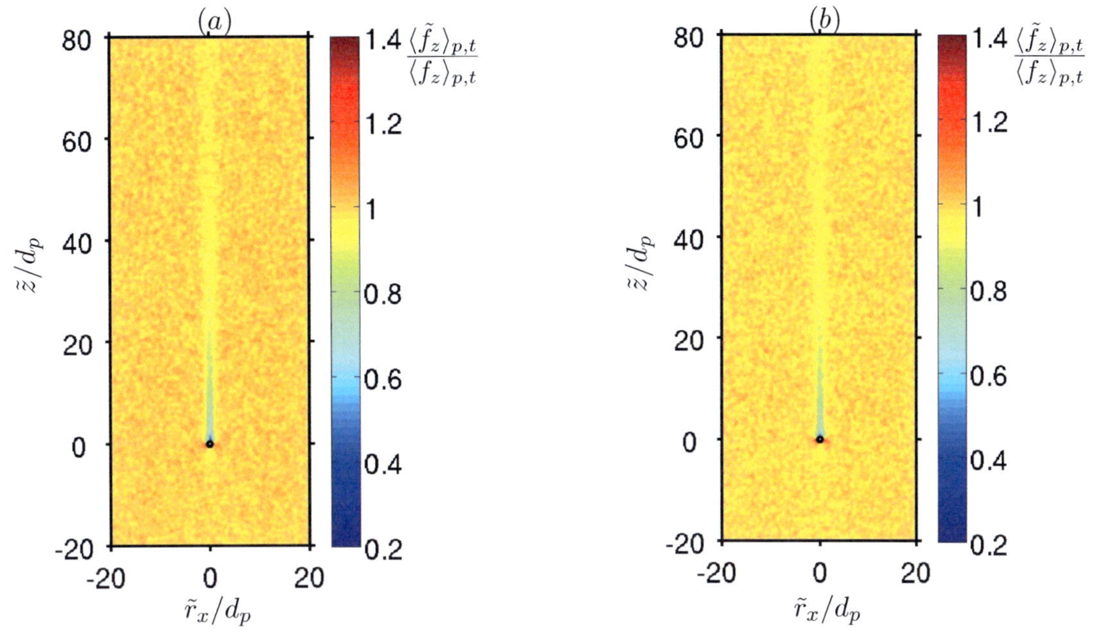

Figure 4.10: Map of the particle-conditioned average value of the vertical hydrodynamic force component $\langle \tilde{f}_z \rangle_{p,t}$ for (a) case $F121$ and (b) case $F178$. The force $\langle \tilde{f}_z \rangle_{p,t}$ is plotted as function of the position $(\tilde{r}_x, \tilde{z})$ relative to the reference particle (vertical plane passing through the center of the reference particle).

the center of the reference particle which are shown in figure 4.11. In the horizontal direction the profile in both cases increase close to the reference particle after which it decreases and fluctuates around the mean force $\langle f_z \rangle_{p,t}$, indicating that the presence of the test particle has only minor effect on the hydrodynamic force of the particles located in the horizontal direction from the reference particle. The maximum in both cases was observed at approximately the same distance of 1.3 particle diameter from the surface of the test particle. The increase of the drag coefficient in the horizontal direction close to the reference particle was also observed in a numerical study of Kim et al. (1993), where the interaction between two stationary spheres placed side by side in a uniform flow was investigated for particle Reynolds numbers similar to the Reynolds number in case $F121$. The authors observed that the drag coefficient increases for distances below 4 particle diameters and tends to the drag coefficient of a single particle for distances larger than $21d_p$. They also observed that the maximum of the drag coefficient is located in immediate distance to the reference particle at approximately $1.8d_p$. On the other hand, in the vertical direction, as expected, the force immediately behind the test particle experience its minimum value after which it increases slowly and reaches unity at a vertical distance of approximately $\tilde{z} = 75d_p$. It is interesting to note that in both cases the rate at which the force increases towards unity is approximately the same, which indicates that in both cases the wake region behind the particle has on average similar properties, despite the differences in the wakes of the single particle from case $S121$ and case $S178$ (the average wake of the particles for case $F121$ and case $F178$ will be discussed in section 4.3.4).

For the purpose of relating the hydrodynamic force of the reference particle with the force exerted on the neighboring particles, the residual force $\tilde{f}_{ij}$ between the test particle i and any neighboring particle j was evaluated. The resultant force is evaluated as the difference of the

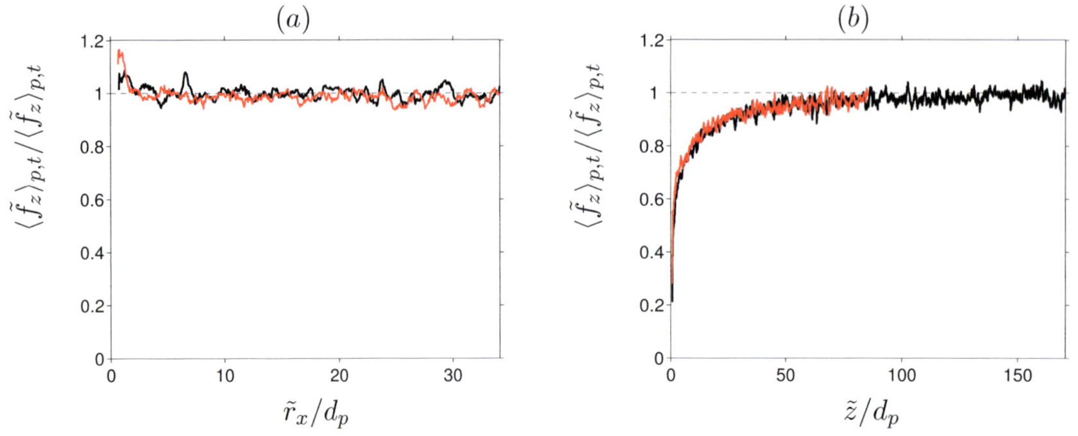

Figure 4.11: Variation of the particle conditioned average value of $\tilde{f}_z$ from figure 4.10 along the two axis passing through the center of the reference particle, i.e. $(\tilde{x}, \tilde{z}) = (0, 0)$. (a) in the horizontal direction $(\tilde{x}, \tilde{z}) = (\tilde{r}_x, 0)$. (b) in the vertical direction $(\tilde{x}, \tilde{z}) = (0, \tilde{z})$. $\tilde{r}_x$ denotes the radial distance from the particle surface in the horizontal direction. Case $F121$ (——). Case $F178$ (——).

hydrodynamic force of the two particles projected on the line connecting the particles:

$$\tilde{f}_{ij} = (\tilde{\boldsymbol{f}}_i - \tilde{\boldsymbol{f}}_j) \cdot \tilde{\boldsymbol{e}}_{ij}, \tag{4.8}$$

where $\tilde{\boldsymbol{e}}_{ij}$ is the unit normal vector of the particle pointing from particle i in direction of particle j. With this definition, negative values imply that the resultant force points in the direction of the reference particle and that, if released to move freely, the particle pair would initially accelerate towards each other. On the other hand, positive values would result in the opposite behavior. Here we would like to stress that this behavior can be expected only in the immediate time after the particles are released , i.e. $t \to 0$.

Figure 4.12 depicts the particle-conditioned average force $\langle \tilde{f}_{ij} \rangle_{p,t}$ in case $F121$. The averaging is performed in the same way as described above. Please note that in contrast to $\langle \tilde{f}_z \rangle_{p,t}$, the average of the quantity $\tilde{f}_{ij}$ is symmetric with respect to the horizontal axis $\tilde{x}$. The average force $\langle \tilde{f}_{ij} \rangle_{p,t}$ has been normalized by the absolute value of the hydrodynamic force obtained in case $S121$ and case $S178$. It can be observed that the force $\langle \tilde{f}_{ij} \rangle_{p,t}$ downstream of the reference particle exhibits negative values, whereas on the horizontal axis crossing the reference particle positive values are observed. The affected region behind the test particle extends approximately up to $40d_p$. From the close up view in figure 4.12b it is evident that the area with positive values of the force $\langle \tilde{f}_{ij} \rangle_{p,t}$ is localized horizontally to the reference particle and due to the symmetry of the flow in the horizontal direction is symmetric with respect to the vertical axis $\tilde{z}$. This region is as well substantially smaller than the region with negative values behind the particle, extending up to roughly $1.7d_p$ on each side of the test particle, after which it assumes values close to $|\langle f \rangle_{p,t}|$.

Figure 4.13 shows the corresponding plot for case $F178$. Overall, similar observations can be made as in case $F121$. However, the region with negative values of $\langle \tilde{f}_{ij} \rangle_{p,t}$ appear to be slightly shorter than in case $F121$. This can be best observed in figure 4.14b where the profile of the force in vertical direction is depicted. The graph reveals that the force in case $F121$ increases faster to zero than in case $F178$. In the horizontal direction, the force steeply increases in the vicinity of the test particle reaching positive value of $0.21|f_{ref}|$ $(0.23|f_{ref}|)$ in case $F121$ $(F178)$ at $\tilde{x} = 1.3d_p$ after which it quickly declines towards zero. In both cases the

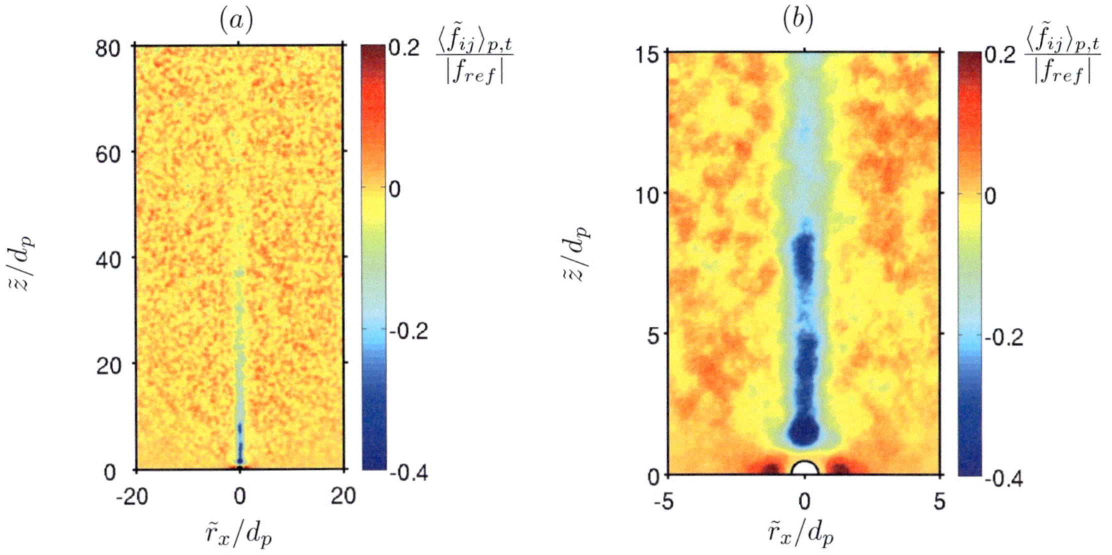

Figure 4.12: Spatial map of the particle-conditioned average value of the resultant hydrodynamic force $\tilde{f}_{ij}$ as function of the position $(\tilde{x}, \tilde{z})$ relative to the reference particle (vertical plane passing through the center of the reference particle) for case $F121$. The force $\tilde{f}_{ij}$ is defined in equation 4.8. The graph in subfigure (b) shows close up view to the reference particle.

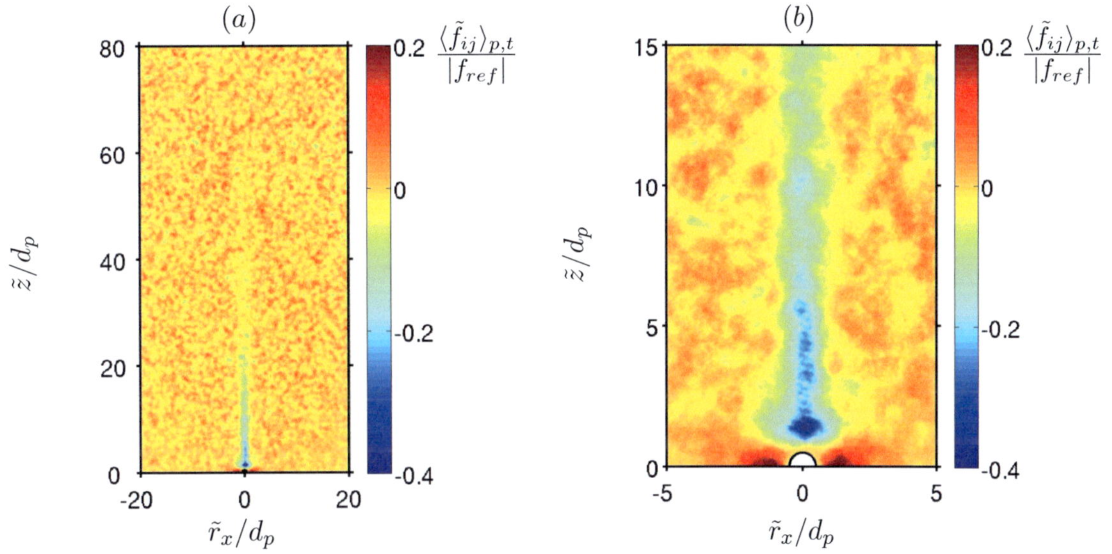

Figure 4.13: Same as in figure 4.12 but for case $F178$.

region with positive values of the force $\langle f_{ij} \rangle_{p,t}$ extends up to 4 particle diameters. Very similar results were obtained in Kim et al. (1993). The authors observed that the flow between the particle pair accelerates differently than the flow on the opposite side of the particles leading to a negative pressure gradient which would contribute to a lift force between the two spheres. The lift force would tend to repel the particles from each other. The authors reported the length of the "repulsion" region to be approximately 3.4 particle diameter.

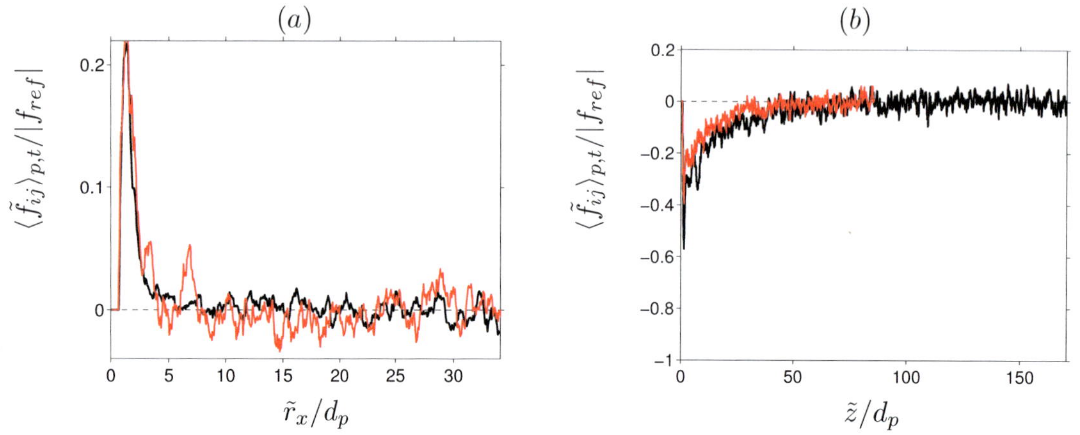

Figure 4.14: Variation of the particle conditioned average values of $\langle \tilde{f}_{ij} \rangle_{p,t}$ from figures 4.12 - 4.13 along the two axis passing through the center of the reference particle, i.e. $(\tilde{x}, \tilde{z}) = (0, 0)$. (a) in the horizontal direction $(\tilde{x}, \tilde{z}) = (\tilde{r}_x, 0)$. (b) in the vertical direction $(\tilde{x}, \tilde{z}) = (0, \tilde{z})$. $\tilde{r}_x$ denotes the distance from the particle surface. Case $F121$ (——). Case $F178$ (——).

4.3.4 Particle wakes

In the following we will present an analysis of the local fluid velocity field in the vicinity of the particles. To this end, the flow field in vicinity of the particles is conditionally averaged with respect to a reference particle. In particular, we employ the above described procedure used to perform the averaging of the hydrodynamic force. Here, instead of using a solid phase indicator function, a indicator function for the continuous phase is applied. This ensures that the averaging is performed over the actual volume occupied instantaneously by the fluid. The hydrodynamic force is also replaced by the Eulerian velocity field relative to the velocity of the test particle. Please note that for the present simulations $F121$ and $F178$ the particle velocity is identical zero. The averaging can be then defined as:

$$\langle \tilde{\boldsymbol{u}}_{f,rel}(\tilde{\boldsymbol{x}}, t) \rangle_{p,t} = \langle \boldsymbol{u}_f(\tilde{\boldsymbol{x}}, t) - \boldsymbol{u}_p^i(t) \rangle_{p,t}, \qquad i = 1 \dots N_p. \tag{4.9}$$

Please recall that the position $\tilde{\boldsymbol{x}}$ is defined as $\tilde{\boldsymbol{x}} = \boldsymbol{x} - \boldsymbol{x}_p^i$ (section §3.4.2, p. 47), thus the coordinates of the position of the center of the i-th particle are given by $\tilde{\boldsymbol{x}} = 0$. The velocity $\tilde{\boldsymbol{u}}_{f,rel}(\tilde{\boldsymbol{x}}, t)$ is related to the relative velocity from equation 3.3 by simply replacing the characteristic fluid velocity $\boldsymbol{u}_f$ with the velocity $\boldsymbol{u}_f(\tilde{\boldsymbol{x}}, t)$. The reader is referred to appendix A for detailed definition of the above defined averaging operator.

Figure 4.15 shows contours of the vertical velocity component of the average relative velocity, $\tilde{w}_{f,rel}/U_\infty$, on a plane through the center of the test particle. The plane is aligned with the vertical coordinate $\tilde{z}$. In analogy with section 3.4.2 the region enclosed by the red contour line corresponds to the recirculation region of the wake, i.e. $\tilde{w}_{f,rel} \leq 0$. Due to the axi-symmetry of the mean wake, the recirculation length of the wake can be defined as the maximum extension of the recirculation region on the vertical axis $\tilde{z}$, i.e. the point at which the velocity $\tilde{w}_{f,rel}/U_\infty$ changes sign. This can be easily examined in the profile of the average relative velocity $\tilde{w}_{f,rel}/U_\infty$ on the vertical axis through the center of the particle, which is shown in figure 4.16. The recirculation length in case $F121$ measures $l_e = 0.955 d_p$ and in case $F178$ $l_e = 1.05 d_p$. When compared with the values obtained for the single particles in section 3.4.2, the recirculation length for the assembly of fixed particles is significantly smaller.

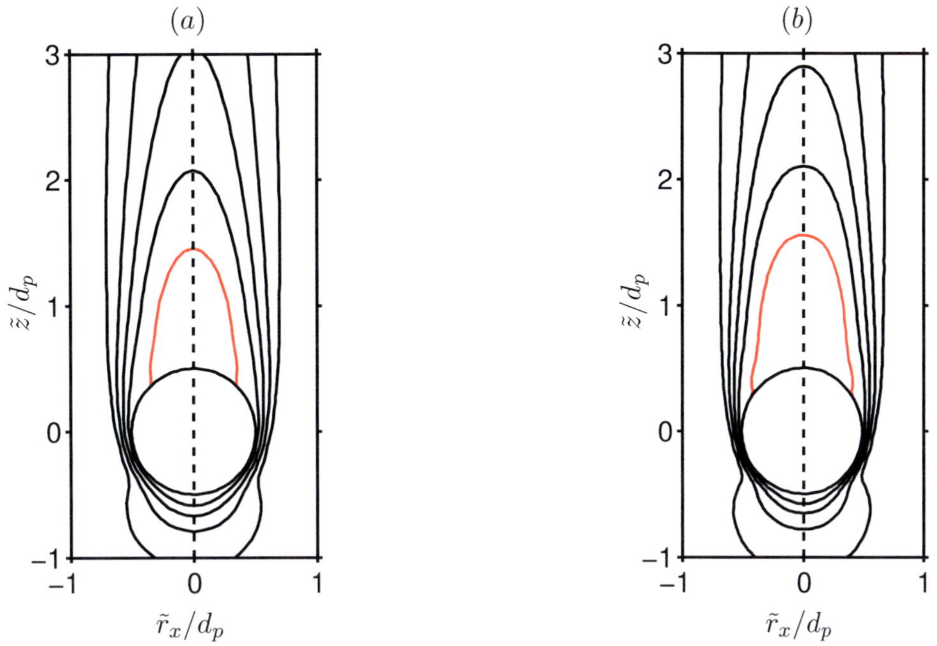

Figure 4.15: Contour plot of the projected relative velocity $\tilde{w}_{p,rel}/U_\infty$ in the plane parallel to the vertical axis passing trough the particle center $\boldsymbol{x}_p$. (a) case $F121$. (b) case $F178$. Contours are shown for values (0:0.2:0.8). The red line (——) marks the extent of the recirculation region, i.e. $\tilde{w}_{p,rel}/U_\infty = 0$. The position of the z-axis is denoted by the black dashed line (- - - -). $\tilde{r}_x$ denotes the radial distance from the reference particle in the averaging domain.

In case $F121$ l_e is 20% smaller than in case $S121$, whereas in case $F178$ the recirculation length of the mean wake is smaller by as much as 33%.

The profile of the velocity $\tilde{w}_{f,rel}/U_\infty$ on the vertical axis $\tilde{z}$ experience similar evolution as in the single particle simulations. The uniform flow upstream of the particle decays quickly for distances between $\tilde{z}/d_p = [-4, -0.5]$ and becomes negative immediately behind the particle after which it increases slowly towards its asymptotic value. The recovery of the velocity downstream of the particle is in both cases consistent with the recovery in the single particle simulations. The velocity $\tilde{w}_{f,rel}$ in case $F178$ increases faster than in case $F121$ and reaches 97% of its asymptotic value after approximately 33 particle diameter downstream of the particle surface, while in case $F121$ this distance was calculated to $46d_p$. This indicates that the mutual interaction of the particle wakes in case $F178$ accomplish faster decorrelation of the fluid velocity than in case $F121$.

It is interesting to note that the relative velocity $\tilde{w}_{f,rel}$ for the fixed assembly of particles is slightly smaller than the mean velocity driving the flow U_∞. In case $F121$ the velocity reaches up to 93% of the velocity U_∞, while in case $F178$ only 90% are obtained. The difference between the upstream velocity at $\tilde{z}/d_p = -20$ and the velocity at $\tilde{z}/d_p = 100$ downstream of the test particle was measured to 1.5% in case $F121$ and to 0.64% in case $F178$, confirming the observation made above that the velocity in case $F178$ recovers over shorter distance than in case $F121$. This also evidences the long-range effect of the particles on the flow field and is in line with the observed pattern of the hydrodynamic force behind the test particle.

The reduced recirculation length as well the faster recovery of the asymptotic velocity in the wake of the particle assembly than in the wake of a single particle confirm the observations

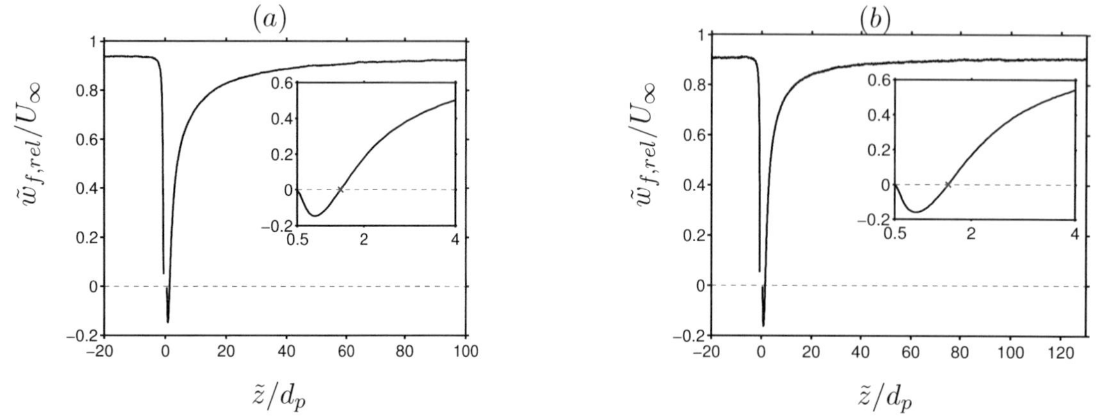

Figure 4.16: Vertical component of the relative velocity on the center line passing trough the center of the reference particle $\boldsymbol{x}_p$ and parallel to the vertical axis. (a) Case $F121$. (b) Case $F178$. The coordinate system is particle centered, i.e. origin at $\boldsymbol{x}_p$. The inset-figures show close-up view of the recirculation region. The blue cross $\times$ marks the position at which $\tilde{w}_{f,rel}/U_\infty$ changes sign. Negative values of the x-axis correspond to upstream positions in front of the particle and positive values represent the domain downstream of the test particle. The distance from the particle was normalized by the particle diameter d_p. The velocity is made non-dimensional with the mean velocity driving the flow U_∞.

made by Risso and Ellingsen (2002) and Risso et al. (2008) that the wake of the particles in the assembly is attenuated, when compared to the wake of a single particle. The wake attenuation can be best evaluated by considering the velocity defect $\tilde{u}_{d0}$ behind the particles. The velocity $\tilde{u}_{d0}$ is defined as in equation 3.11 using the coordinate system $(\tilde{x}, \tilde{y}, \tilde{z})$ instead of the system $(\hat{x}, \hat{y}, \hat{z})$. Figure 4.17 shows the evolution of $\tilde{u}_{d0}$ as function of the vertical distance $\tilde{r}_z$ downstream from the particle surface. The velocity deficit was evaluated in both cases for distances up to 100 particle diameters downstream from the particle surface. As in case $S121$ and case $S178$ the velocity deficit is unity at the particle surface after which it increases and reaches a global maximum at $0.9d_p$ in both cases $F121$ and $F178$. While in case $F121$ this value correspond to the value observed for the single particle in case $S121$, the value in case $F178$ is smaller than in case $S178$ by as much as 17%. After reaching its maximum value the velocity $\tilde{w}_{f,lag}$ decays experiencing regions with different power laws.

When comparing the velocity deficit for the network of particles with that obtained for the single particle simulation from chapter 3, figure 4.17 reveals that the velocity $\tilde{w}_{f,lag}$ experience slightly different decay rates. Similar to the single particle simulations, immediately after reaching its maximum value, the velocity $\tilde{w}_{f,rel}$ decays with $-2/3$ power law, resembling the characteristics of a turbulent wake. This and the location of the maximum of $\tilde{w}_{f,rel}$ indicates that the flow close to the particle ($\tilde{z} \leq 2.6d_p$) is similar to that induced by a singe particle. The region with $-2/3$ power law is followed by a region, where the velocity deficit decays as a laminar wake with -1 power law. However, for case $F121$ and case $F178$ the change in slope from $\tilde{r}_z^{-2/3}$ to $\tilde{r}_z^{-1}$ takes place much closer to the particle as in the case for the single particle. In case $F121$ the change in slope was observed at $2.63d_p$, while in case $F178$ at $2.1d_p$. This values are an order of magnitude smaller than for the single particle in case $S121$ and case $S178$. It is worth noticing that the -1 power law in case $F178$ is somehow more prominent than in case $F121$, which is in contrast to the results for the single particle. In the region of the final decay, the velocity deficit for the particle network does not experiences similar cut-off as in the case of single particle. Overall, the velocity deficit in the wake of the particle assembly

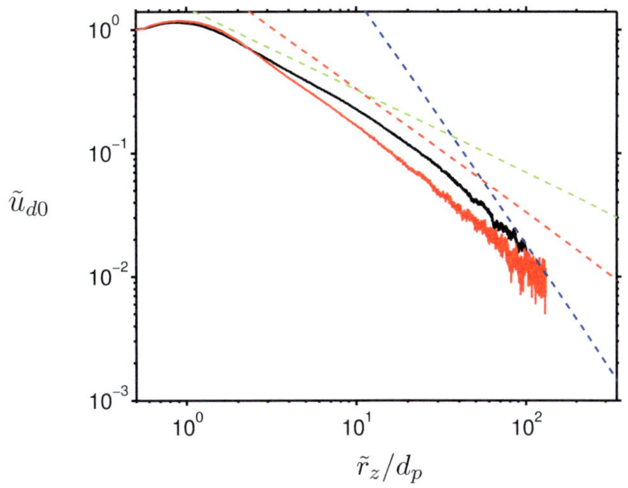

Figure 4.17: Streamwise velocity deficit $\tilde{u}_{d0}$ defined as in equation 3.11 along the vertical axis $\tilde{z}$ as function of the downstream distance $\tilde{r}_z$ from the reference particle surface. Case $F121$ (——), case $F178$ (——). The downstream distance $\tilde{r}_z$ from the particle surface is defined as $\tilde{r}_z = \tilde{z} - z_p$. The dashed straight lines indicate decay rates proportional to $\tilde{r}_z^{-1}$ (- - - -), $\tilde{r}_z^{-2}$ (- - - -) and $\tilde{r}_z^{-2/3}$ (- - - -).

support the above results that the mean wake is strongly affected by the collective presence of the particles with the consequence of being significantly attenuated with respect to the wake of a single particle.

In large number of studies the velocity deficit behind the particles in *a priori* turbulent flows is observed to decay approximately as $\tilde{r}_z^{-2}$ (Wu and Faeth, 1994; Legendre et al., 2006; Amoura et al., 2010). The smaller decay rate observed in our simulations indicates that the uniform flow past a random assembly of particles has a different nature. As result, for uniform flows the presence of the particles affects the flow over longer distances downstream of the particles.

Now let us turn to the question of self-similarity of the mean particle wake. To this end, we consider the cross-stream profiles of the velocity deficit $\tilde{u}_d$ in the particle wake, which is depicted in figure 4.18 as function of the radial distance $\tilde{x}$ from the wake centerline. The velocity deficit $\tilde{u}_d$ is calculated as in equations 3.11, 3.13. As for the single particle the wake is considered to be self-similar, when by appropriate scaling the cross-profiles of the velocity deficit $\tilde{u}_d$ follow a Gaussian distribution. As can be observed in figure 4.18 the mean particle wake in both cases $F121$ and $F178$ do indeed follow a Gaussian function. For the normalization of the velocity deficit in the cross-profile the value of $\tilde{u}_d$ at $\tilde{x} = 0$ was used and for the normalization of the radial distance the half-width h_{wz} of the particle wake was used. Thereby the half-width of the particle wake is defined as in equation 3.12.

For laminar wakes it is known that the spreading of the wake obey $h_{wz} \sim \tilde{x}^{1/2}$, whereas for turbulent wakes the half-width scales as $h_{wz} \sim \tilde{x}^{1/3}$. The streamwise evolution of the wake half-width is shown in figure 4.19. It can be observed, that in the present simulations the half-width of the wake appear to follow over large range a linear expansion, i.e. $h_{wz} \sim \tilde{x}$:

$$h_{wz}/d_p = \tilde{a}\tilde{z}/d_p + \tilde{b}. \tag{4.10}$$

The coefficients in $\tilde{a}$ and $\tilde{b}$ were evaluated in case $F121$ to $\tilde{a} = 0.011$, $\tilde{b} = 0.42$ and in case $F178$ to $\tilde{a} = 0.017$, $\tilde{b} = 0.405$. A linear spreading of the particle wakes was also observed in

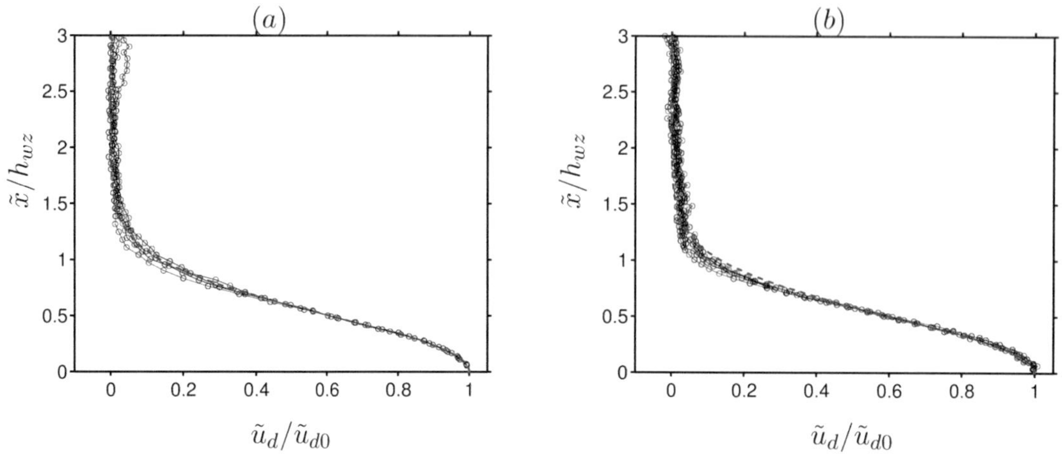

Figure 4.18: Profiles of the velocity deficit in the average particle wake as function of the cross-stream coordinate $\tilde{x}$. The profile was taken at four different downstream locations, $\tilde{z}/d_p = [6, 10, 13, 17]$. The blue dashed (----) line corresponds to a Gaussian function. The velocities $\tilde{u}_d$ and $\tilde{u}_{d0}$ are defined as in equations 3.11, 3.13. (a) Case $F121$. (b) Case $F178$. The coordinate system $(\tilde{x}, \tilde{y}, \tilde{z})$ is chosen as in figure 3.5. The distance from the particle center was normalized by the wake half-width h_{wz}.

the studies of Bagchi and Balachandar (2004); Legendre et al. (2006); García-Villalba et al. (2012). However, this studies considered turbulent flows. The evolution of the wake half-width also illustrates that the wake in case $F178$ experience larger values than in case $F121$.

In the context of particle induced turbulence, the question of turbulence generation by the particles can be indirectly addressed by focusing on the turbulence statistics in the particle wake. For this reason we study the evolution of the total turbulent kinetic energy $\tilde{k}_f$ in the particle wake. The energy $\tilde{k}_f$ is defined as follow:

$$\tilde{k}_f = \frac{1}{2}\langle \tilde{u}'_{fi,rel}\tilde{u}'_{fi,rel}\rangle_{p,t}, \tag{4.11}$$

where the velocity fluctuations $\tilde{u}'_{fi,rel}$ are defined with respect to the average $\langle \tilde{\boldsymbol{u}}_{f,rel}\rangle_{p,t}$ as:

$$\tilde{\boldsymbol{u}}'_{f,rel}(\tilde{\boldsymbol{x}}, t) = \tilde{\boldsymbol{u}}_{f,rel}(\tilde{\boldsymbol{x}}, t) - \langle \tilde{\boldsymbol{u}}_{f,rel}(\tilde{\boldsymbol{x}}, t)\rangle_{p,t}. \tag{4.12}$$

Following García-Villalba et al. (2012) the covariance of the fluctuations $\langle \tilde{u}'_{fi,rel}\tilde{u}'_{fi,rel}\rangle_{p,t}$ can be decomposed in the following three contributions:

$$\langle \tilde{u}'_{fi,rel}\tilde{u}'_{fi,rel}\rangle_{p,t}(\tilde{\boldsymbol{x}}, t) = \langle u'_{f_i}(\tilde{\boldsymbol{x}}, t)u'_{f_i}(\tilde{\boldsymbol{x}}, t)\rangle_{p,t} + \langle u_p^{(i)\prime}(t)u_p^{(i)\prime}(t)\rangle_{p,t} - 2\langle u_{f_i}(\tilde{\boldsymbol{x}}, t)u_p^{(i)\prime}(t)\rangle p, t. \tag{4.13}$$

In the above equation the first term on the right hand side represents the covariance of the fluctuating fluid velocity field, conditioned on the particle position. The second term express the contribution from the particle velocity fluctuations and the third term stands for the contribution of the mixed covariance between the particle velocity fluctuations and the particle-conditioned fluid velocity fluctuations. Please note that this equation holds for flow configurations with fixed and with mobile particles. For the present two cases of fixed particles the last two contributions are identically zero and the total turbulent kinetic energy can be calculated as:

$$\tilde{k}_f = \frac{1}{2}\langle u'_{f_i}(\tilde{\boldsymbol{x}}, t)u'_{f_i}(\tilde{\boldsymbol{x}}, t)\rangle_{p,t}. \tag{4.14}$$

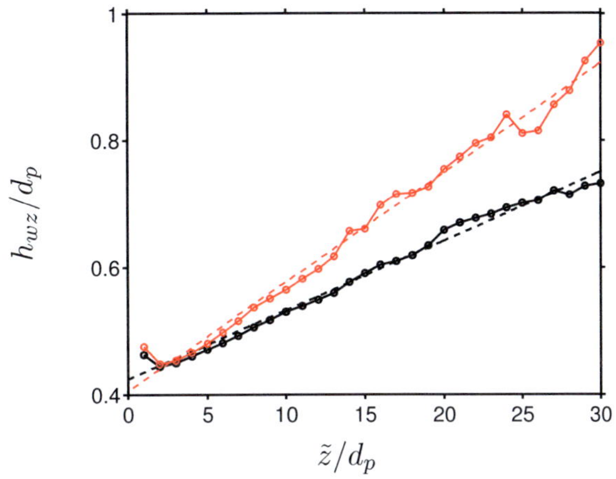

Figure 4.19: Cross-stream half-width of the average particle wake as function of the downstream distance $\tilde{z}$. Case $F121$ (—•—, - - - -). Case $F178$ (—•—, - - - -). Dashed lines represent a linear fit of the spreading of the wake, $h_{wz}/r_p = \tilde{a}\tilde{z}/d_p + \tilde{b}$, with $\tilde{a} = 0.011$, $\tilde{b} = 0.42$ for case $F121$ and $\tilde{a} = 0.017$, $\tilde{b} = 0.405$ for case $F178$.

The variation of the kinetic energy $\tilde{k}_f$ along the vertical axis $\tilde{z}$ in the particle wake is shown in figure 4.20. The turbulent kinetic energy has been normalized by a reference energy defined as:

$$\tilde{k}_{ref} = \frac{1}{2}\tilde{u}^{(i)\prime}_{ref}\tilde{u}^{(i)\prime}_{ref}, \tag{4.15}$$

where the velocity $\tilde{u}^{(i)\prime}_{ref}$ is defined as:

$$\tilde{u}^{(i)\prime}_{ref} = (\langle u'_f u'_f \rangle_{\Omega_f,t} + \langle u_p^{(i)\prime}(t) u_p^{(i)\prime}(t) \rangle_{p,t})^{1/2}. \tag{4.16}$$

Since the particles in the present two cases $F121$ and $F178$ are fixed in space, the second term in the equation 4.16 becomes identically zero and the reference velocity becomes identical to the unconditioned box-averaged covariance $\langle u'_f u'_f \rangle_{\Omega_f,t}$.

In both cases the energy $\tilde{k}_f$ increases in the near vicinity of the particle, reaching a local maximum at $1.1d_p$ in case $F121$ and at $1.3d_p$ in case $F178$. Then the turbulent kinetic energy decreases reaching a local minimum after which increases slowly towards its asymptotic value. In case $F121$ the recovery of the turbulent kinetic energy is approximately 80% at 100 particle diameter downstream of the reference particle. On the other hand, in case $F178$ the recovery of $\tilde{k}_f$ is more complete reaching 90% of the reference energy at the same distance. This again strengthen the observation that the velocity field in the mean particle wake in case $F178$ recovers over shorter distances than in case $F121$. Similar evolution of the turbulent kinetic energy in the particle wake was observed in Zeng et al. (2010) and more recently in García-Villalba et al. (2012). In the first work the authors numerically studied the effect of a stationary finite size particle in a channel flow, while in the second the channel flow of finite-sized mobile particles was investigated. Nevertheless, for the regions of the flow where the flow is dominated by the particle wakes, the observations appear to be in good agreement with the here observed evolution of the turbulent kinetic energy. The only visible difference appear to be in the vicinity of the particle, where in the above two studies the local maximum, respectively minimum is not as pronounced as in the present case. A reason for this could be the turbulent background flow, which would tend to diminish the effect of the particle wake in vicinity of the particle surface.

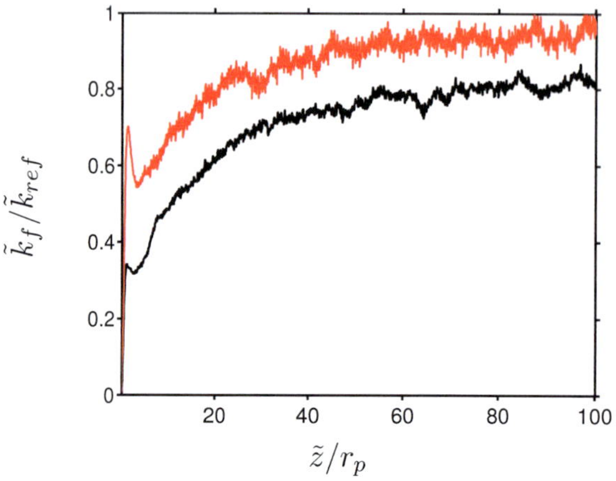

Figure 4.20: Evolution of the energy $\tilde{k}_f$ as function of the downstream distance $\tilde{z}$ form the reference particle for case $F121$ (——) and case $F178$ (——). The energy $\tilde{k}_f$ is calculated according to equation 4.14 and normalized by the reference kinetic energy $\tilde{k}_{ref}$ defined in equation 4.15.

4.4 Conclusion

The particle-induced flow field was elaborated in detail. It was observed that in both cases the fluctuations of the fluid velocity were highly anisotropic with the vertical velocity component being the dominant one. The anisotropy of the flow field was attributed to the dominance of the particle wakes in the flow field. As a result, the probability distribution function of the flow velocity was strongly affected by the presence of the particle wakes, especially the p.d.f. of the vertical velocity component. Nevertheless, the horizontal component of the fluid velocity was found to experience higher intermittency than the vertical component. In order to better characterize the velocity fluctuations, the fluid velocity fluctuations were decomposed in their spatial and temporal contributions. The results led to the conclusion that the flow field is unsteady and that large-scale wake interactions are responsible for the fluctuations of the flow field. The spatial fluctuations were found to be more prominent than the temporal fluctuations accounting for the major contribution to the total fluctuations. Furthermore, a spectral analysis of the flow field was performed. The spectral density function in both cases was found to experience a κ_α^{-3} sub-range for both the horizontal and vertical velocity components. As a consequence, it was concluded that the mobility of the particles is not required for the energy spectrum to exhibit -3 decay power law. The above results were found to be in good agreement with the available literature. Additional to the energy spectrum, the corresponding two-point correlation function for the flow field was evaluated, revealing that the flow field in the vertical direction was not fully decorrelated over the half height of the computational box.

The hydrodynamic force exerted by the flow upon the particles was analyzed. The mean particle drag coefficient in both cases proved to be close to the drag coefficient of a single particle. This circumstance implies that the local flow field acting on the individual particles resembles very closely the flow field around a single settling particle. The hydrodynamic force was also conditionally averaged on the presence of the particles. This led to a map of the force experienced by the neighboring particles in vicinity of a test particle. The map revealed a substantial decrease of the drag immediately downstream of the test particle and that the presence of the particle was experienced over long distances. Analysis of the resultant force of a particle pair consisting of the reference particle and any other neighboring particle at certain distance showed that the test particle and the particles located downstream of it would be more prone to attract each other (as result of the reduced drag behind the particles), whereas particles located sideways are more prone to repel each other. This features of the particle conditioned hydrodynamic force support the results for the wake induced aggregation of particle pairs (drafting-kissing-tumbling (Hu et al., 1993)).

Finally, we have performed detailed analysis of the particle-conditioned averaged flow field in the vicinity of the particles, i.e. the average particle wake. It was found that the average length of the recirculation length was significantly attenuated by the presence of multiple particles (up to 33% reduction in case $F178$), supporting the results from the literature (Roig and De Tournemine, 2007; Risso et al., 2008). The streamwise velocity deficit on the axis through the center of the particles was found to experience decay regions with two different power laws. In the near-wake the velocity deficit was observed to decay as $\tilde{z}^{-2/3}$ and as $\tilde{z}^{-1}$ for distances larger than five particle diameters. This led to the conclusion that close to the particle the flow is similar to that induced by a single particle. The vertical velocity component in the particles' wakes was found to be reasonably correlated with a Gaussian function, evidencing a self-similarity of the average particle wake. This was observed in both cases, thus being independent of the Galileo number. The half-width of the mean particle wake in both cases evolved linearly with the downstream distance from the reference particle. Thereby, the spreading of the wake in case $F178$ was observed to be higher than in case $F121$. The energy of the velocity fluctuations with respect to the particle-conditioned average, revealed that the kinetic energy in case $F178$ recovered faster than that of case $F121$. The results for the turbulent kinetic energy for the present simulations reasonably correlates with the results from Zeng et al. (2010); García-Villalba et al. (2012). This again confirms that in case $F178$ the flow field in the wake of the particles recovers faster than in case $F121$.

The overall conclusion is that the velocity field in the mean wake for both cases was affected over long distances by the presence of the particles. Thereby the velocity recovery in the particle wake showed different properties in both flow cases, with the velocity in case $F178$ recovering faster (over shorter distance) than in case $F121$.

Part II

Settling of many finite-size particles in an initially quiescent flow field

Chapter 5

Introduction and computational set-up

In this part of the thesis the dynamics of freely moving heavy finite-size particles settling under the influence of gravity in an initially ambient flow field is investigated. This study extends the investigation from chapter §4 by allowing the particles to move freely within the computational domain. This will allow the investigation of the effects of mobility on both the dispersed and the carrier phase. As introduced in chapter §1 the present flow configurations involve large number of complex coupling mechanisms between the dispersed and the carrier phase, which are most prominently manifested by quantities such as inhomogeneous particle distribution, turbulence generation or modulation by the particles and the mean particle settling rate.

Particle clustering

A remarkable feature of such flow configurations is the ability of the particles to agglomerate into clusters. In the past a large number of literature was accumulated on the topic of particle clustering. However, the most of the studies considered flows which are *a priori* turbulent (cf. chapter §1). In case of pseudo-turbulent flows, i.e. flow generated solely by the presence of the particles, surprisingly little is known about the mechanisms responsible for the clustering of the particles. To our knowledge Kajishima and Takiguchi (2002) were the first to observe clustering of heavy sedimenting particles in pseudo-turbulence. The authors performed direct numerical simulation of heavy particles settling under the influence of gravity in an initially ambient flow with particle Reynolds number of $\mathcal{O}(100)$. They observed that the particles agglomerated in highly anisotropic clusters, forming columnar like structures and concluded that clusters are formed due to particle wakes. Similarly structured particle clusters were found in the experimental work of Nishino and Matsushita (2004). The flow in the work of Nishino and Matsushita (2004) was considered to be turbulent with turbulent intensities of the order of 3%. However, in both works the relative turbulence intensities in the flows were comparable. The relative turbulence intensity is a measure of the influence of the fluid (pseudo) turbulence on the particle motion and is defined as the ratio between the intensity of the surrounding fluid and the apparent slip velocity. The clustering of light particles (e.g. bubbles) in pseudo-turbulence has received comparably more attention in the past. Although bubble flows poses some features which set them apart from dispersed flows with heavy particles, it is nevertheless instructive to consider the present results in view of the literature for bubbles, which appear to be more profound than the one for heavy particles. Bubble clustering in pseudo-turbulence has been studied numerically for example in Sangani and Didwania (1993); Mazzitelli and Lohse (2009) and experimentally in Cartellier and Riviére (2001); Zenit et al. (2001); Risso and Ellingsen (2002); Esmaeeli and Tryggvason (2005); Roig and De Tournemine (2007); Martínez-Mercado et al. (2010). This studies were however not fully conclusive with some of the authors reporting clusters being oriented in the horizontal direction and some in the vertical. Risso and Ellingsen (2002) reported, on the other hand, no clustering and suggested that for the low volume fraction considered in their study ($\Phi_s = 0.5\% - 1.05\%$), the hydrodynamic interactions between the bubbles only weekly influenced the bubble distribution.

Mean particle settling rate

It is well known that the presence of particles in the fluid will disturb the flow and as we showed in the previous chapter instabilities in the flow are generated by the presence of the particles, which lead to hydrodynamic interactions between the particles. For mobile particles these interactions could significantly affect the hydrodynamic force exerted by the fluid on the particles as well as the mean settling rate of the particles. The mean settling rate of the particles as well as the hydrodynamic force were identified in section §1.2.1 as one of the most important quantities responsible for the dynamics of the particles. As noted in §1.2.1 the common observation is that the presence of multiple particles could lead to net increase of the drag coefficient and to a decrease of the average particle settling rate. In this context, the fluctuations of the flow filed induced by the particles are known to be responsible for the so called non-linear drag effect. On the other hand, when the particles are inhomogeneously distributed, i.e. they are organized in clusters, it is known that the particle clusters could have the opposite effect on the settling velocity of the particles. Kajishima and Takiguchi (2002); Kajishima (2004a) observed that the velocity of the particles inside clusters are on average higher than outside and that particle clusters behave as objects settling with higher velocity. The review of the literature in section §1.2.1 revealed that the relative importance of the different mechanisms, which affect the average particle settling velocity, is incompletely defined. Thus, one of the main objectives of this work is to contribute to the understanding of these effects.

One of the open questions with respect to the particle settling rate is its dependence with the solid volume fraction. To this date there is no reliable theoretical approach to estimate the effect of the solid volume fraction on the particle settling rate. Commonly, this effect is estimated based on empirical correlations such as Richardson and Zaki (1954). Thus, it will be highly beneficial to estimate the reliabilty of such correlations and if possible to contribute to their improvement.

Regarding the regimes of settling (cf. chapter §3), the real impact of the transition of the particle motion on many particle systems is expected to considerably modify the dispersion characteristics of the particles. The individual particles will tend to drift horizontally by themselves without interactions with the remaining particles. The dispersion of the particles in pseudo-turbulent flow field was studied experimentally and theoretically by Parthasarathy and Faeth (1990a). The authors considered dilute flow of heavy particles settling under the influence of gravity with particle Reynolds numbers of the order of $\mathcal{O}(100)$. They observed that the particles exhibited self-induced lateral motion due to eddy shedding. The fluctuations of the particle velocities were reported to be highly anisotropic with the vertical component being the dominant one.

Particle-induced flow field

As aforementioned, the fluctuations of the flow field play a crucial role in the dynamics of the particles. In pseudo-turbulence the fluctuations in the flow field are introduced solely by the relative motion of the particles to the fluid and, as we saw in chapter §4, are mainly controlled by the wakes of the particles. In such flow configurations the decay of the velocity disturbances generated by the particles play an important role. In the limit of vanishing solid volume fraction and very small Reynolds numbers ($Re_p \ll 1$), the Stokes solution predicts that the flow perturbations behind the particles decay as $\tilde{r}_z^{-1}$, where $\tilde{r}_z$ is the relative distance downstream from the particle surface. In an theoretical work Caflisch and Luke (1985) showed that due to the slow decay the superposition of the flow induced by multiple settling particles will lead to the divergence of the fluid velocity variance with the size of the container. At finite Reynolds numbers a wake develops behind the particles, wherein the velocity decays as $\tilde{r}_z^{-1}$ in the case of laminar wake and as $\tilde{r}_z^{-2/3}$ in case of turbulent wake, which was confirmed in the simulations reported in chapters §3 and §4. In an experimental study Parthasarathy and Faeth (1990a) showed that even in the case of finite Reynolds numbers and very low volume concentrations ($\Phi_s < 0.01\%$) the variance of the fluid velocity diverges as well. In their study they used the principle of superposition to predict the fluid velocity fluctuations. They overcome the divergence issue, by assuming that the wake vanishes for distances larger than 175 particle diameters. Moreover, the authors showed that the particle wake has to be taken into account for the estimation of the velocity fluctuations. The experimental work of Parthasarathy and Faeth (1990a,b) will be considered in the present study as an important reference case, whose parameters match closely with the flow parameters in the present work, except for the solid volume fraction, which was in their work considerably lower. In an experimental work Risso and Ellingsen (2002) investigated the velocity fluctuations in a homogeneous dispersion of bubbles. They identified two regions in the flow in which the properties of the fluid velocity fluctuations are different. In the vicinity of each bubble, the flow was found to be similar to that generated by a single bubble. Far from the bubbles the fluid velocity fluctuations were found to be controlled by a non-linear interactions between the wakes of all bubbles and showed to be self-similar when scaled with $\Phi_s^{0.4}$. The self-similarity of the fluid velocity fluctuations was confirmed in following studies of Roig and De Tournemine (2007); Risso et al. (2008); Riboux et al. (2010, 2013). However, the review of the literature was not conclusive to what extent the self-similarity of the fluid velocity fluctuations applies for dispersed flows with heavy particles.

Regarding the energy spectrum of the flow induced by a large number of freely moving particles, the question arise whether the wakes of the particles will have the same contribution to the fluctuations of the flow field as when the particles are fixed. Particularly, it is interesting to observe whether the mobility and the spatial distribution of the particles will affect the shape of the power spectral density function of the vertical fluid velocity component.

To summarize, the main objective of the present part of the thesis is to qualitatively assess the effects of particle mobility on both the carrier and the dispersed phase as well the effects of the solid volume fraction by addressing the following open questions with respect to the principal two-way coupling phenomena:

- Particle clustering:
    - Do particles agglomerate in clusters in these flows?
    - What is the exact mechanism of cluster formation?

  - What is the structure of particle clusters in pseudo-turbulence?
  - What is the temporal dynamics of the clusters?
- The effect of the particle-induced flow field on the particle motion:
  - How does the pseudo-turbulence affect the tendency to form wake-induced clusters?
  - How does the pseudo-turbulence affect Lagrangian dynamics of the particles?
  - What are the consequences of particle cluster on the settling rate and the hydrodynamic forces of the particles?
- Characteristics of the particle-induced flow field:
  - What are the characteristics of the fluid velocity fluctuations? Are they self-similar?
  - What is the probability distribution function of the fluid velocity?
  - What is the shape of the energy spectrum?
  - What are the characteristics of the mean particle wake?

This questions are approached in this and the following chapters as follows. In the present chapter the relevant parameters, the computational set-up and the simulation start-up procedure will be presented. Additionally we will give an overview of the general evolution of the dispersed flow and we will provide a discussion of the initial transient of the flow from the moment when the particles are released until the moment at which the flow is considered to be at statistically steady state. The analysis in the following chapters §6 - §8 will be based on the data accumulated during the time at which the flow is considered statistically steady. In chapter §6 results for the spatial distribution of the dispersed phase will be presented. Here the focus will be on whether the particles agglomerate in clusters or not and on the physical mechanisms responsible for the clustering of the particles. Thereby, in case of clustering, the global and local characteristics of the particle clusters will be provided. Moreover, the temporal dynamics of the clusters will be discussed in detail. In chapter §7, a comprehensive description of the particle motion will be provided. The main objective in this chapter is to provide a deep insight into the effect of the particle-generated flow field on the particle motion, especially the effects of possible particle clustering on the particle motion will be elucidated in detail. To this end, a detailed discussion of various particle statistical quantities, such as statistics of the mean and the fluctuating particle velocities and hydrodynamic forces will be performed. From particular interest is the average particle settling velocity relative to the motion of the fluid and the drag force exhibited by the particles. Furthermore, the contribution of the local flow field on the particle velocity and forces is identified and analyzed. Finally, in section §8 the effect of the particles and the particle clusters on the flow field is discussed. Here the statistical description of pseudo-turbulence is elaborated in detail. The focus in this chapter will be on the fluctuations of the fluid velocity and the shape of the energy spectrum. A detailed description of the average particle wake will be presented as well.

5.1 Flow configuration, initial- and boundary conditions

The simulations introduced in this chapter consider the motion of mobile heavy finite-size particles settling under the influence of gravity in an initially ambient fluid. For this purpose we performed in total three simulations. As introduced in section §2.1.4, dimensional analysis shows that under such conditions the system is characterized by three non-dimensional parameters: (i) the particle-to-fluid- density ratio ρ_p/ρ_f, (ii) the Galileo number Ga and (iii) the solid volume fraction Φ_s. The first two simulations considered here, correspond to the two simulations considered in chapter §4 with the only difference that the particles are allowed to move freely in the computational domain adding six additional degrees of freedom for the particles, i.e. the three components of the particle translational and angular velocities. Here the first two parameters are kept the same as in case $S121$ and case $S178$ from chapter §3, i.e. $\rho_p/\rho_f = 1.5$ and Ga $= 121$ and Ga $= 178$. The third parameter, the solid volume fraction, was set in both simulations to $\Phi_s = 0.5\%$, as in case $F121$ and case $F178$. For this value of the solid volume fraction the dispersed flow is considered to be dilute. Hereafter we will refer to the simulation with the triplet ($\rho_p/\rho_f = 1.5, \text{Ga} = 121, \Phi_s = 0.005$) as $M121$ and with the triplet ($\rho_p/\rho_f = 1.5, \text{Ga} = 178, \Phi_s = 0.005$) as $M178$ (M for mobile particles). The third simulation is derived from the simulation $M178$ by reducing the solid volume fraction by order of ten, when compared with $M178$. Consequently, this simulation is characterized by the triplet ($\rho_p/\rho_f = 1.5, \text{Ga} = 178, \Phi_s = 0.0005$) and will be referred to as $ML178$.

The simulations are performed as in section §4 in triply periodic computational domains with cuboid shape elongated in the vertical direction in order to account for the anisotropic length scales of the particle wakes. The extensions of the computational domain in case $M121$ are the same as in case $F121$. Similarly, in case $M178$ and case $ML178$ the size of the computational domain is as in case $F178$. The flow in all three cases is driven by the constant mean velocity U_∞ introduced in section §4.2. This implies that the average vertical velocity of the particles will be close to zero, assuming that the particles are homogeneously distributed and neglecting any two-way coupling and collective effects. The particle Reynolds number based on the velocity U_∞ and the particle diameter can be evaluated to $Re_\infty = U_\infty d_p/\nu = 141$ in case $M121$ and to $Re_\infty = 245$ in both cases $M178$ and $ML178$. A homogeneous body force is added to the vertical component of the fluid momentum equation in order to assure that the systems in all three cases can attain a steady state (Uhlmann, 2005). In case particles approach closely each other, an artificial repulsion force (cf. section §2.3.3) is applied in order to prevent unphysical overlapping of particles. Since in the present work all flow cases are considered in the dilute regime, the effect of the collision model upon the statistics is expected to be marginal.

The particles in case $M121$, $M178$ and $ML178$ are resolved with the same number of points per particle diameters as in case $F121$ and case $F178$, i.e. $d_p = 15\Delta x$ in case $M121$ and $d_p = 24\Delta x$ in cases $M178$ and $ML178$. Consequently the total number of grid points remain $1024^2 \times 5120$ in case $M121$ and $2048^2 \times 4096$ in cases $M178$ and $ML178$. The particle-fluid interface in case $M121$ is resolved by 708 and in cases $M178$ and $ML178$ by 1188 Lagrangian marker points.

The simulations $M121$ and $M178$ are initialized with the last computed flow field from case $F121$ and case $F178$ respectively. At this time the particles are randomly distributed and the particle wakes are already fully developed and in a statistically stationary state. For a detailed description of the initial flow field in case $M121$ and $M178$ the reader is referred to the result section §4.3 of the previous chapter. The simulation in case $M121$ was run for total of $3424\tau_g$, while case $M178$ was evolved 1672 gravitational time units. On the other hand, the simulation $ML178$ is initialized from the last computed flow field in case $M178$ and the number of particles was reduced by randomly taking particles out of the domain. Please note that the wakes of the remaining particles in this case are also fully developed and in statistically steady state, however, the initial flow field differs significantly from that in case $M178$, due to the occurrence of strong particle clustering in case $M178$, as we will see in chapter §6. Thus, the flow field at this time is strongly correlated with the distribution of the particles at this time. Nevertheless, we believe that simulation $ML178$ will provide valuable insight into the effects of the solid volume fraction on the particles' clusters. The simulation in case $ML178$ was run for 1756 gravitational time units. The computational domain and the initial particle position in case $M121$ and $M178$ are visualized in figure 4.1 (page 62), whereas the computational domain and particle initial position in case $ML178$ are depicted in figure 5.1. Hereafter, in order to better analyze and compare the time evolution of the three systems, the time at which the simulations are started is set arbitrary to zero, i.e. $t/\tau_g = 0$.

The physical and numerical parameters for the three simulations $M121$, $M178$ and $ML178$ are summarized in tables 5.1 and 5.2. The simulation $M121$ was conducted on 640 processor cores on the IBM system at the LRZ München supercomputing center, requiring approximately $1.8 \cdot 10^6$ CPU core hours, whereas the simulation $M178$ was performed on 8192 processor cores of the Blue Gene/p system at JSC Jülich consuming approximately $25 \cdot 10^6$ CPU core hours and the simulation in case $ML178$ was performed on 8192 processor cores at the LRZ München and the JSC Jülich supercomputing centers. In this case approximately $20 \cdot 10^6$ CPU core hours were consumed.

	Φ_s	ρ_p/ρ_f	Ga	Re_∞	$L_x \times L_y \times L_z$	N_p
$M121$	0.005	1.5	121.24	141	$68d_p \times 68d_p \times 341d_p$	15190
$M178$	0.005	1.5	178.46	245	$85d_p \times 85d_p \times 171d_p$	11867
$ML178$	0.0005	1.5	178.46	245	$85d_p \times 85d_p \times 171d_p$	1130

Table 5.1: Physical parameters of the flow of heavy particles settling under the influence of gravity in an initially quiescent fluid. Solid volume fraction Φ_s, density ratio ρ_p/ρ_f, Galileo number Ga, Reynolds number $Re_\infty = U_\infty d_p/\nu$ based on the imposed mean fluid velocity in the vertical direction, the particle diameter d_p and the fluid viscosity ν, extensions of the computational box L_x, L_y, L_z and number of particles N_p.

	$d_p/\Delta x$	$N_x \times N_y \times N_z$	N_l	N_{proc}
$M121$	15	$1024 \times 1024 \times 5120$	708	640
$M178$	24	$2048 \times 2048 \times 4096$	1188	8192
$ML178$	24	$2048 \times 2048 \times 4096$	1188	8192

Table 5.2: Numerical parameters for the flow cases $M121$, $M178$ and $ML178$. Particle resolution $d_p/\Delta x$, where Δx denotes the grid spacing. Number of grid nodes N_i in the i-th coordinate direction. Number of Lagrangian marker points used for the discretization of the fluid-particle interface, N_l. Number of processors used for the present simulations, N_{proc}.

5.2 Temporal evolution of the system

Here the evolution of the dispersed flows in the above introduced three cases $M121$, $M178$ and $ML178$ from the moment when the particles are released will be discussed. We would

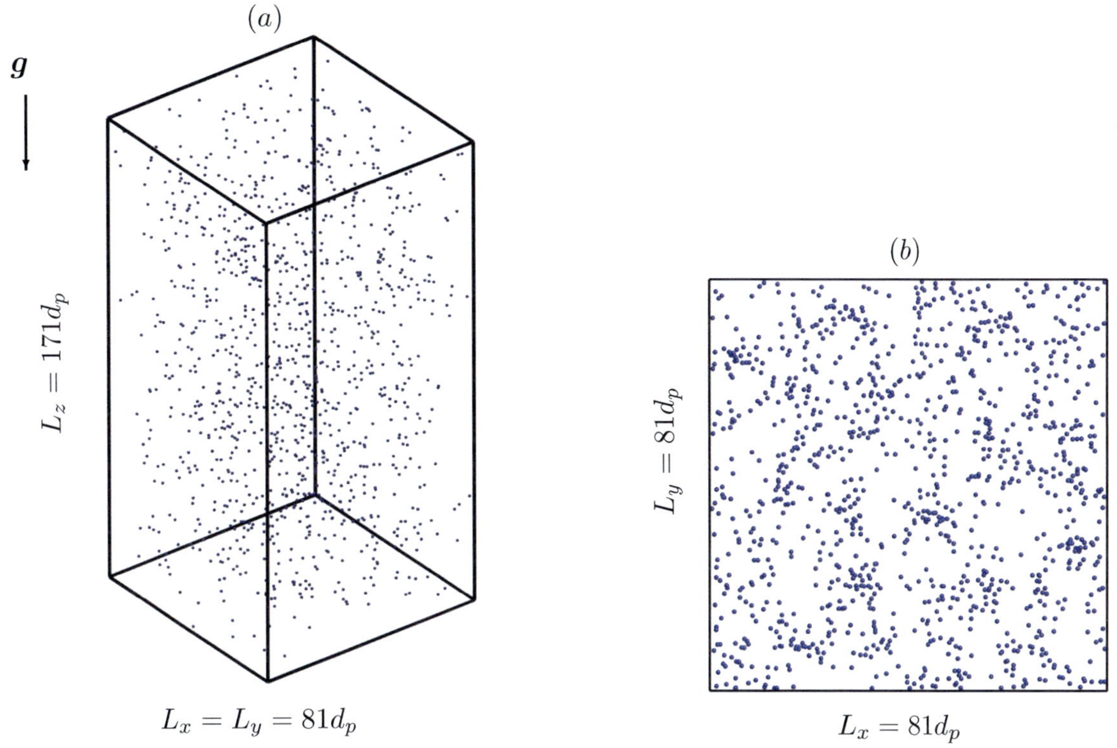

Figure 5.1: Dimensions of the computational domain with the particle distribution for case $ML178$ at the begin of the simulation. (a) side view (b) top view showing the initial particle distribution projected upon the horizontal plane. d_p denotes the particle diameter. Periodic boundary conditions are applied in all three directions in both cases. The corresponding plot for case $M121$ and case $M178$ are depicted in figure 4.1 on page 62.

like to stress that a detailed analysis of the here presented quantities will be provided in the following chapters §6 - §8. The main concern of this section is to provide an overview over the general evolution of the particulate flow system. To this end, we discuss the temporal evolution of the mean particle settling rate and the fluctuations in both the fluid and the particle phase. This will allow us to define a time interval over which the flow is considered as statistically steady. The temporal statistics in the following chapters are consequently computed in this time interval.

The instantaneous particle relative velocity with respect to the fluid-phase-averaged velocity is defined as:

$$\boldsymbol{u}_{p,rel}(i,t) = \boldsymbol{u}_p^i(t) - \langle \boldsymbol{u}_f \rangle_{\Omega_f}(t), \quad \forall i = 1 \ldots N_p. \tag{5.1}$$

Then the mean particle settling velocity can be defined as the particle ensemble average of the vertical component of the particle relative velocity $\boldsymbol{u}_{p,rel}$, viz.:

$$w_s(t) = \langle w_{p,rel}(t) \rangle_p. \tag{5.2}$$

Figure 5.2 depicts the temporal evolution of the average of the particle settling velocity w_s for all three cases. The velocities have been scaled with the viscous velocity scale based on the particle diameter, i.e. $w_s d_p/\nu$. The magnitude of the velocity w_s normalized by the same viscous scale is often referred to as the average particle Reynolds number and will be denoted

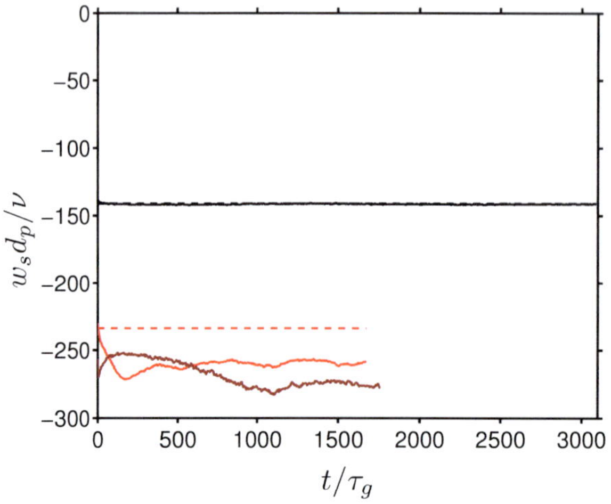

Figure 5.2: Average of the relative particle settling velocity, w_s, for case $M121$ (——), case $M178$ (——) and case $ML178$ (——). as function of time. The velocity w_s has been normalized by the ratio of the particle diameter d_p and the kinematic viscosity ν. Dashed lines indicate the terminal settling velocity of the single particle from case $S121$ (- - - -) and case $S178$ (- - - -).

in the present work by:

$$Re_{p,rel} = |w_s| d_p/\nu. \tag{5.3}$$

The corresponding terminal settling velocities of the single particles from the simulations $S121$ and $S178$ are shown as well in figure 5.2. It can be observed that the particles in case $M121$ settle on average approximately at the same rate as the corresponding single particle in case $S121$. The average particle settling velocity does not deviate substantially from the velocity in case $S121$ and remains over the entire course of the simulation at an approximate value of $Re_{p,rel} = 141.6$. This indicates that the collective effects, i.e. the presence of multiple particles, as well the hydrodynamic interactions between the particles have only a marginal effect on the average settling velocity in the present configuration ($\rho_p/\rho_f = 1.5, \mathrm{Ga} = 121, \Phi_s = 0.005$). On the other hand, in case $M178$ it can be observed that the mean particle settling velocity deviates substantially from the settling velocity of the single particle in case $S178$. The average particles velocity initially increases for approximately $200\tau_g$ and reaches a maximum value of $Re_{p,rel} = 271$, after which the velocity levels off and fluctuates around a value of 260.6. The collective effects in case $M178$ lead to an overall enhancement of the particle settling rate, when compared to the terminal settling velocity of a single particle. The difference between the two velocities for the present simulation measures approximately 12%. This will be investigated in greater detail in chapter §7. Here we would like to note that the only physical difference between the two cases $M121$ and $M178$ is the value of the Galileo number and as we will see in the next chapter (§6) the particles in case $M178$ agglomerate into clusters, while the particles in case $M121$ remain homogeneously distributed. The one of the consequences of the particle clusters in case $M178$ is an increase of the mean particle settling rate (cf. 7.1.1).

Concerning case $ML178$, the graph in figure 5.2 reveals that the mean settling rate of the particles initially decreases and reaches a minimum value of $Re_{p,lag} = 256$ approximately $100\tau_g$ after the simulation was initiated. Please note that due to the removing of the majority of the particles from the system, the remaining particles in case $ML178$, at the begin of the simulations, are more probably exposed to the mean flow, which would lead to a net

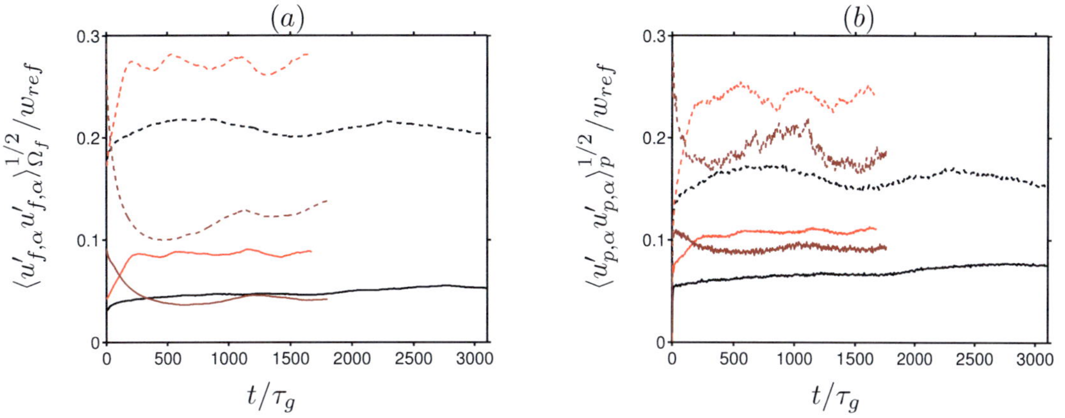

Figure 5.3: Intensity of the velocity fluctuations of (a) the fluid phase and (b) the dispersed phase as function of time. Case $M121$ (——, - - - -). Case $M178$ (——, - - - -). Case $ML178$ (——, - - - -). Solid lines correspond to horizontal velocity components ($\alpha = 1, 2$), dashed lines show the vertical velocity component ($\alpha = 3$). The velocity fluctuations have been normalized with the absolute value of the particle settling velocity $w_{ref} = |w_\infty|$ of the corresponding single particle cases $S121$ and $S178$.

reduce of the average particle settling velocity. After reaching its minimum, the mean particle settling velocity increases slowly for approximately $1000\tau_g$ and reaches a maximum value of $Re_{p,rel} = 282.6$ at $t/\tau_g = 1100$. The velocity then appear to fluctuate at the mean value of $Re_{p,rel} = 280$. As we will see in chapter §6, the main reasons for these evolution of the particle settling velocity in case $ML178$ is the spatial distribution of the particles.

Let us now turn to the time evolution of the velocity fluctuations in the fluid and the dispersed phase, which are depicted in figure 5.3. In all three cases the common observation is that the fluctuations of the velocity fields in both phases are anisotropic with the vertical components being highly dominant. In case $M121$ after a short initial transient, the fluctuations of the vertical velocity component in both phases exhibit slow fluctuations with time scale of the order of $1000\tau_g$. However, as we saw in figure 5.2, the mean settling rate of the particles appears to be uncorrelated (unaffected) by these long time oscillations. On the other hand, the fluctuations of the vertical velocity component in case $M178$ increase more slowly as in case $M121$ and similar to the average settling rate of the particles, reach a statistically steady state after approximately 200 gravitational time units. In contrast to the velocity fluctuations in case $M121$, in case $M178$ the fluctuations oscillate around their asymptotic value with comparably shorter period. Moreover, the temporal evolution of the mean settling rate of the particles in case $M178$ seems to correlate with that of the fluid velocity fluctuations. These feature merits a more detailed investigation, which would be conducted in the following three chapters. Similar to the mean settling rate, the fluctuations in case $ML178$ for both phases exhibit long transitional period of roughly $1100\tau_g$ in which they decrease significantly. The difference between the two cases $M178$ and $ML178$ at the end of the simulated time intervals was calculated to be as high as 50.9% for the vertical component of the fluid velocity fluctuations and 22.2% for the vertical component of the particle velocity fluctuations. For the horizontal components this difference was calculated to 52.7% and 17.6% for the fluid and the particle velocity fluctuations respectively.

The fluctuations of the cross-stream components of the fluid and the particle velocity have in all three cases comparable values, which indicates that turbulent dispersion may play an important role in the present numerical experiments. As we will see later, the particles do not settle on a straight trajectories oriented in streamwise direction, rather they settle on curved or oblique paths. As result the particle wakes are not oriented perfectly with the vertical direction (in case $M178$ and case $ML178$ this is true even for a single particle in undisturbed flow), which causes momentum along the wake axis to be deposited in the lateral direction. This is in line with the observations made in Parthasarathy and Faeth (1990a,b). However, in their work the relative turbulent intensities are one order of magnitude smaller than in the present work, thus the turbulent dispersion of the particles is expected to be more prominent in the present simulations. It is interesting to note that the velocity fluctuations in the dispersed phase in all three cases are entirely due to collective effects, since a single particle at this Galileo numbers is in steady motion.

Before we conclude, let us recall (section §4.3.1) that the initial flow field in case $M121$ and case $M178$ was fully developed at the time the particles were released. Due to the strong anisotropy of the particle wakes, the flow field in both cases evidenced a small positive correlations of the vertical velocity component in the stream-wise direction (figures 4.7 - 4.8). Please note that the particles at this time were fixed and randomly distributed, thus the flow field would contain a three dimensional time average (fixed particles) contribution which will affect the spatial correlation of the fluid velocity. A preliminary coarse grid simulations at $Re_{p,lag} = 230$ (data provided by courtesy of Markus Uhlmann) showed that even an extended domain size in the vertical direction of $L_z = 680d_p$ does not assure full decorrelation of the flow field. Thus, in case of streamwise elongated particle clusters, as in case $M178$, the necessary domain size required to obtain full decorrelation of the flow field would be even more restrictive. On the other hand, in the horizontal direction the velocity signals decorrelated for separations smaller than the extension of the computational domain in the horizontal direction, concluding that the domain size in the horizontal direction is sufficiently large in order to accommodate the largest flow structures. The same considerations apply for case $ML178$, where, as aforementioned, the initial flow field is strongly correlated with the distribution of the particles at this time (cf. chapter §8).

5.3 Conclusion

Here we have introduced the computational set-up and the initial flow configurations for the three direct numerical simulations performed in this part of the thesis. This chapter was not meant to provide a detail analysis of the results of the present simulations, rather to provide a brief introduction to the open questions which are going to be addressed in this part of the thesis. Additionally, the initial transient of the flow from the moment when the particles are released was discussed. The main concern was to identify the time interval in which the dispersed flow is considered to be in statistically steady state. This time interval is then consequently used in the following chapters §6 - §8 to accumulate the temporal statistics. The main observation is that while the particles and the flow in case $M121$ evolved in a statistically stationary state shortly after the particles were released, the particles and the flow field in case $M178$ required approximately 200 gravitational time units to reach statistically steady state. For consistency between the two cases, the dispersed flow in case $M121$ and case $M178$ is considered as fully developed after an initial transient of $200\tau_g$. On the other hand, the flow in case $ML178$ experienced longer initial transient due to the strong large scale correlations in the initial flow field. Therefore, the statistics in case $ML178$ will be accumulated after the initial transient of $1100\tau_g$. A comparison of the fluid (particle) velocity fluctuations in case

$M178$ and $ML178$ evidenced, as expected, a strong dependency with solid volume fraction, higher solid volume fraction led to higher fluctuations in the flow. Moreover, the mean settling rate of the particles showed to correlate with the fluctuations of the fluid velocity field.

Finally, a brief discussion of the two-point correlation of the initial flow field in case $M121$ and $M178$ lead to the conclusion that due to anisotropy of the particle wakes an even larger domain extensions in the vertical directions are required in order to obtain a full decorrelation of the flow field.

Chapter 6

Spatial structure of the dispersed phase

As introduced in sections §1.2.3 and §5 an interesting feature of dispersed particulate flows is the ability of the particles to sample the flow in an inhomogeneous manner, i.e. to agglomerate into clusters. The flows under consideration in this chapter consist of multiple heavy freely moving finite-size particles settling under the influence of gravity in an initially ambient flow field. Further the flows are considered to be homogeneous in all three directions and all of the fluctuations in the flow field are generated by the relative motion of the fluid to the particles. Particularly, in the present chapter the spatial distribution of the particles in case $M121$, $M178$ and $ML178$ will be studied in detail. For detailed description of the configuration of these cases, the reader is referred to chapter §5. The main question addressed here is whether the particles in such flow configurations agglomerate into clusters or not. In the context of particle clustering, the distribution of the particles is implicitly compared with the Poisson distribution of uniformly distributed particles. Due to the homogeneity of the flows, any deviation of the particle distribution from the random (homogeneous) distribution can be attributed to the collective effects and to the coupling effects between the particles and the particle-induced flow field. In case of particle clustering a characterization and detailed description of the clusters will be provided. The discussion will focus on questions such as what is the characteristic shape, size and life-time of the particle clusters.

6.1 Qualitative cluster analysis

The question of whether particles cluster or not is often approached qualitatively by visualizations of the instantaneous particle positions. Visualizations often have been used to provide an initial reference for further quantitative analysis.

Figure 6.1 depicts the instantaneous particle positions in case $M121$ at two instants, at the time the particles are released to move freely, showing the initial particle distribution and at some instant when the system is in statistically stationary state. As can be seen, there is no significant difference between the particle distribution at this two instants. The corresponding visualization for case $M178$ is shown in figure 6.2. It is clearly observable that the particle distribution at time when the flow is in statistically stationary state deviates significantly from the initial particle distribution. One can easily identify localized regions in the projection upon

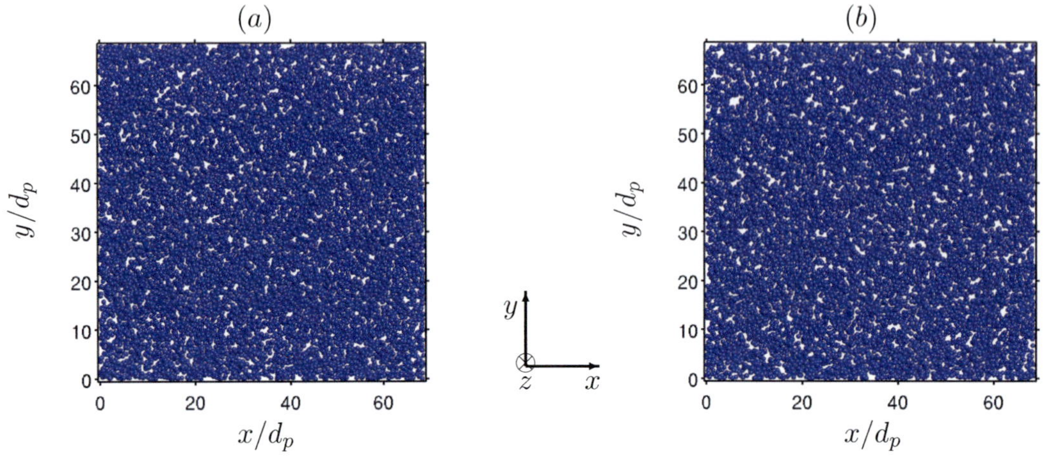

Figure 6.1: Instantaneous particle position in case $M121$ viewed from the top: (a) initial state $t/\tau_g = 0$; (b) in the statistically stationary regime at $t/\tau_g = 1200$.

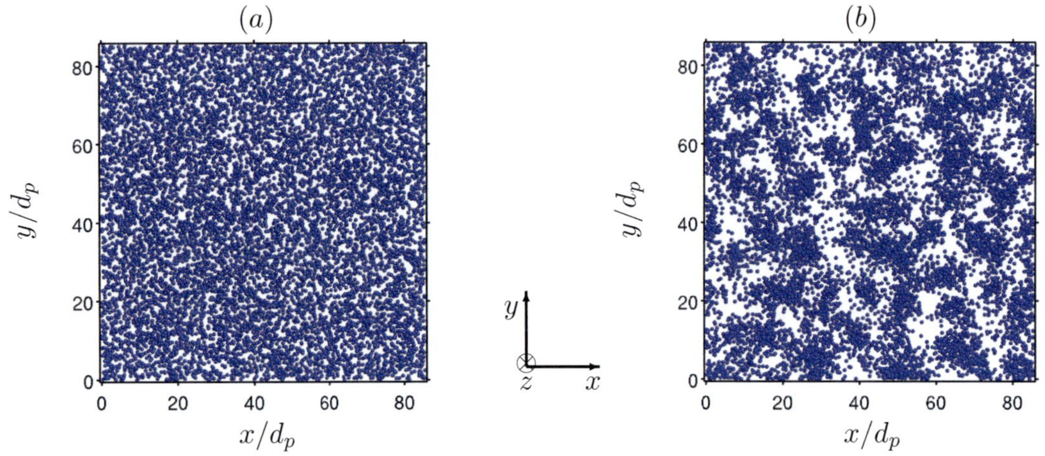

Figure 6.2: The same as in figure 6.1 but for case $M178$: (a) $t/\tau_g = 0$; (b) $t/\tau_g = 820$.

the horizontal plane with a high particle density and regions with low particle concentration, or even void regions with complete absence of particles.

Hereafter, regions with high particle concentration are referred to as *clusters* and regions with low particle concentration as *voids*, where the definition of high- and low- particle concentration will be discussed and defined more precisely later in this section. Seen from this vertical perspective the particles appear to form structures which are elongated in the vertical direction. More precise inspection reveals that the clusters and the voids extend throughout the entire height of the computational domain. On the other hand the characteristic size of the particle clusters in the horizontal direction is much smaller than the extensions of the computational domain and they appear to be randomly distributed in the horizontal plane. This initial qualitative description of the clusters in case $M178$ is in line with previous experimental (Nishino and Matsushita, 2004) and numerical (Kajishima and Takiguchi, 2002; Kajishima, 2004b,a) findings, where similar *columnar particle accumulation* was observed. Time sequences of such visualization shows that these structures persist over long time intervals.

6.2 Voronoï Analysis

In the following we quantify the tendency of the particles to agglomerate into clusters by performing Voronoï analysis of the particle distribution. A number of techniques have been established for the characterization of the spatial structure of the dispersed phase. Besides the conventional tools for detecting preferential concentrations such as box counting methods (Fessler et al., 1994), pair correlation function estimation (Sundaram and Collins, 1997) and genuine clustering detection algorithms (Melheim, 2005), a technique based upon Voronoï tessellation has recently been proposed in the context of particulate flows (Monchaux et al., 2010). The Voronoï tessellation or the so called Voronoï diagrams is a decomposition of the space into independent cells, which are associated to a single particle p_i, $\mathcal{V}_i = \mathcal{V}(p_i)$. The Voronoï cells $\mathcal{V}_i$ possess the property that all points inside the cell are closer to the particle p_i than to any other particle p_j, $j \neq i$ (Okabe et al., 1992). Example of a Voronoï diagrams in two- and three-dimensional space can be found in figure 6.3. The cells in figure 6.3a fill the entire space by periodical extension of the cell boundaries. As consequence the local concentration of the particles can be related to the inverse of the volume of the Voronoï cells. That is, Voronoï cells with small (large) volume indicate areas with high (low) particle concentration. A main advantage of the Voronoï analysis in comparison to the previously mentioned approaches is the absence of any *a-priori* choice of length scales and its computational efficiency. Here we would like to stress that the Voronoï analysis provides only a general tendency of clustering at some instant and does not provide any information about the size or shape of the clusters. Additionally, since Voronoï volumes are evaluated around each individual particle, one can track the local particle concentration along the particle trajectories in a Lagrangian manner. This kind of cluster analysis was successfully applied in the past in number of studies in both the two- and three-dimensional space, e.g. Monchaux et al. (2010, 2012); García-Villalba et al. (2012); Tagawa et al. (2012); Dejoan and Monchaux (2013).

In order to investigate the spatial distribution of the dispersed phase, we have performed a three dimensional Voronoï analysis by computing the Voronoï diagrams and the corresponding volumes V_i of the Voronoï cells for all particles in the computational domain. The Voronoï cell volumes are normalized to be of unit mean. This normalization allows a qualitative comparison of the p.d.f.s for different particle number densities, since the so normalized p.d.f.s are independent of the particle number density (Ferenc and Néda, 2007). Before we proceed with the results, let us briefly clarify the procedure applied here to quantify the general tendency for clustering. As pointed out by Monchaux et al. (2010, 2012) one can study the probability distribution function of the normalized Voronoï volumes and their standard deviation. A tendency of clustering is given, when the p.d.f. of the normalized Voronoï volumes deviates from the p.d.f. for random distribution and the standard deviation is larger than for randomly distributed particles. The standard deviation of the p.d.f. is then directly linked to the level of clustering. By definition of the Voronoï tessellation, for particles distributed in a more regular manner than random, the corresponding p.d.f. of the Voronoï cell volumes should be less intermittent than the p.d.f. for randomly distributed particles. Eventually for particles on regular lattice all Voronoï volumes will become identical and the p.d.f. will be resembled by a Dirac function. However, for particles with more inhomogeneous distribution than random, the p.d.f. of the Voronoï cell volumes is expected to be more intermittent than the one for randomly distributed particles with higher probability of finding cells in the “extremes”, i.e. the probability of finding Voronoï cells with very large or very small volumes will become larger than for randomly distributed particles.

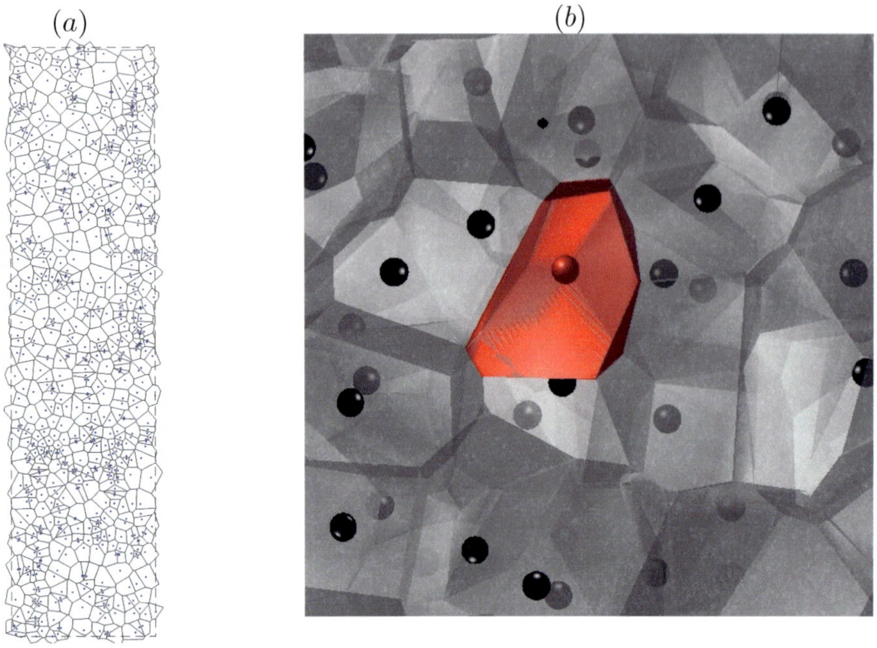

Figure 6.3: (a) Two dimensional Voronoï diagram for case $M178$. The dashed line shows the domain boundaries. (b) Close-up of a three dimensional Voronoï diagram.

Figures 6.4a and 6.5a depict the p.d.f. of the normalized Voronoï volumes for case $M121$ and case $M178$ computed at different times during the course of the simulations. Except for the initial state, the tessellations are performed at several instants covering short time intervals of length less than $30\tau_g$. As aforementioned, in the three dimensional space particles with random distribution are usually implicitly assumed to be distributed as a random Poisson process. In this case a random Poisson process is well represented by a Gamma function (Ferenc and Néda, 2007). As can be observed, the p.d.f. for both cases at time the particles were released in the computational domain is well represented by the Gamma function, confirming that the initial particle distribution was indeed random. It can be observed that in case $M121$ (figure 6.4a) the occurrence of very small and very large Voronoï cells is less probable than in the case of randomly distributed particles. Additionally the standard deviation of the normalized Voronoï volumes at all instants was found to be smaller than in the random case, indicating that the deviation from the average cell volume in case $M121$ is less probable than in the case of randomly distributed particles. This indicates that the distribution of the particles in case $M121$ becomes more ordered than randomly distributed particles. Regarding case $M178$ (figure 6.5a), as time advances the p.d.f. of the Voronoï volumes begins to deviate more and more from the one at initial time. Finally, the p.d.f.s in time saturate at their final position, indicating the reaching of a stationary state. Furthermore, the p.d.f.s of the present data exhibit tails with higher probability of finding Voronoï cells with very large or very small volumes than in the random case with uniform probability, indicating that cluster and respectively void formation takes place.

As we showed in section §5 the flow in the present simulations is highly anisotropic due to preferential spatial direction due to gravity. Furthermore, a columnar like structures are observed in case $M178$, viz. figure 6.2, indicating the highly anisotropic shape of the particle clusters in case $M178$. This kind of non-isotropic clusters can sometimes be “invisible” to the

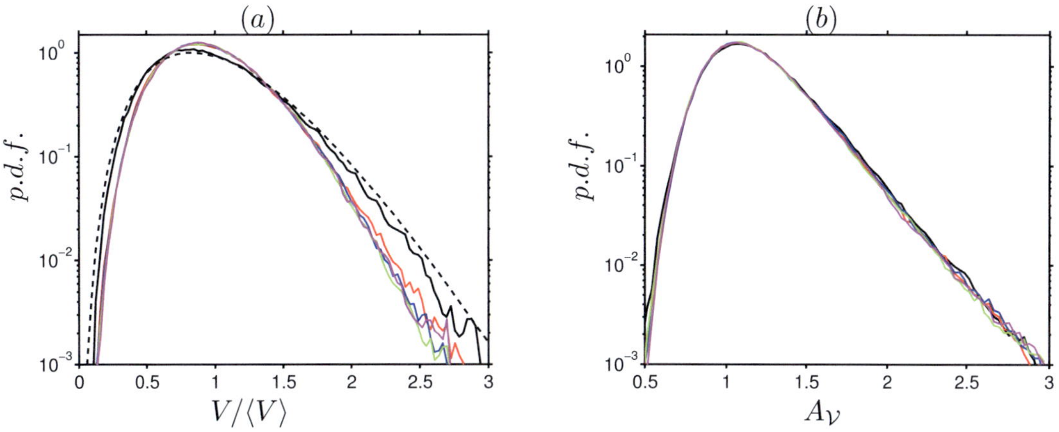

Figure 6.4: Case $M121$: Probability density function of (a) the normalized Voronoï cell volumes and (b) the aspect ratios $A_{\mathcal{V}}$, where $A_{\mathcal{V}}$ is defined as in equation 6.1. Different lines represent data assembled over different time intervals. The initial random distribution is represented by (——). $t/\tau_g = [279.7,\ 307.6]$ (——). $t/\tau_g = [559.4,\ 587.3]$ (——); $t/\tau_g = [1216.7,\ 1244.5]$ (——); $t/\tau_g = [1496.4,\ 1524.2]$ (——). The dashed black line (----) represents an analytical Gamma function fit (Ferenc and Néda, 2007).

Voronoï analysis, purely based upon the cell volumes. Therefore, we have analyzed the shape of these structures by computing the aspect ratio (AR) of each Voronoï cell, defined as the ratio of the largest horizontal extension $l_x(\mathcal{V}_i)$ to the largest vertical extension $l_z(\mathcal{V}_i)$ of the Voronoï cell $\mathcal{V}_i$:

$$A_{\mathcal{V}_i} = l_x(\mathcal{V}_i)/l_z(\mathcal{V}_i), \qquad \forall i = 1...N_p. \tag{6.1}$$

The so defined aspect ratio $A_{\mathcal{V}}$ provides a qualitative measure of the anisotropy of the particle clusters and can be seen as measure for the stretching of the cells. For the present simulations, large values of $A_{\mathcal{V}}$ indicate a “flat” cells squeezed (stretched) in the vertical (lateral) direction and small values of $A_{\mathcal{V}}$ indicate cells which are squeezed (stretched) in the horizontal (vertical) direction. Figures 6.4b and 6.5b depict the distributions of the AR $A_{\mathcal{V}}$ of the Voronoï cells for case $M121$ and case $M178$. The p.d.f.s are calculated at the same instants as the p.d.f.s of the Voronoï cell volumes. It can be seen that the p.d.f.s in case $M121$ does not show any deviation from the p.d.f. of the randomly distributed particles, while in case $M178$ an appreciable difference can be observed. Finding Voronoï cells with large values of $A_{\mathcal{V}}$ is more probable than in case of randomly distributed particles. This is an indication that the majority of the Voronoï cells in case $M178$ are squeezed (stretched) in the vertical (horizontal) direction and that the particle agglomerations are more likely to be oriented in the vertical direction. This confirms the observations made in figure 6.2b, where particle columnar like structures were observed, which is in line with the results from the literature (Kajishima and Takiguchi, 2002; Kajishima, 2004a).

As mentioned above, in case of particle clustering the p.d.f. of the normalized Voronoï volumes deviate from the p.d.f. for randomly distributed particles. Furthermore the second moment of the distribution should exceed the one for randomly distributed particles. Contrarily, for particle with more homogeneous spatial distribution, the standard deviation of the normalized Voronoï volume p.d.f. is expected to be smaller than in the random case. Therefore, the standard deviation of the p.d.f. of the normalized Voronoï volumes can be used to characterize the level of clustering (Monchaux et al., 2010, 2012). Figure 6.6 depicts the

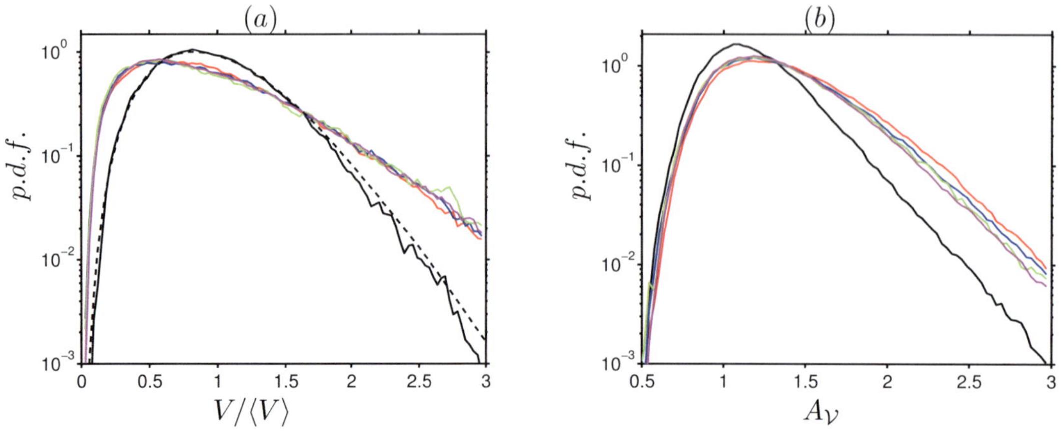

Figure 6.5: As figure 6.4 but for case $M178$. Different lines represent data assembled over different time intervals. Initial random distribution is represented by (——). $t/\tau_g = [199.1,\ 213.7]$ (——). $t/\tau_g = [345.3,\ 359.9]$ (——); $t/\tau_g = [572.5,\ 576.9]$ (——); $t/\tau_g = [794.1,\ 810.9]$ (——). The dashed black line (- - - -) represents an analytical Gamma function fit (Ferenc and Néda, 2007).

temporal evolution of the standard deviation of the p.d.f. of the normalized Voronoï volumes, $\sigma_{\mathcal{V}}$, for both cases $M121$ and $M178$. Since there is no analytical expression for the second moment of the distribution of the Voronoï volumes (Ferenc and Néda, 2007; Monchaux et al., 2010), the standard deviation is compared with respect to the standard deviation of the initial particle distribution ($\sigma_{\mathcal{V}} = 0.411$). As can be seen from figure 6.6, the standard deviation in case $M121$ decreases after particles have been released to settle freely and reaches quickly a statistically steady value of approximately 0.345. Contrary, the standard deviation in case $M178$ shows the opposite behavior: as the particles organize in clusters, the standard deviation increases with time and reaches a statistically steady value of approximately $\sigma_{\mathcal{V}} = 0.616$. From the figure it is clearly observable, that for the particles in case $M178$ the transient time, that the particle require in order to reach statistically stationary state is considerably longer than in case $M121$. Moreover, the time at which an approximately steady state is reached represents approximately the time at which the fluctuations of the particle- and fluid velocity field reached steady state.

The Voronoï diagrams can be used to define an intrinsic definition of cluster regions with high particle concentration and voids with low particle concentration by determining the intersection points between the p.d.f. of the DNS data and that of the random distribution. Particles whose Voronoï cell volume is smaller than the lower intersection points V^c are considered to be located in a regions with high particle concentration and particles whose Voronoï cell volume is larger than the upper intersection point V^v are attributed to void regions (Monchaux et al., 2010). By tracking the particle positions in time, the Voronoï tessellation allows us to determine the instantaneous local concentration of the particles. Combining this with the above intrinsic definitions of clusters and voids we can determine instantaneously for each particle, if it reside in regions with high or low particle concentration. We calculated the mean residence time of the particles in clusters (voids), by determining the temporal intervals during which the volume of the Voronoï cells of each particle is below (above) the threshold V^c (V^v). It turns out that on average the particles in case $M178$ reside for approximately $24.9\tau_g$ in regions with high particle concentration and approximately $87.7\tau_g$ in regions with low particle concentration. This indicates, that once the particles enter cluster or void regions

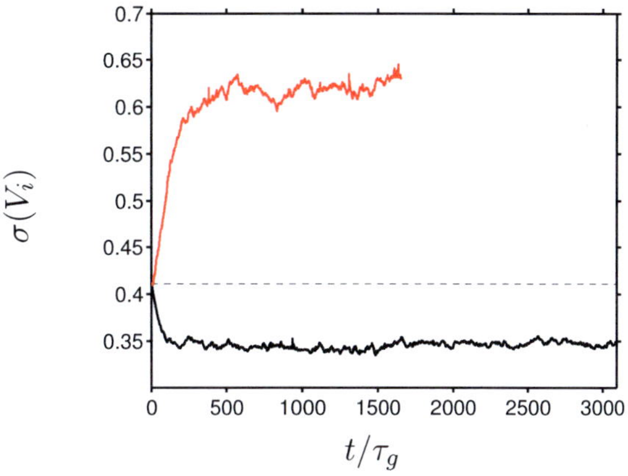

Figure 6.6: Standard deviation of Voronoï cell volumes as function of time for case $M121$ (——) and case $M178$ (——). The corresponding value for randomly distributed particles with uniform probability (- - - -).

they remain trapped therein for relatively long times, with a bias towards voids. Please note that the Voronoï tessellation would attribute rather large cell volumes to particles located on the periphery of clusters. With the above intrinsic definition of clusters and voids, the particles located inside clusters or voids can be identified at each instant. Figure 6.7 shows the temporal evolution of the total number of particles incorporated in clusters and voids in case $M178$ evidencing that once statistically stationary state was reached the total number of particles residing in voids or clusters remains approximately constant over time with only minor temporal fluctuations. A total of approximately 30% of the particles are attributed to clusters and 15% to voids. Moreover, we have calculated the frequencies with which the particles enter clusters and voids to $0.016/\tau_g$ and $0.006/\tau_g$ respectively. The higher frequency of particles entering clusters stems most probably from the geometrical origin of the Voronoï tessellation and could be attributed to particles located on the periphery of clusters.

Since each Voronoï cell can be uniquely assigned to a single particle, the derived cell quantities (volume and AR) of each particle can be tracked in time. This allows us to study the behavior of these quantities in a Lagrangian manner. To this end we have studied the Lagrangian autocorrelation function related to a cell quantity $Q_\mathcal{V}(t)$, which is calculated as function of the separation time τ_{sep} as:

$$R_{Q_\mathcal{V}Q_\mathcal{V}}(\tau_{sep}) = \frac{\langle Q'_\mathcal{V}(t)Q'_\mathcal{V}(t+\tau_{sep})\rangle_{p,t}}{\langle Q'_\mathcal{V}(t)Q'_\mathcal{V}(t)\rangle_{p,t}}, \tag{6.2}$$

where $Q'_\mathcal{V}(t) = Q_\mathcal{V}(t) - \langle Q_\mathcal{V}\rangle_{p,t}$ is the fluctuation with respect to the mean over the particle ensemble and time. For detailed definition of the averaging operator $\langle\cdot\rangle_{p,t}$ the reader is referred to appendix A.

The Lagrangian autocorrelation of the Voronoï cell volume and AR for both cases $M121$ and $M178$ are depicted in figure 6.8. It can be observed that in both cases the correlation time of the Voronoï volume is appreciably larger than the correlation time of the Voronoï cell aspect ratio. Comparison of both cases reveals that the Voronoï volume in the clustering case $M178$ is somewhat correlated over longer times than in case $M121$ where no clustering takes place. The corresponding integral time scale were calculated to $43.7\tau_g$ in case $M121$ and to $73.7\tau_g$ in case $M178$. Regarding the Voronoï cell aspect ratio, it features in both cases

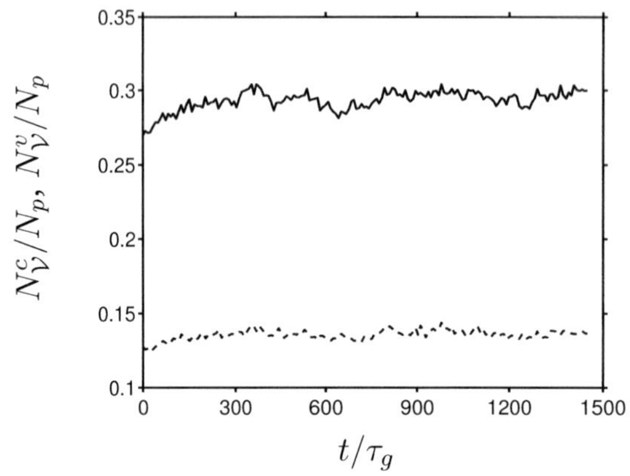

Figure 6.7: Time evolution of the number of particles with high probability to be in clusters $N^c_\mathcal{V}$ (——) and in voids $N^v_\mathcal{V}$ (- - - -) for case $M178$. The plotted data has been thinned out for clarity purposes. Note, that the data for case $M121$ is not shown, since in case $M121$ no clustering is observed.

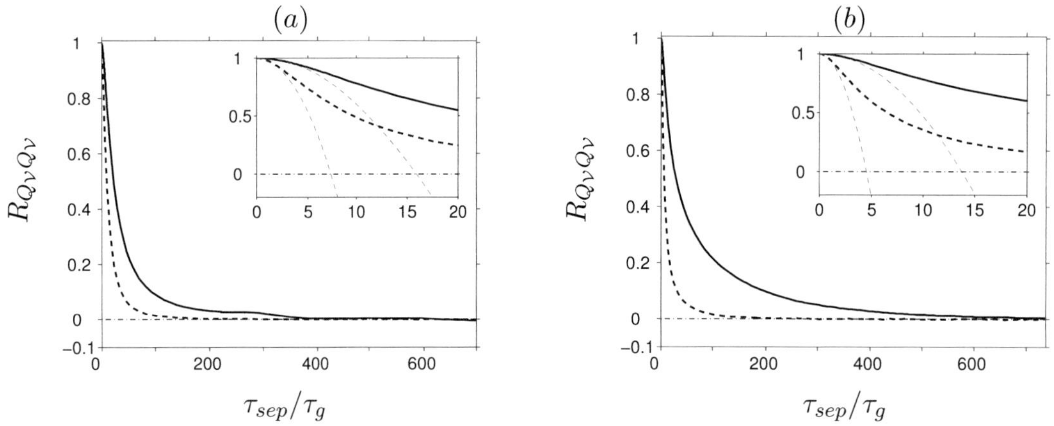

Figure 6.8: Lagrangian autocorrelation of Voronoï volumes (——) and Voronoï cell aspect ratio $A_\mathcal{V}$ (- - - -) as function of the separation lag τ_{sep} for (a) case $M121$ and (b) case $M178$. The inset show closeup of the same data, but for small separation lags τ_{sep}. The blue dashed lines (- - - -) represent the osculating parabola at $\tau_{sep}/\tau_g = 0$.

considerably faster decorrelation than the cell volume, practically vanishing at the same time after approximately $200\tau_g$. The corresponding integral time scale measures $17.2\tau_g$ and $14.2\tau_g$ in case $M121$ and case $M178$ respectively. Please note that the aspect ratio of the Voronoï cells is a geometrical property of the shape of the cells and it may be significantly modified while the particle dynamically rearrange in space without necessarily changing the volume of the corresponding cells.

Concerning the short time behavior of the particle structures, it can be studied by considering the Taylor micro-scale of the Voronoï cell properties. The Taylor micro-scale is defined as the osculating parabola at zero separation and is depicted in the insets of figure 6.8. In contrast to the integral time scale, the Taylor micro-scale in both cases have similar values: 15.8 in case $M121$ and 13.7 in case $M178$. The corresponding values for the Voronoï cell aspect ratio measure 7.3 and 4.5 in case $M121$ and $M178$ respectively. The comparable values of the Taylor micro-scale in both cases suggests that the short time behavior of the particle structure in both cases is very similar.

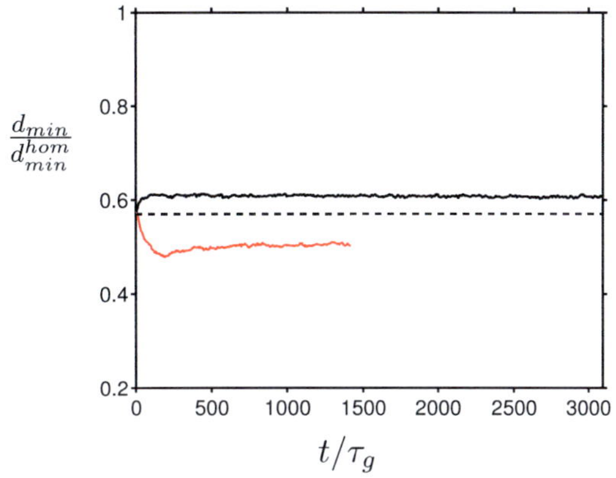

Figure 6.9: Temporal evolution of the average distance to the nearest neighbor for cases $M121$ and $M178$, normalized by the value for a homogeneous distribution on a regular cubical lattice. Case $M121$ (———). Case $M178$ (———). Random particle distribution with the same volume fraction (- - - -).

An alternative way of characterizing the spatial structure of the dispersed phase is by performing a nearest-neighbor analysis (Kajishima, 2004b). Figure 6.9 depicts the time evolution of the average distance to the nearest particle neighbor d_{min}, which is calculated as:

$$d_{min} = \frac{1}{N_p} \sum_{i=1}^{N_p} \min_{\substack{j=1 \\ j \neq i}} d_{i,j}, \tag{6.3}$$

where $d_{i,j} = |\boldsymbol{x}_p^i - \boldsymbol{x}_p^j|$ is the distance between the centers of i-th and the j-th particle. The distance d_{min} has been normalized by its value for a homogeneous distribution with the same solid volume fraction, $d_{min}^{hom} = (\|\Omega\|/N_p)^{1/3}$. The lower limit of d_{min}/d_{min}^{hom} corresponds to the minimum value of the function, when all particles are in contact with a neighboring particle. The upper limit of the function d_{min}/d_{min}^{hom} has a value of unity and arises for homogeneously distributed particles, i.e. particles are positioned on a regular lattice. As expected, the average distance to the nearest neighbor at initial time is very close to the value for a randomly distribute particles. With time the value of d_{min}/d_{min}^{hom} in case $M121$ increases quickly and reaches a value of 0.61 in the time interval where the flow is considered as statistically steady. This is in line with the above findings, that the particles in case $M121$ are more ordered than randomly distributed particles. Contrarily, the value of d_{min}/d_{min}^{hom} in case $M178$ decreases for approximately $200\tau_g$ and undulates at approximate value of 0.5. This again corroborates the results obtained above, that the particles in case $M178$ tend to form agglomerations.

We can preliminarily conclude that the particles in case $M121$ are distributed in more regular way, whereas the particle distribution in case $M178$ exhibits strong inhomogeneities, i.e. particle form clusters. Further we showed that the shape of the particle structures is anisotropic and these structures are oriented in vertical direction. Lagrangian analysis revealed that the particle structures maintain coherency over substantial time scales.

6.3 Conditionally averaged particle concentration

The Voronoï analysis in the previous chapter showed the tendency of the particles in case $M178$ to agglomerate in clusters as a whole, whereas in case $M121$ no clustering was evidenced.

In this section we study the local particle distribution in the vicinity of a test particle. This kind of analysis will provide more insight into the micro-structure of the dispersed phase. The micro-structure is a result of the migration of the particles induced by particle-particle interactions and is affected by features such as particle wake interactions, vortex shedding or hydrodynamic forces (Hu et al., 2001). For this purpose we performed particle-conditioned averaging of the particle locations. The averaging procedure is very similar to that introduced in section §4.3.3. First we define an averaging volume Ω_A with vertical axis coinciding with the vertical direction and a discrete solid phase indicator function, which is unity for computational nodes occupied exclusively by the dispersed phase and zero otherwise. The indicator function is then applied to the computational domain identifying the grid nodes located in the domain occupied by the dispersed phase Ω_p. The origin of the domain Ω_A is then shifted to the location of each particle, taking into account the periodicity of the domain, and added up. The domain Ω_A is discretized by the same grid spacing as the corresponding cases $M121$ or $M178$. A detailed definition of the averaging procedure can be found in appendix A. This procedure was repeated for 16816 and 26180 particle snapshots in case $M121$ and $M178$ respectively. By normalization of the so defined quantity with the number of samples one obtains an averaged field ϕ_s^{cond} with values in the interval $\phi_s^{cond} \in [0, 1]$, which can be interpreted as the particle conditioned local volume fraction (Kidanemariam et al., 2013). The so defined quantity can be also interpreted as the probability that a point of the averaging domain is located inside the domain occupied by the solid phase Ω_p. When normalized with the global solid volume fraction Φ_s, the quantity ϕ_s^{cond}/Φ_s can be related to the pairwise distribution function, which provides the probability of finding a particle at certain distance away from a given reference particle (Sundaram and Collins, 1997; Shotorban and Balachandar, 2006; Sardina et al., 2012). By definition, for mono-dispersed flows the quantity ϕ_s^{cond} is a symmetric function, viz.:

$$\phi_s^{cond}(\tilde{\boldsymbol{x}}) = \phi_s^{cond}(-\tilde{\boldsymbol{x}}). \tag{6.4}$$

Figures 6.10 and 6.11 show the two dimensional maps of ϕ_s^{cond}/Φ_s for both flow cases. As can be clearly seen on macroscopic level the dispersed phase in case $M121$ (figure 6.10a) does not show any signs of accumulation and particles appear to be randomly distributed ($\phi_s^{cond}/\Phi_s \simeq 1$). The local solid volume fraction differs from that for randomly distributed particles only in vicinity to the reference particle (figure 6.10b), where the probability of finding another particle is significantly larger than the global average sold volume fraction. As can be seen the local particle concentration exhibits anisotropic spatial distribution, i.e. the region with higher particle concentration is clearly oriented in vertical direction. The maximum can be found at approximately one particle diameter downstream of the reference particle on the vertical axis. In the horizontal direction ϕ_s^{cond} is observed to be below the average at small horizontal separations from the reference particle. The average value is adopted for distances larger than one particle diameter.

In case $M178$ (figure 6.11a), the distribution of the local solid volume fraction clearly exhibits a strongly anisotropic spatial structure even at the macroscopic level. The distribution evidences high values of the local concentration located in a vertical stripe along the vertical direction. As can be seen the region of high local particle density extends throughout the entire height of the computational domain. This is in line with the observation that the particles in case $M178$ agglomerate in columnar like structures (section §6.2). In contrast to the micro-structure in case $M121$, the local sold volume fraction in case $M178$ shows larger than the average values in a small region in the horizontal direction with maximum at distances around $1.5d_p$.

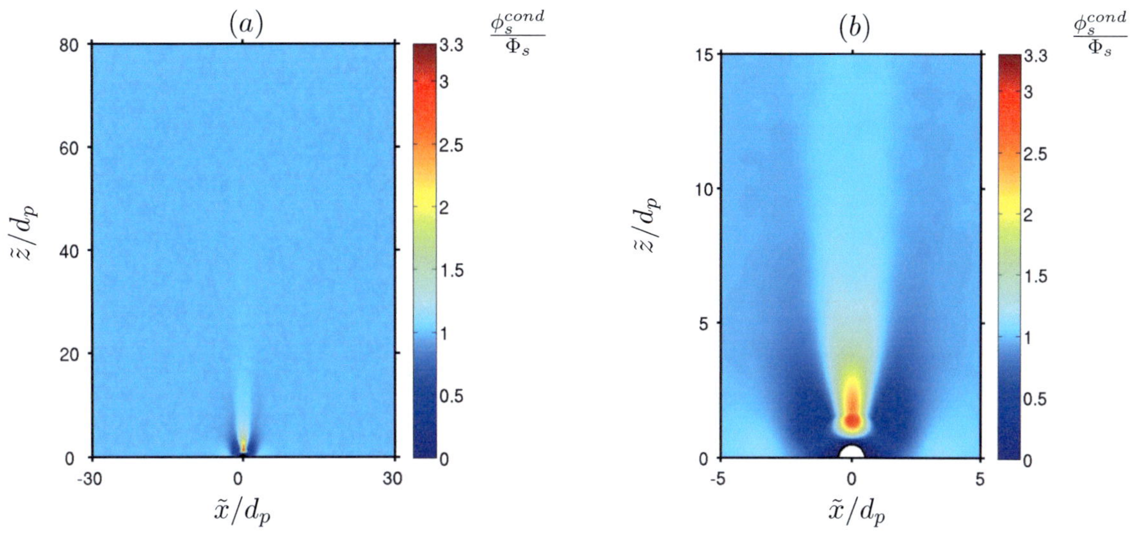

Figure 6.10: (a) Averaged local solid volume fraction ϕ_s^{cond} conditioned on the particle positions for case $M121$, normalized by the global solid volume fraction Φ_s. (b) Close-up of the same data. The coordinates are relative to the reference particle position (section §3.4.2, p. 47).

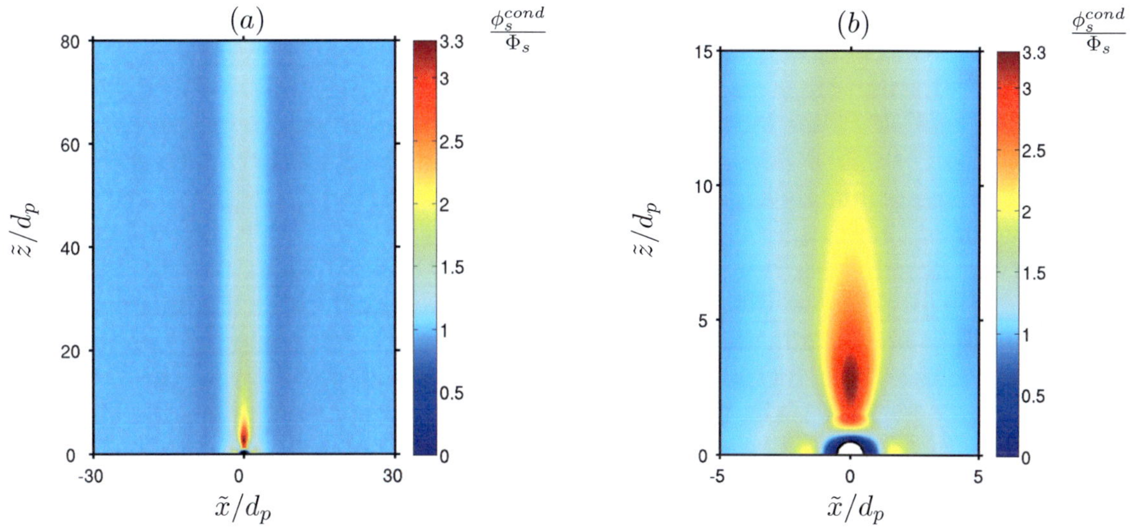

Figure 6.11: As in figure 6.10, but for case $M178$.

The evolution of ϕ_s^{cond}/Φ_s with the distance from the test particle can be better examined by considering the axis profiles of ϕ_s^{cond}/Φ_s through the center of the reference particle. Figure 6.12 depicts the profile of particle conditioned solid volume fraction for both flow cases through the center of the test particle in both the horizontal and the vertical direction. In the horizontal direction (figure 6.12a) the local particle concentration exhibit in both cases different behavior. In case $M121$ the quantity ϕ_s^{cond}/Φ_s increases slowly and reaches unity at a horizontal distance of approximately $\tilde{x} = 2.5d_p$ from the surface of the reference particle. Beyond that distance the solid volume fraction remains close to unity. On the other hand, in case $M178$ the local particle concentration increases more quickly with the horizontal distance and crosses unity at $\tilde{x} \approx 1d_p$ after which it increases further until a maximum value of $\phi_s^{cond} = 2\Phi_s$ located at $\tilde{x} = 1.5d_p$. After reaching maximum ϕ_s^{cond}/Φ_s decreases and converges

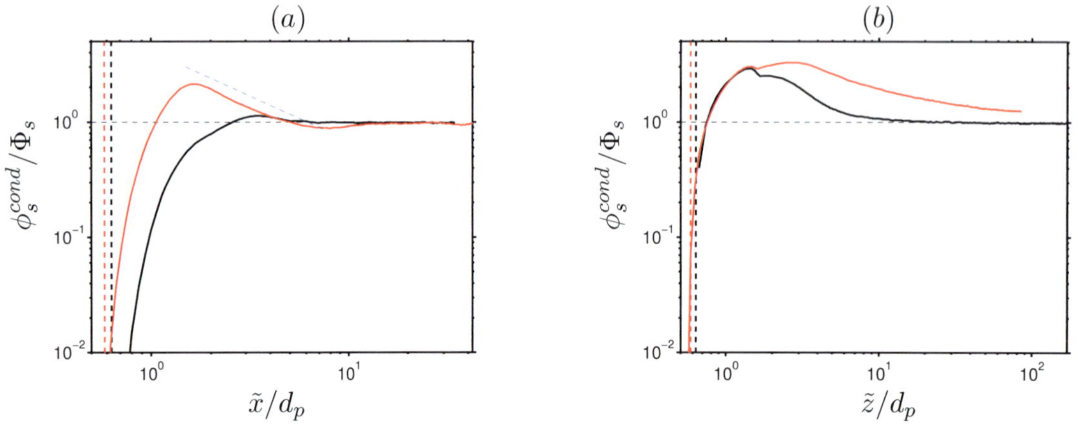

Figure 6.12: Evolution of ϕ_s^{cond} along two perpendicular axis through the reference particle: (a) $\tilde{z} = 0$. (b) $\tilde{x} = 0$. Case $M121$ (——) and case $M178$ (——). The vertical dashed lines (- - - -, - - - -) represent the distance $|\tilde{\boldsymbol{x}}_{ijk}| = d_p/2 + 2\Delta x$, which marks the limit of the range of action of the collisional force. The dashed blue line, (- - - -), indicate decay rate proportional to $\tilde{x}^{-0.65}$.

towards unity at a rate approximately proportional to $\tilde{x}^{-\alpha}$ with $\alpha = 0.65$. As can be seen the region in the horizontal direction which is depleted of particles is much smaller in the clustering case $M178$ than in case $M121$.

In vertical direction (figure 6.12b), the profile of ϕ_s^{cond}/Φ_s in both cases exhibit similar evolution. In case $M121$ the local particle concentration is observed to increase steeply for distances up to $\tilde{z} = 1.5d_p$. Beyond this distance ϕ_s^{cond}/Φ_s decays slowly and reaches unity at distance of approximately $8d_p$. In case $M178$, on the other hand, the graph increases for distances up to $3d_p$, after which it decays slowly, not reaching unity within the present domain size. Thus, the long distance behavior of ϕ_s^{cond}/Φ_s reflects directly the macroscopic clustering structure of the particles, i.e. the particles in case $M178$ form columnar like structure which expends throughout the entire domain in vertical direction. It is interesting to note that while on macroscopic level the particles distribution in both cases evidence difference behavior: (no) clustering in case ($M121$) $M178$, the microscopic structure is remarkably similar in both cases. It appears that the mechanism responsible for this behavior is "robust", which does not depend on the Galileo number (the only difference between the two flow cases).

6.4 Effect of the solid volume fraction

Here the effect of the solid volume fraction upon the particle clustering is discussed. From particular interest are the general shape and extension of the particle cluster as well as the properties of the micro-structure of the dispersed phase. For this purpose we performed a comparison of various statistics related to the particle distribution in case $M178$ and case $ML178$. Please recall, that the only physical difference between the two cases is the solid volume fraction Φ_s (cf. chapter §5). Beside this, the initial flow field and the particle distribution in case $ML178$ corresponds to the latest available state of case $M178$ where nearly 90% of the particles in case $M178$ were randomly removed from the computational domain. Visualization of the initial particle distribution as well at the end of the simulated time interval projected upon the horizontal plane can be found in figure 6.13. As can be seen the initial particle distribution inherits the distribution features of the particles in case $M178$. However, the

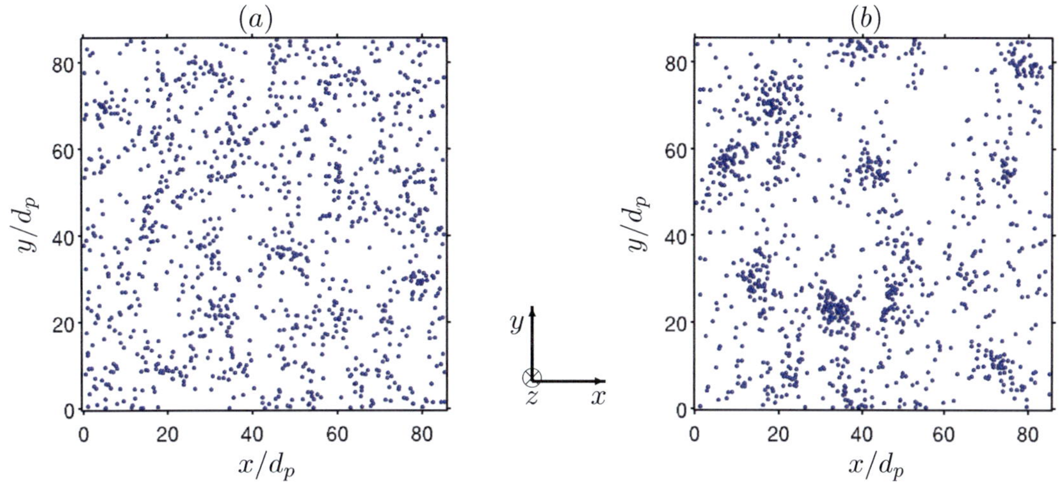

Figure 6.13: The same as in figure 6.2 but for case $ML178$: (a) $t/\tau_g = 0$; (b) $t/\tau_g = 1756$.

particle distribution appear to be slightly more homogeneous and as we will see below, Voronoï analysis confirms that the particles are more randomly distributed than at the end of the simulation in $M178$. On the other hand, the final particle agglomeration in case $ML178$ (figure 6.13) indicate that strong particle clustering takes place revealing large void regions accompanied by localized regions with high number density of particles.

Similar to the analysis in section §6.2, a more quantitative results are obtained by means of Voronoï analysis. Figure 6.14 depicts the corresponding p.d.f.s of the volume and the aspect ratio of the Voronoï cells in case $ML178$. Please note that due to the sufficiently smaller number of particles in case $ML178$ the curves of the p.d.f.s exhibit more noise than in case $M178$. It can be observed that the p.d.f. of the Voronoï cell volume of the initial particle distribution in case $ML178$ indeed reassembles closely the Gamma distribution corresponding to a random Poisson distribution. However, the p.d.f. of the aspect ratio deviates from that for randomly distributed particles with bias towards the p.d.f. in case $M178$, indicating the columnar like organization of the particles. At later times the p.d.f.s of both the Voronoï cell volume and the aspect ratio experience similar shape as in case $M178$ indicating that particles agglomerate in columnar like clusters as in case $M178$. The clustering, however, appear to be more pronounced in case $ML178$ than in case $M178$. This can be best observed in figure 6.15 where the temporal evolution of the standard deviation of the Voronoï volume for case $M178$ and case $ML178$ are depicted. The figure reveals that in fact at begin of the simulation, the standard deviation of the Voronoï cell volume in case $ML178$ is close to the value for randomly distributed particles. As time advances the standard deviation increases steadily for approximately $1100\tau_g$ (note that, this time correlates strongly with the temporal evolution of the particle and fluid quantities shown in section §5.2) after which it levels off and fluctuates around $\sigma(V_i) \approx 0.8$, which is approximately 25% larger than the value in case $M178$ at statistically stationary state. This observations confirms that the initial particle distribution in case $ML178$ is close to random and that in the course of the simulation strong clustering takes place with substantially higher clustering levels than in case $M178$. An alternative nearest-neighbor analysis (figure omitted) verifies this findings as well.

By tracking the particle position in time and applying the intrinsic definition of clusters (voids), we computed the mean residence time of the particles in case $ML178$ in cluster

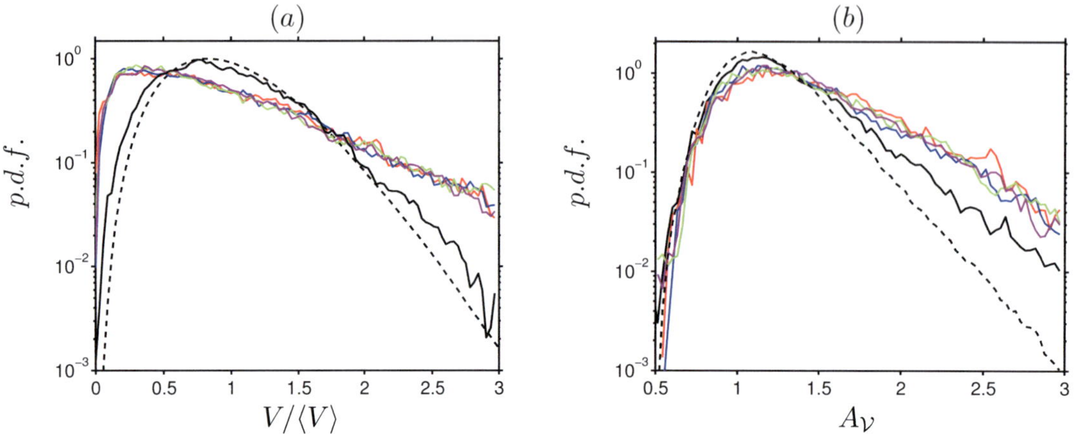

Figure 6.14: As figure 6.4 but for case $ML178$. Different lines represent data assembled over different time intervals. $t/\tau_g = [0,\ 101.2]$ (——). $t/\tau_g = [1205,\ 1250]$ (——). $t/\tau_g = [1367,\ 1431]$ (——); $t/\tau_g = [1586,\ 1653]$ (——); $t/\tau_g = [1699,\ 1756]$ (——). The dashed black line (----) represents: (a) an analytical Gammma function fit (Ferenc and Néda, 2007); (b) Voronoï cell aspect ratio for randomly distributed particles;

and voids respectively. The residence time in clusters measures $90.3\tau_g$ and in voids $166.4\tau_g$. This indicates that the cluster respectively void structures in case $ML178$ are somehow more "robust" and persist over longer times than in case $M178$ and evidence that once the particles enter one of the either regions they remain trapped therein for relatively longer times than in case $M178$. However the number of particles located in cluster and void regions increases only slightly when compared to case $M178$. We evaluated the amount of particles residing in clusters (voids) to approximately 35% (18%).

In order to obtain more insight into the temporal behavior of the particle clusters in case $ML178$ we have calculated the corresponding Lagrangian autocorrelation function of the Voronoï cell volume and the cell aspect ratio (figure omitted). In both cases the autocorrelation function experienced similar behavior, unfortunately, the considered time interval was not sufficient for full decorrelation of the autocorrelation function. Nevertheless, the short time behavior can be addressed by considering the Taylor micro-scale, which calculates for the Voronoï cell volume to $37.1\tau_g$ and for the aspect ratio to $6.9\tau_g$. Comparison with the values obtained for case $M178$ reveals that both values are larger in case $ML178$. While the difference for the Voronoï aspect ratio is relatively small, the values for the Voronoï cell volume in case $ML178$ a nearly three times larger than in case $M178$.

Now let us turn to the conditionally average particle concentration. Figure 6.16 shows the two dimensional map of ϕ_s^{cond}/Φ_s for case $ML178$. It can be clearly seen, that while the overall structures of the particle agglomeration in case $M178$ and case $ML178$ are very similar, the clustering intensity in case $ML178$ is significantly higher than in case $M178$. As in case $M178$, on macroscopic level the particles in case $ML178$ are organized in highly anisotropic columnar like structures. However the region where the maximum of the local particle concentration is observed is substantially larger in case $ML178$ extending in the vertical direction up to $20d_p$ from the surface of the reference particle (recall that in case $M178$ this distance was measured to be $3d_p$). Its extensions in the horizontal direction increased substantially as well, now covering region of $10d_p$ on each side of the reference particle. As in case $M178$ the local solid volume fraction in the horizontal direction experiences values larger than the average

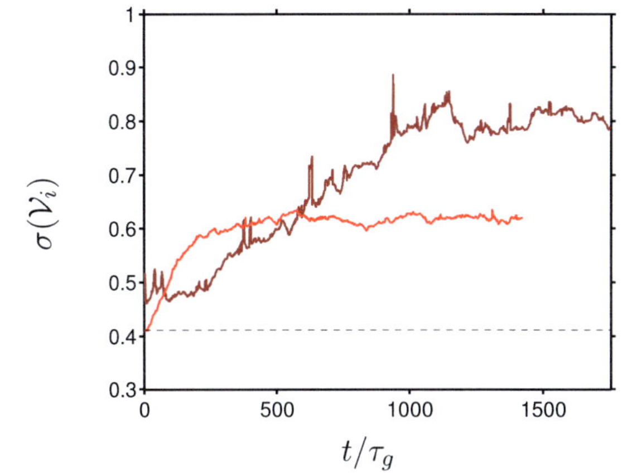

Figure 6.15: Standard deviation of Voronoï cell volumes as function of time for case $M178$ (——) and case $ML178$ (——). The corresponding value for randomly distributed particles with uniform probability (- - - -).

value, however, the region experiencing maximum values covers larger area and the maximum value can be observed at distances up to $4.6d_p$. This can be best observed in the close-up in figure 6.16b showing the distribution of the local particle concentration in vicinity of the test particle.

The spatial variation of the particle conditioned solid volume fraction can be more conveniently examined by considering the horizontal and the vertical profiles through the center of the particle, as shown in figure 6.17. In the horizontal direction (figure 6.17a) the profile of ϕ_s^{cond}/Φ_s in case $ML178$ increases more quickly than in case $M178$ and reaches unity approximately at $\tilde{x} = 0.83d_p$ after which it increases further until a maximum value of $\phi_s^{cond} = 5.1\Phi_s$ located at $1.6d_p$. The local particle concentration then decays at higher rate than in case $M178$ with values of $\alpha = 0.9$. In contrast to case $M178$ the profile of ϕ_s^{cond}/Φ_s converges towards a slightly negative value $\Phi_s^{cond} = 0.85\Phi_s$, indicating the deficit of particles caused by the strong clustering. Furthermore, the region depleted of particles in direct vicinity of the reference particle is much smaller compared to case $M178$. In the vertical direction (figure 6.17b), the profile of ϕ_s^{cond}/Φ_s in case $ML178$ evidences similar evolution with $\tilde{z}$ as in case $M178$. However, the maximum value in case $ML178$ is approximately three times larger than in case $M178$, i.e. $\phi_s^{cond} = 9.1\Phi_s$. In contrast to case $M178$, the profile of ϕ_s^{cond} remains flat over a distances ranging from $\tilde{z} = 1.3d_p - 3.87d_p$. The particle conditioned concentration then decays slowly towards a value $\phi_s^{cond} = 2.01\Phi_s$, which is about 38% larger than in case $M178$.

6.5 Conclusion

In the present chapter we have investigated the spatial structure of the dispersed phase in flows of heavy finite-size particles settling under the influence of gravity in an initially ambient fluid. Here we considered the simulations introduced in chapter §5, which allowed us to investigate (although limited to the number of parameters in the parameter space) the effects of the Galileo number and the solid volume fraction.

Concerning the effect of the Galileo number, a strong clustering was observed for the particles with the higher Galileo number of Ga $= 178$, while no evidence of clustering was

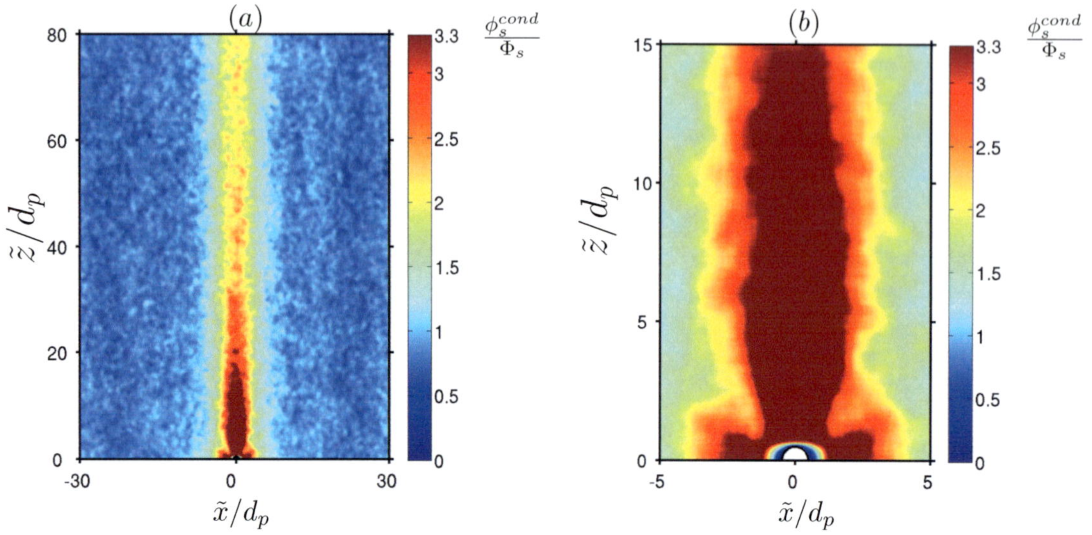

Figure 6.16: As in figure 6.10, but for case $ML178$.

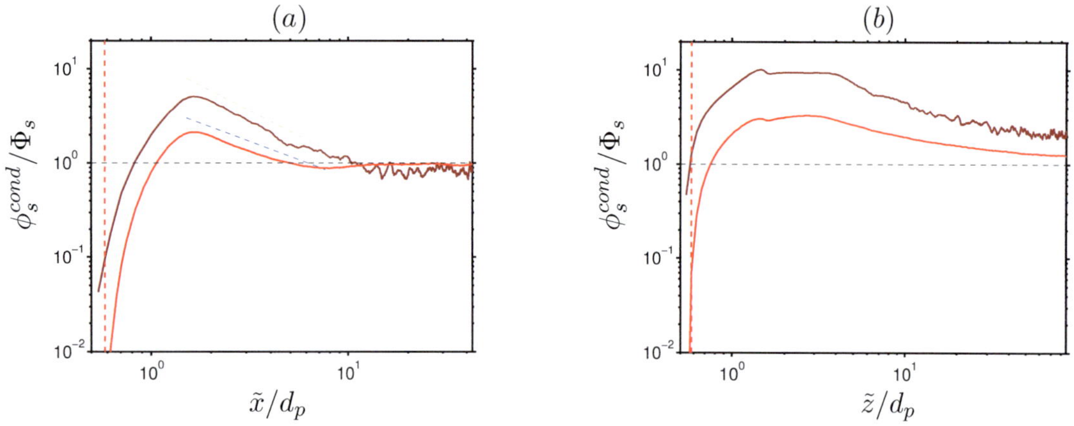

Figure 6.17: Evolution of ϕ_s^{cond} along two perpendicular axis through the reference particle: (a) $\tilde{z} = 0$. (b) $\tilde{x} = 0$. Case $M121$ (——) and case $M178$ (——). The vertical dashed lines (----, ----) represent the distance $|\tilde{\boldsymbol{x}}_{ijk}| = d_p/2 + 2\Delta x$, which marks the limit of the range of action of the collisional force. The dashed blue line, (----), indicate decay rate proportional to $\tilde{x}^{-0.65}$. The dashed blue line, (----), indicate decay rate proportional to $\tilde{x}^{-0.9}$.

observed for Ga $= 121$, on the contrary the particles rather organized in slightly more regular manner than a randomly distributed particles. The particles in the clustering case were observed to agglomerate in highly anisotropic columnar like structures extending in the present simulation in the vertical direction throughout the entire domain length of 170 particle diameters. On the other hand, in the horizontal direction the particle cluster extended up to equivalent diameter of 10 particle diameters.

Voronoï analysis revealed that the residence time of the particles inside cluster or void regions is relatively long compared to the gravitational time scale. An intrinsic cluster (void) definition revealed that nearly 30% (15%) of the particles reside in clusters (voids). Lagrangian analysis of the Voronoï cell volume revealed that the particle structures maintained coherency over substantially long time. Study of the short time behavior of the particle structures showed that the particles in both cases experienced similar behavior.

Similarity between both flow cases was also found in the micro-structure of the dispersed phase, which was studied by means of the particle-condioned average of the solid volume fraction. This indicated that while on macroscopic level the two cases exhibit completely different spatial structure of the dispersed phase, on microscopic level they appear to feature similar micro-structure, indicating that under this conditions the mechanism responsible for this behavior does not depend on the Galileo number and that the short range particle-to-particle interaction is remarkably "robust".

Regarding the effect of the solid volume fraction upon the cluster characteristics, we have performed a third simulation with the same physical and numerical parameters as the simulation with Ga $= 178$, except for the solid volume fraction, which was reduced by the factor of ten. Voronoï analysis and study of the conditionally averaged particle concentration showed that while the overall cluster structure was maintained, the level of clustering increased with decrease of the solid volume fraction which manifested itself in the micro-structure of the dispersed phase. The micro-structure revealed that for the smaller value of the solid volume fraction the extensions of the region with higher local particle concentration than the average increased substantially. The higher clustering levels and the lower solid volume fraction (small number of particles) led to a net deficit of particles in the horizontal direction, when compared to the average concentration.

The present results are in line with the observed cluster structure made by Kajishima and Takiguchi (2002); Kajishima (2004a). When the clustering of bubbles in pseudo turbulence is considered, similar vertical orientation of the bubble clusters was found by Esmaeeli and Tryggvason (2005) and more lately by Martínez-Mercado et al. (2010). In the later work the authors reported robust vertical alignment of the bubbles at both small and large scales. The particle respectively the bubble Reynolds number in all of this works was of the order of $\mathcal{O}(100)$. For smaller particle (bubble) Reynolds numbers of the order of $\mathcal{O}(10)$ the works of Cartellier and Riviére (2001); Cartellier et al. (2009); Yin and Koch (2007) investigated the micro-structure of the dispersed phase and found a deficit of neighboring particles downstream of a reference particle and increased probability of finding a particle located horizontally from the test particle. Yin and Koch (2007) attributed this to the wake induced "drafting-kissing-tumbling" mechanism (Hu et al., 1993). In a numerical work of bubbly flows Sangani and Didwania (1993); Smereka (1993) suggested that, when potential flow is assumed, the bubbles form horizontally aligned rafts. This leads to the question of how important is the local flow around the particles, in particular the recirculation in the wake of the particles for the orientation of the particle agglomerations, since for particle Reynolds number below $Re_p < 20$ no, or negligible separation is observed (Clift et al., 1978).

In view of the results in case $M121$ and $M178$ the question arise what is the exact mechanism responsible for the clustering of the particles. Kajishima (2004a) identified the wake attraction in the sense of "drafting-kissing-tumbling" as the main reason for the clustering of the particles. However, this does not account for the difference in the particle distribution in both cases investigated here. As we saw in chapter §4 an isolated particle at the Galileo numbers and particle-to-fluid density ratio considered in the present study settle steadily on a vertical and on an oblique path, respectively. Thus, the question arise to what extent the regime of particle settling is responsible for the particle clustering and it merits a further investigation. Unfortunately due to the high computational cost, at the present it was not possible to perform a detailed parameter study in order to determine an exact value of the critical Galileo number demarcating the onset of clustering. A possible critical value for the Galileo number could be the bifurcation point of the wake of an isolated particle, $\mathrm{Ga} = 155$ (cf. chapter §3). In a future work the identifying of the critical Galileo number would allow more detailed analysis of the mechanism of wake-induced particle attraction.

Chapter 7

Particle motion

In this chapter the motion of the finite-size particles in case $M121$, $M178$ and $ML178$, introduced in chapter §5, is investigated. Particularly the coupling phenomena related to the influence of the particle-induced flow field upon the mean particle settling rate and the hydrodynamic forces exerted by the fluid on the particles are elucidated in detail. The main question addressed here is whether the presence of multiple particles would lead to a net increase or decrease of the mean particle settling rate. In the context of the particle settling rate, commonly the relative velocity between the particles and the fluid is considered. This leads to the question of the most appropriate definition of the fluid velocity "seen" by the particles with respect to which the relative particle velocity is computed. This is addressed here by providing a definition for the fluid velocity "seen" by the particles. Also of great interest are the (statistical) consequences of the particle clustering on the motion of the particles. For this reason a comprehensive analysis of the correlation between various particle and cluster quantities is performed. Likewise, the contributions of the local flow field on the particle velocity field and forces is identified and analyzed. Finally a limited study of the effect of the solid volume fraction on the particle motion is investigated. For more detailed introduction to the topic the reader is referred to sections §1.2.1 and §5. The chapter is organized as follows. In section §7.1 analysis of the mean particle properties such as the particle settling velocity and hydrodynamic forcesis performed. Here also a definition of the velocity "seen" by the particles is provided and discussed. Section §7.2 is devoted to the analysis of the fluctuations of various particle quantities. In particular the probability density functions and the Lagrangian statistics are discussed. The chapter then closes in section §7.3 with a brief conclusion.

7.1 Mean particle settling velocity and hydrodynamic force

Before we proceed with the results, let us first provide a definition of the local fluid velocity seen by the particles. As we will see the choice of the reference fluid velocity used for the calculation of the particle relative velocity is vital for the understanding of the particle motion in the present flow configurations.

7.1.1 Fluid velocity seen by the particle

In the context of the mean particle settling rate, commonly the relative velocity of the particles to the fluid is considered. Thus, a reference fluid velocity that is used for the calculation of the relative velocity is required. For complex flows (multiple mobile particles, (pseudo) turbulence, etc.) usually the free-stream fluid velocity, to which each particle is subjected, is unknown. In such flow configurations usually the particles are assumed to be smaller than the smallest relevant flow structures. This leads to the so called point-particles models, where, in case of one-way coupling between the two phases, the relative velocity of the particles is often calculated with respect to the fluid velocity at the particle's center. For finite-size particles, as in the present case, this approach is, however, not justified, due to two-way coupling phenomena (cf. section §1.2.1). Since for such configurations there is no separation of scales, the main challenge in determining the relative velocity is the calculation of the undisturbed free stream fluid velocity, which is "seen" by the particles.

There have been several attempts to define the free stream fluid velocity "seen" by the particles. Bagchi and Balachandar (2003) investigated the effect of homogeneous isotropic turbulence on the drag and lift of a single stationary particle subjected to isotropic turbulent flow field. They examined two different approximations of the undisturbed fluid velocity "seen" by the particle. In their study this velocity was alternatively defined (i) as the fluid velocity at the position of the particle in an otherwise identical particle-free simulation, or (ii) as the fluid velocity averaged over a spherical volume around the particle in the corresponding particle-free simulation. Later Zeng et al. (2008) simulated the flow over a fixed sphere in a non-periodical channel. Similarly to the work of Bagchi and Balachandar (2003) they used the fluid velocity at position of the particle center. Merle et al. (2005) investigated the forces acting on a bubble in a turbulent pipe flow. A Taylor's hypothesis was invoked to approximate the undisturbed fluid velocity "seen" by the bubble without recurring to a particle-free companion simulation. The authors of the just mentioned studies highlighted, however, that in case of finite-size particles the use of the fluid velocity at the position of the particles is questionable. In more recent work Lucci et al. (2010) investigated the modification of isotropic turbulence by particles with size of the order of the Taylor length-scale in the absence of gravity. They proposed two definitions of the fluid velocity "seen" by the particles: (i) the fluid velocity in direction of the particle velocity vector taken at distance comparable to the Taylor micro-scale and (ii) the average of the fluid velocity over a small spherical cap along the particle velocity vector taken at the same distance. In both cases the definition of the fluid velocity "seen" by the particles is biased by the choice of the direction in which the fluid velocity is considered. As noted by Kidanemariam et al. (2013), the use of directional information from the instantaneous particle velocity is rather questionable. Recently Bellani and Variano (2012a) defined a stochastic slip velocity for particles of Taylor length-scale size. The definition of the slip velocity proposed by Bellani and Variano (2012a) does not require any filtration of the flow phase velocity field. However, the authors noted that this approach is applicable to cases where no preferential concentration of the particles is present, otherwise they suggest once again the use of conditional averaging of the flow field in vicinity of the particle.

Since inertial finite-size particles tend to "filter" the flow field (Qureshi et al., 2007) we believe that the fluid velocity "seen" by the particles should be comprised from the local flow field. Therefore, in the present work we use a definition of the fluid velocity seen by the particles similar to that proposed by Kidanemariam et al. (2013), where the instantaneous fluid velocity "seen" by the particles is approximated by the average of the fluid velocity located on a spherical surface $\mathcal{S}$ of radius R_S centered at the particle's center location. As discussed in Kidanemariam et al. (2013) the choice of the radius R_S has to fulfill the following two requirements: (i) the radius R_S should be chosen large enough in order to avoid any possible influence of the flow field, due to the presence of the particle and (ii) R_S should be chosen sufficiently small for the result to be of relevance to the motion of the respective particle. Please note that for sufficiently large number of particles and when the shell radius R_S tends to infinity the fluid velocity $\boldsymbol{u}_f^S$ tends to the globally averaged fluid velocity $\boldsymbol{u}_f$. In the present work we chose the same averaging radius as Kidanemariam et al. (2013), i.e. we set for all simulations $R_S = 1.5d_p$. In their work they used the result of three different direct numerical simulations of type $S121$ and $S178$ for the calibration of the radius R_S, thus we believe that the choice of R_S is justified in the present flow cases. For detailed definition of the fluid velocity seen by the particles and a discussion of the chosen value for R_S the reader is referred to Kidanemariam et al. (2013). Hereafter, the so defined fluid velocity corresponding to the i-th particle at instance t will be denoted by $\boldsymbol{u}_f^S(i, R_S = 1.5d_p, t)$. In the present work the radius R_S is kept constant in all three cases, thus, for sake of simplicity the parameter R_S will be omitted from the list of arguments of $\boldsymbol{u}_f^S$, i.e. $\boldsymbol{u}_f^S(i,t) = \boldsymbol{u}_f^S(i, R_S = 1.5d_p, t)$.

By definition the fluid velocity $\boldsymbol{u}_f^S$ samples only a subset of the fluid velocity $\boldsymbol{u}_f$ and accounts for the contribution of the local flow field in vicinity of the particles. Thus, it is interesting to compare the statistical description of the particle conditioned fluid velocity $\boldsymbol{u}_f^S$ with the unconditioned fluid velocity field $\boldsymbol{u}_f$. In order to do so, we compared the first and second moments (mean value and standard deviation) of the vertical component of the fluid velocities $w_f^S(i,t)$ ($\forall i = 1 \ldots N_p$) and $w_f(\boldsymbol{x}, t)$, which can be found in table 7.1. It can be seen, that in all flow cases the average value of the particle conditioned fluid velocity is smaller than the average value of the unconditioned fluid velocity. Moreover, the smallest difference was observed in the case with the lower Galileo number and accounts for approximately 3% of the absolute value of the average particle settling velocity $|w_s|$, indicating that the local fluid velocity in vicinity of the particles is only marginally affected by the presence of the particles and resembles closely the unconditioned average of the fluid velocity. On the other hand, the difference becomes larger for the flow cases with the higher Galileo number and accounts in case $M178$ for about 11% and in case $ML178$ for nearly 20% of the velocity $|w_s|$. This implies that the particle in case $M178$ and $ML178$ are on average surrounded by fluid with significantly smaller vertical velocity than the box averaged value, i.e. they are located inside regions with a downward fluid motion when compared to the globally averaged fluid velocity. This is in line with the observation made in the previous chapter that the particles in case $M178$ and case $ML178$ are organized in a clusters with columnar like shape. It appears also that the values increase with the clustering level (cf. chapter §6; in case $ML178$ a higher clustering level was observed than in case $M178$).

Regarding the standard deviation of the velocities $\boldsymbol{u}_f^S$ and $\boldsymbol{u}_f$, it can be seen that the values increase with the Galileo number and the standard deviation of the particle conditioned

case	$(\langle w_f^S \rangle_{p,t} - \langle w_f \rangle_{\Omega_f,t})/w_{ref}$	$\sigma(w_f^S)/w_{ref}$	$\sigma(w_f)/w_{ref}$
$M121$	-0.0315	0.1262	0.2081
$M178$	-0.1155	0.1972	0.2407
$ML178$	-0.1965	0.2852	0.102

Table 7.1: Comparison of the first and second moments (mean value and standard deviation) of the vertical component of the particle conditioned fluid velocity $w_f^S(i,t)$ ($\forall i = 1 \ldots N_p$) and the unconditioned velocity field $w_f(\boldsymbol{x}, t)$ for all three flow cases $M121$, $M178$, $ML178$. All values are normalized with the absolute value of the time averaged mean particle settling velocity $|\langle w_{p,rel} \rangle_{p,t}|$.

local fluid velocity is significantly smaller than for the unconditioned velocity, except for case $ML178$. Most probably the smaller value of the standard deviation of the unconditioned fluid velocity in case $ML178$ can be originated in the smaller solid volume fraction. Please recall that all the fluctuations in the carrier phase are due to the presence of the particles and the solid volume fraction in case $ML178$ is factor of ten smaller than in case $M178$. That is, for decreasing solid volume fraction and otherwise identical parameters it can be expected that the overall intensity of the fluid velocity fluctuations in the computational box would steadily decrease. On the other hand, it is interesting to note that the standard deviation of the local fluid velocity $\boldsymbol{u}_f^S$ in case $ML178$ remains of the same order as in case $M178$, although with higher value, which once again stems most probably from the higher clustering level observed in case $ML178$.

Preliminarily, we can conclude that the particle conditioned averaging over a spherical shell of diameter $R_S = 1.5d_p$ reflects the fact that the particles do not sample the flow field randomly and appears to acts like a spatial filter which tends to suppress small scale fluctuations. However, as we will see later in chapter §8 the distribution of the particle conditioned fluid velocity in case $M121$ and $M178$ are very similar to those of the unconditioned field, while in case $ML178$ a significant difference between the two distributions can be observed.

7.1.2 Particle average settling velocity

Now let us turn to the average settling velocity of the particles. With the above definition of the fluid velocity in vicinity of the particles, we can now define the instantaneous relative velocity based upon the local reference value of the fluid velocity as:

$$\boldsymbol{u}_{p,rel}^S(i,t) = \boldsymbol{u}_p^i(t) - \boldsymbol{u}_f^S(\boldsymbol{x}_p^i, t), \quad \forall i = 1 \dots N_p. \tag{7.1}$$

The definition of the particle relative velocity in equation 7.1 is expected to provide better representation of the local flow conditions which affect the particle motion, since it accounts for the local flow conditions of each individual particle. Please note that the definition of the relative velocity in equation 7.1 differ from the conventional definition of the relative velocity in equation 5.1 only in the choice of the reference fluid velocity. Thus, the difference in the particle relative velocity reduces to the difference in the values of $\boldsymbol{u}_f^S$ and $\boldsymbol{u}_f$. The instantaneous mean settling velocity of the particles with respect to the local fluid velocity can be now simply computed by applying an average over the particle ensemble to the vertical component of the velocity $\boldsymbol{u}_f^S$, viz.:

$$w_s^S(t) = \langle w_{p,rel}^S(i,t) \rangle_p. \tag{7.2}$$

Hereafter, the particle Reynolds number based on the velocity w_s^S will be denoted by $Re_{p,rel}^S$ and is defined as:

$$Re_{p,rel}^S = |w_s^S| d_p / \nu. \tag{7.3}$$

Figure 7.1 depicts the temporal evolution of the mean particle settling velocity w_s^S along with the particle average of the conventional settling velocity w_s, which was already briefly discussed in section §5.2. The data is normalized as in figure 5.2 with the viscous velocity scale based on the particle diameter, i.e. $w_s d_p / \nu$. It can be observed that the temporal evolution of the mean particle settling velocity w_s^S in case $M121$ resembles closely the temporal evolution of the velocity w_s. The velocity w_s^S remains essentially constant over the entire course of

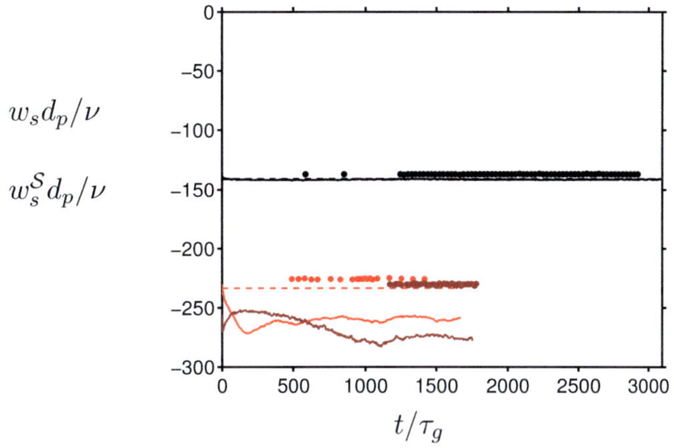

Figure 7.1: Average of the relative particle settling velocity, w_s (equation 5.1) and w_s^S (equation 7.2), for case $M121$ (——), case $M178$ (——) and case $ML178$ (——) as function of time. The velocities w_s and w_s^S have been normalized by the ratio of the particle diameter d_p and the kinematic viscosity ν. Dashed lines indicate the terminal settling velocity of the single particle from case $S121$ (- - - -) and case $S178$ (- - - -). The settling velocity with respect to the fluid velocity in the vicinity of the particles as defined in equation 7.2 and is depicted by the circular symbols with corresponding color (•, •, •).

the simulation at absolute values slightly smaller than the values of $|w_s|$. The corresponding particle Reynolds number was evaluated to $Re^S_{p,rel} = 137.1$ and the difference between the two velocities was measured to be approximately 3%. Please recall, that the difference between the two values stems only from the difference in the values of the velocities $\langle \boldsymbol{u}_f^S \rangle_p$ and $\langle \boldsymbol{u}_f \rangle_{\Omega,f}$, which was measured to be approximately 3% (cf. table 7.1). On the other hand, due to the significant difference between the velocities $\boldsymbol{u}_f^S$ and $\boldsymbol{u}_f$ in case $M178$ and $ML178$ the mean particle settling rate w_s^S deviates significantly from the conventional average settling velocity w_s. In both cases the mean settling rate of the particles was measured to be smaller, but close to that of the single particle from case $S178$. In case $M178$ the difference between the terminal settling velocity in case $S178$ and the mean particle settling rate w_f^S was evaluated to be 3.3% and, surprisingly, in case $ML178$ even smaller difference of 1.4% was measured. Please recall that the absolute value of the mean particle settling rate based on the unconditional average of the fluid velocity was substantially larger than the terminal settling velocity of the single particle from case $S178$. This results implies that the particles are preferentially located (on average) in regions with negative fluid velocity fluctuations and mean values of the fluid velocity smaller than the box-average. The graphs in figure 7.1suggest that the choice of the reference fluid velocity "seen" by the particles could play a significant role in the determination of the mean apparent velocity lag, especially when the particles exhibit inhomogeneous spatial distribution. Further, this strengthen the remarks in the referenced literature that an inhomogeneous particle distribution can lead to large spatial variation in the particle relative velocity and that the local flow field should be incorporated in the calculation of the mean particle settling rate.

As we saw in the previous chapter §6 the particles in case $M178$ and case $ML178$ agglomerate into clusters, i.e. regions experiencing high particle concentrations. Combined with the above observations for the particle relative velocity it is tempting to relate the higher particle relative velocity to the particle located inside clusters. In order to quantify this, we have calculated the joint p.d.f. of the fluctuations of the particle settling velocity w_s (equation 5.1)

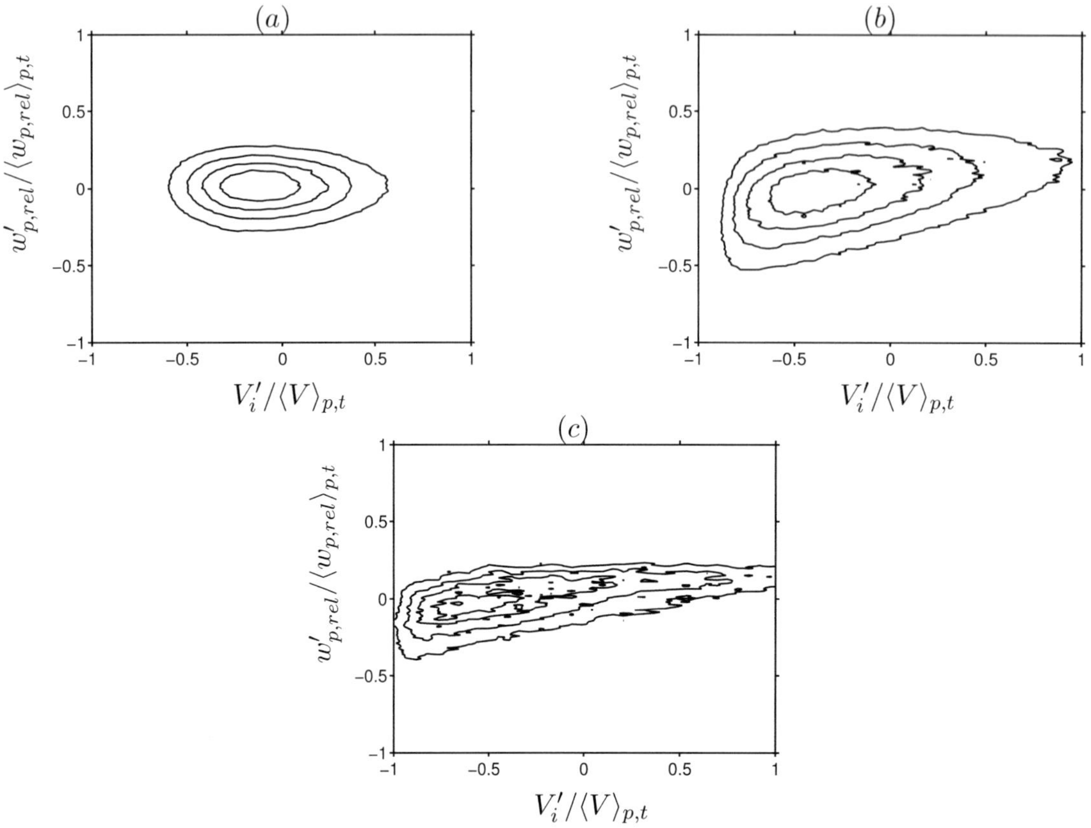

Figure 7.2: Joint p.d.f.s of the vertical component of the relative particle velocity fluctuations $w'_{p,rel} = w_{p,rel} - \langle w_{p,rel}\rangle_{p,t}$ and the fluctuations of the Voronoï cell volumes $V'_i = V_i - \langle V\rangle_{p,t}$. (a) Case $M121$, (b) case $M178$ and (c) $ML178$. In all three plots the lines indicate contours at $0.2, 0.4, 0.6, 0.8$ times the maximum number of occurrences.

and the fluctuations of the Voronoï cell volume associated to the corresponding particle at the corresponding time. Figure 7.2 depicts the corresponding p.d.f.s for all three cases. It can be observed that in case $M121$, as expected, the two quantities correlates only weekly with a correlation coefficient of 0.002. On the other hand, the correlation in the two flow cases, where clustering was observed ($M178$ and $ML178$) a positive correlation between the Voronoï volumes and the particle relative velocity can be observed. The graphs in figure 7.2b,c evidence that particles associated with small Voronoï cells tend to poses faster than the average settling velocities (in absolute value) and vice versa particles located in large Voronoï cells own smaller-than-the-average settling velocities. Thereby, the correlation case $ML178$ appears to be more pronounced than in case $M178$, as evidenced by the larger value of the correlation coefficient which were evaluated to 0.347 and 0.4531 in case $M178$ and case $ML178$, respectively. This confirms the above suggestion, that particles located inside clusters settle on average faster than particles located in void regions.

As aforementioned, the particle in case $M178$ and $ML178$ are surrounded on average by fluid with significantly smaller vertical velocity than the box averaged value, i.e. they are located inside regions with a downward fluid motion when compared to the globally averaged fluid velocity. Additionally the clusters in case $M178$ and case $ML178$ were observed to be oriented in the vertical direction and to extend throughout the entire vertical length of the computational domain with clearly visible patterns of the horizontal projection (figures 6.1, 6.2 and 6.13). Thus, one could expect that the downward fluid motion correlates with the higher velocity of the particles located inside clusters. This can be qualitatively illustrated in figure 7.3, where the instantaneous position of the particles (viewed from the top) and the instantaneous vertical average of the vertical component of the fluid velocity are illustrated. The particles in figure 7.3 are colored by the fluctuations of their respective settling velocity $w'_{p,rel}(i,t)$ and the vertical average of the fluid velocity fluctuations is calculated as:

$$\hat{w}'_f(x,y,t) = \langle w_f(\boldsymbol{x},t)\rangle_z - \langle w_f(\boldsymbol{x},t)\rangle_{\Omega_f}, \tag{7.4}$$

where the definition of the averaging operator $\langle\cdot\rangle_z$ can be found in appendix A. It can be observed that the vertical component of the relative velocity in case $M178$ and $ML178$ strongly correlates with the spatial distribution of the particles. At locations with high (low) particle concentration, particles with larger (smaller) absolute values than the average of the vertical relative velocity are observed. Moreover, the companion figure for the vertically averaged vertical component of the fluid velocity shows strong correlations with the regions with high (low) particle concentration. A significantly smaller than the average fluid velocities are observed at locations with notable particle agglomerations and vice versa for voids. This indicates the presence of coherent flow structures occupying regions in the horizontal plane of the order of the size of the particle clusters (cf. chapter §8, figures 8.1 - 8.2). The effect of the solid volume fraction is clearly visible by a comparison of the two clustering cases. Although the overall features of the flows in both cases are very similar, it can be observed that, as expected, the void regions in case $ML178$ are significantly larger than in case $M178$ and the deviation from the average value of both, the particle relative velocity and the fluid velocity $\hat{w}'_f$ is smaller than in case $M178$. Time sequences of the plots for case $M178$ and case $ML178$ in figure 7.3 (not shown here) reveal that the flow structures (and the particle clusters) do not change significantly with time.

In case $M121$, on the other hand, such strong correlations between the particle position and the relative velocity are not clearly visible. On the contrary, fast and slowly settling particles appear to be much less clearly segregated in the horizontal direction. Concerning the vertically average fluid velocity, in case $M121$ significantly smaller values of $\hat{w}'_f$ than in the cases $M178$ and $ML178$ are observed. Nevertheless, in case $M121$ a small but finite correlation between the instantaneous particle settling velocity an the vertically averaged fluid velocity can be observed, indicating the presence of, although weak, large scale fluid structures in the background flow.

In order to better quantify the impact of the particle clustering on the particle settling velocity we performed a particle-conditioned averaging of the vertical component of the particle relative velocity $w_{p,rel}$. This allows us to provide a local (macroscopic) view on the relative distribution of the particle relative velocity in vicinity of the particle surface. The averaging procedure corresponds to the particle-conditioned averaging of the hydrodynamic force introduced in section 4.3.3. The number of samples used for the averaging is the same as in the calculation of the particle-conditioned solid volume fraction in section 6.2. For more details the reader is referred to appendix A. The particle-conditioned value of the settling velocity $w_{p,rel}(i,t)$ will henceforth be denoted as $\tilde{w}_{p,rel}(\tilde{x},\tilde{z})$, i.e. $\tilde{w}_{p,rel}(\tilde{x},\tilde{z}) = \langle w_{p,rel}(i,t)\rangle_{p,t}$.

Figure 7.4 shows the map of the average velocity $\tilde{w}_{p,rel}$ for all three cases. The velocity $\tilde{w}_{p,rel}$ has been normalized with the absolute value of the long-time average of the particle settling velocity, i.e. $|\langle w_{p,rel}\rangle_{p,t}|$. Note that, as the particle-conditioned average of the hydrodynamic

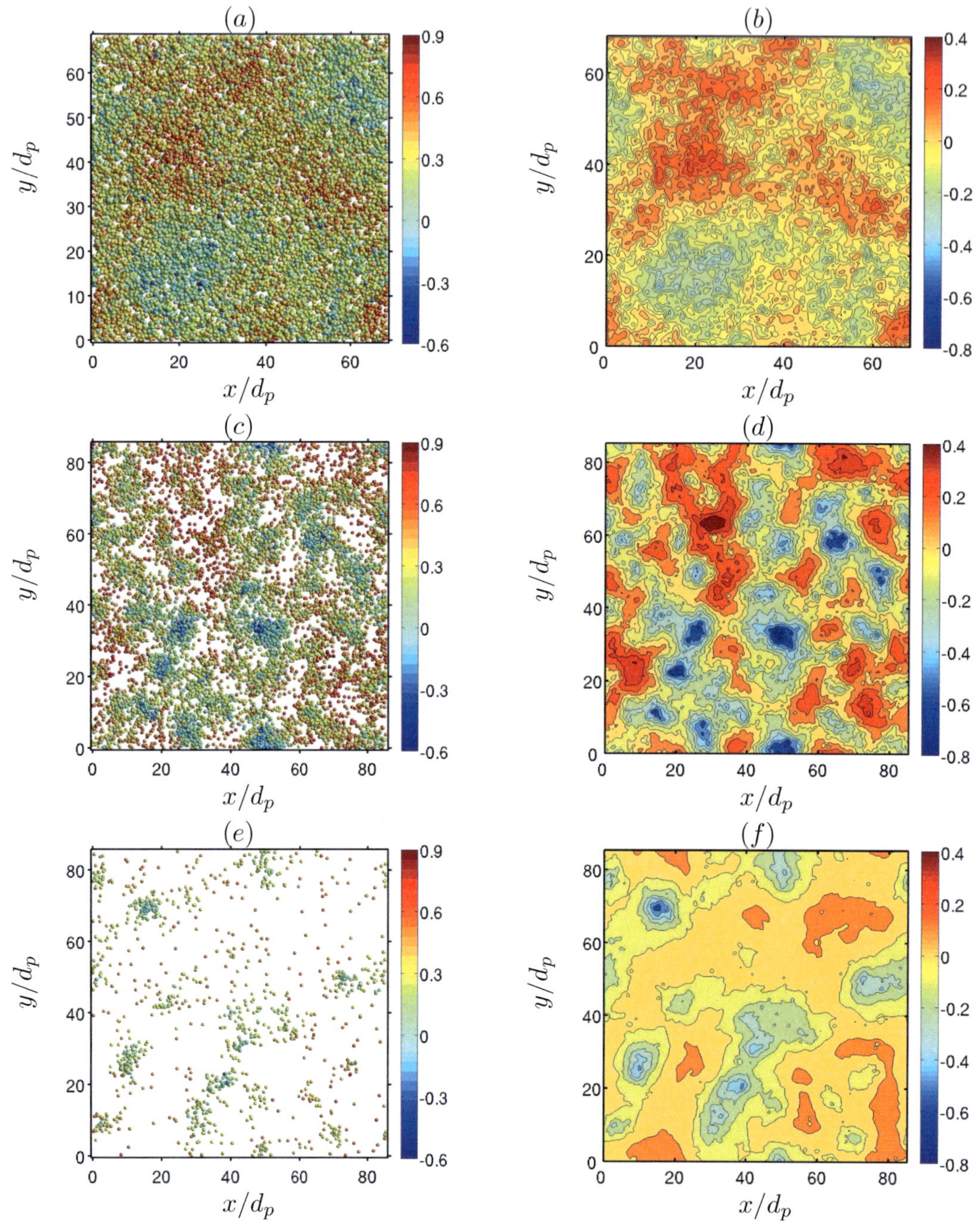

Figure 7.3: (a,c,e) Top view of the particle locations at one instant in time. The particles are colored according to the fluctuation of their settling velocity $w'_{p,rel}/\langle w_{p,rel}\rangle_{p,t}$, where $w'_{p,rel} = w_{p,rel} - \langle w_{p,rel}\rangle_{p,t}$. (b,d,f) Instantaneous, vertically-averaged fluid velocity fluctuations normalized with the instantaneous box-average, $\hat{w}'_f(x,y,t)/\langle w_f(\boldsymbol{x},t)\rangle_{\Omega_f}$ where $\hat{w}'_f(x,y,t) = \langle w_f(\boldsymbol{x},t)\rangle_z - \langle w_f(\boldsymbol{x},t)\rangle_{\Omega_f}$. (a,b) Case $M121$ at $t/\tau_g = 1100.1$, (c,d) case $M178$ at $t/\tau_g = 671.8$ and (e,f) case $ML178$ at $t/\tau_g = 1167.1$.

force, the quantity $\tilde{w}_{p,rel}$ is not invariant with respect to the vertical direction $\tilde{z}$. Thus, the map of the particle-conditioned particle settling velocity is not symmetric with respect to the horizontal axis $\tilde{x}$. Figure 7.4 clearly shows that in all three flow cases the probability of finding a particle with higher than the average settling velocity is confined in the wake region

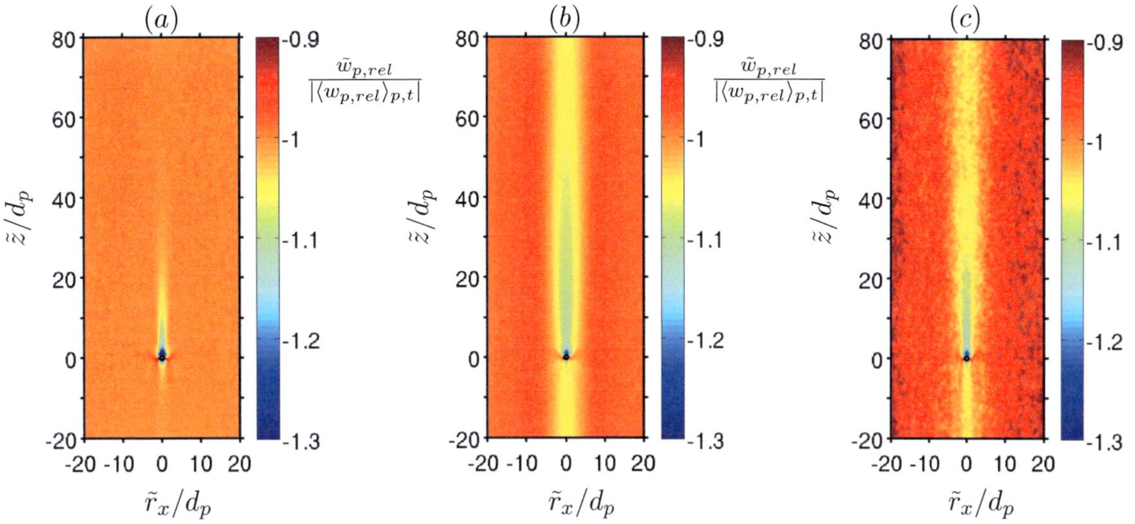

Figure 7.4: Spatial map of the particle-conditioned average value of the settling velocity, $\tilde{w}_{p,rel}$ for (a) case $M121$, (b) case $M178$ and (c) case $ML178$. The velocity $\tilde{w}_{p,rel}$ is plotted as function of the position $(\tilde{x}, \tilde{z})$ with respect to a test particle (vertical plane passing through the center of the reference particle).

of the reference particle, confirming that the trailing particles are settling on average faster than the average. In case $M121$ this effect is limited to a compact region extending up to approximately $40d_p$ downstream of the test particle, after which the velocity $\tilde{w}_{p,rel}$ adopts values close to the average $|\langle w_{p,rel}\rangle_{p,t}|$. It can be observed that the domain of influence of the reference particle in the horizontal direction is confined in a narrow region of the order of the particle diameter. In case $M178$ and case $ML178$ on the other hand, the particle-conditioned average of the particle settling velocity shows features similar to the particle-conditioned solid volume fraction ϕ_s^{cond}, i.e. it exhibits an anisotropic spatial distribution oriented in the vertical direction, which extends downstream from the test particle throughout the entire vertical length of the computational domain. The absolute values of the particle settling velocity along this vertical stripe is significantly larger than the ensemble average. Similar to ϕ_s^{cond} this region extends in the horizontal direction approximately to 10 particle diameter from the test particle.

The evolution of the particle-conditioned velocity $\tilde{w}_{p,rel}$ with the distance from the reference particle can be studied more conveniently by considering the profiles of $\tilde{w}_{p,rel}$ along the vertical and the horizontal axes through the center of the reference particle, which are depicted in figure 7.5. In the horizontal direction (figure 7.5a) the particles in case $M121$ settle on average with smaller settling velocity than the global average for distances up to $5d_p$, beyond this distance the settling velocity is found to be equal to the global average. In case $M178$ the opposite is observed: the particles in horizontal distances up to $5d_p$ (extensions of the average cluster) settle on average with higher velocities than the global average. For larger distances ($\tilde{r}_x > 5d_p$) the settling velocity of the particles is found to be smaller than the average quantity. In case $ML178$ the particle settling velocity is found to be for all distances smaller than the global average.

In the vertical direction (figure 7.5b) it can be seen that in case $M121$ the particle settling velocity upstream from the reference particle decreases quickly with decreasing distance to the reference particle. Downstream from the test particle the particle-conditioned settling velocity increases slowly towards the average global average value, which is reached approximately

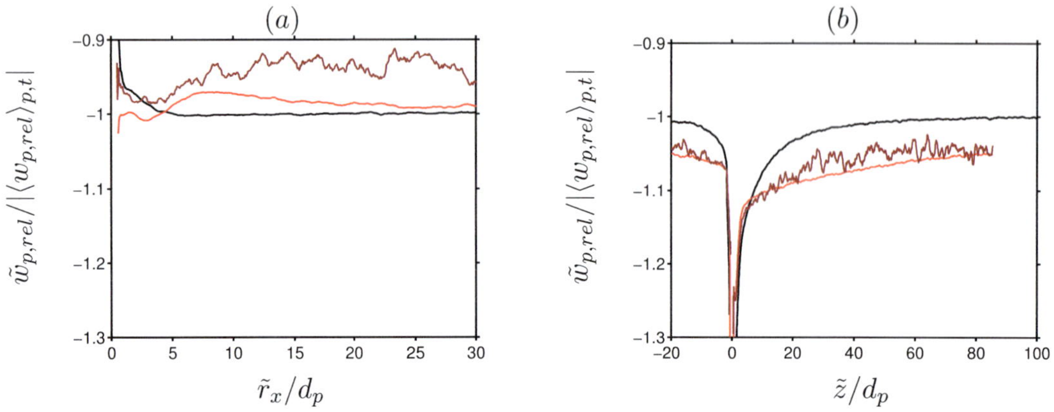

Figure 7.5: Variation of the particle conditioned average value $\tilde{w}_{p,rel}$ from figure 7.4 along the two axis passing through the center of the reference particle, i.e. $(\tilde{x},\tilde{z}) = (0,0)$. (a) in the horizontal direction $(\tilde{x},\tilde{z}) = (\tilde{r}_x,0)$. (b) in the vertical direction $(\tilde{x},\tilde{z}) = (0,\tilde{z})$. $\tilde{r}_x$ denotes the radial distance from the particle surface. Case $M121$ (——). Case $M178$ (——). Case $ML178$ (——).

at $\tilde{z} = 48d_p$ from the particle surface. In the clustering cases $M178$ and $ML178$ the velocity $\tilde{w}_{p,rel}$ experience similar behavior, it increases at much smaller rate than in case $M121$, not reaching the global average value as expected. Regarding the leading particles upstream from the reference particle (i.e. negative values of $\tilde{z}$) it is found that they settle, on average, faster than the global average at any distances.

We can preliminarily conclude that the definition of the velocity "seen" by the particles may play a significant role for the particle motion. We found out that the particles in case $M178$ and $ML178$ settle, locally, at significantly slower rate, close to that of a single particle, than when a global reference value of the fluid velocity is considered. Moreover, the particles are preferentially located in regions with downward fluid motion. The settling velocity of the particles was found to correlate with the particle concentration, with the particles settling on average faster in regions with higher concentration than the global average. Joint p.d.f.s of the particle settling velocity fluctuations and the fluctuations of the Voronoï cell volume quantified this observations. Moreover, it was shown that the fluid velocity correlates with the particle clusters and was found to be on average smaller in regions with higher particle concentration than the global average and vice versa for voids.

Particle-conditioned average of the settling velocity revealed that the settling velocity is spatially distributed in an anisotropic manner, featuring similar pattern as the local solid volume fraction discussed in chapter §6. In all three flow cases it was found that the particles downstream of the reference particle settle at higher rates than the global average. The domain of influence of the reference particle in case $M121$ was found to be substantially smaller than in case $M178$ and $ML178$. The length scales which can be extracted from the particle-conditioned analysis of the settling velocity correspond closely to that obtained from the particle-conditioned average of the solid volume fraction, providing another way of characterizing the geometry of the particle clusters themselves.

7.1.3 Particle Hydrodynamic force

In this section the hydrodynamic force exerted by the fluid on the particles is investigated. For the present configurations the drag force (i.e. the vertical component of the force in the

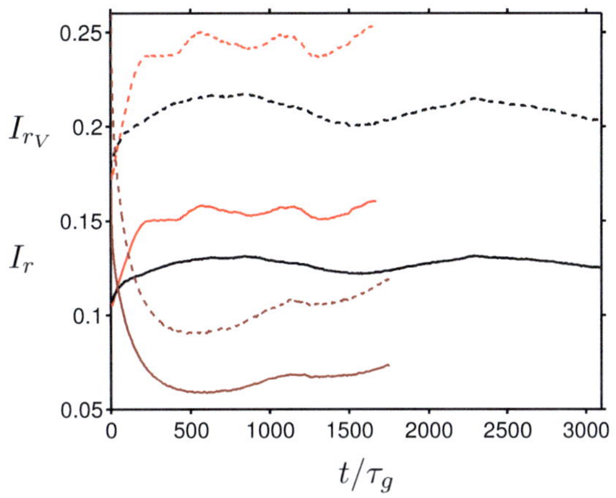

Figure 7.6: Temporal evolution of the relative turbulence intensity as function of time. Solid lines $I_r = \sqrt{\langle u'_{i,f} u'_{i,f} \rangle_{\Omega_f}/3}/w_s$, dashed lines $I_{r_V} = \sqrt{\langle w'_f w'_f \rangle_{\Omega_f}}/w_s$. Case $M121$ (——, - - - -), case $M178$ (——, - - - -) and case $ML178$ (——, - - - -).

present flow cases) was identified in section §1.2.1 as one of the most important hydrodynamic parameters. Thus, the focus in the present section will be on the drag force of the particles. The drag force is often expressed in the form of the particle drag coefficient. Despite the large body of literature (section §1.2.1), the study of the drag coefficient of the particles is by far not trivial, especially in the presence of multiple freely moving particles, which introduce additional complexities due to hydrodynamic interactions between the particles. As introduced in section §1.2.1 a large number of works were devoted to the study of the drag coefficient. In most of the studies the coupling effects of the flow field on the motion and the hydrodynamic forces of the particles were investigated. In this context, the so called relative turbulence intensity is used as parameter to characterize the interaction between the particles and the (pseudo) turbulent flow field. Before we turn to the results of the hydrodynamic force, let us first briefly discuss the relative turbulence intensities of the flow fields observed in the present flow configurations.

The relative turbulence intensity is often defined as the ratio between the intensity of the incoming fluid fluctuations and the apparent slip velocity of the particles. In isotropic flows the definition of the relative turbulence intensity is often based upon the three-component (pseudo) turbulent intensities, viz.:

$$I_r = (\langle u'_{f,i} u'_{f,i} \rangle_{\Omega_f}/3)^{1/2}/w_s, \tag{7.5}$$

whereas in cases with unidirectional mean flow (here in the z-coordinate direction) the following definition is commonly employed:

$$I_{r_V} = (\langle w'_f w'_f \rangle_{\Omega_f})^{1/2}/w_s. \tag{7.6}$$

By definition the relative turbulence intensity measures the influence of the fluid fluctuations on the particle settling and reflects the relative importance of the flow ($I_{r_V} \gg 1$), or the settling of the particles ($I_{r_V} \ll 1$). In the limit of $I_{r_V} \to 0$, the fluid motion is expected to have negligible effects on the particle settling, whereas for $I_{r_V} \to \infty$, the gravity induced relative motion is expected to be irrelevant.

Figure 7.6 depicts the temporal evolution of the relative turbulence intensity I_{r_V} (I_r) for all three flow configurations. Although, the flow in the present simulations is not considered to be turbulent in the general sense, the relative turbulence intensities in the present simulations are showed to be comparable to the turbulence levels often considered in studies of the influence of background turbulence upon the particle motion, e.g. Wu and Faeth (1994); Bagchi and Balachandar (2003, 2004); Yang and Shy (2005); Legendre et al. (2006); Poelma et al. (2007); Snyder et al. (2008); Amoura et al. (2010). It can be observed that the relative turbulence intensity exhibits similar temporal evolution as the fluid velocity fluctuations (cf. figure 5.3). In case $M121$ the values of I_{r_V} (I_r) reach quickly their asymptotic value and as the fluid velocity fluctuations, experience similar low frequent fluctuations with time scale of the order of $1000\tau_g$. In case $M178$ the values of I_{r_V} (I_r) increase more slowly and oscillate around their asymptotic value with comparably shorter period. The relative turbulence intensity clearly shows dependence with the solid volume fraction, the values of I_{r_V} (I_r) in case $ML178$ are significantly smaller than in case $M178$. The temporal evolution of the relative turbulence intensity in case $ML178$ shows an uptrend, suggesting that the flow may still develop. This suggests the presence of low frequent oscillation with time scales of the order of $1000\tau_g$, which is opposite to the temporal evolution in case $M178$ (please note that the simulated time in case $ML178$ is longer than in case $M178$). A much longer simulation times may be required until the flow field reaches a statistically steady state. This needs to be addressed in a future study. The relative turbulent intensities I_{r_V} (I_r) observed in the present work ranges from approximately 0.12 (0.07) in case $ML178$ to 0.2 (0.12) in case $M121$ and 0.25 (0.15) in case $M178$. Therefore, we can assume that in the present flow configurations the fluid motion may significantly affect the motion of the particles.

Now let us turn to the average drag force experienced by the particles. It is common practice to specify the drag force in the form of the drag coefficient. Commonly the mean drag coefficient of the particles is defined as:

$$C_d = \frac{\langle f_z \rangle_p}{\frac{1}{2}\rho_f A_p w_s^2}, \tag{7.7}$$

where ρ_f is the fluid density and $A_p = \pi r_p^2$ is the projection area of the particles. As can be seen the definition of the mean drag coefficient depends on the particle relative velocity, hence the definition of the mean fluid velocity seen by the particle is crucial for the determination of the mean drag coefficient. As shown above, the mean fluid velocity seen by the particles obtained by the global averaging over the entire volume occupied by the fluid differ substantially from that obtained by using the local fluid velocity in vicinity of the particles, especially when the particles do not sample the computational volume homogeneously. Therefore, additionally to the definition of the drag coefficient in equation 7.7 we employ the following definition based on the local relative velocity w_s^S (equation 7.2):

$$C_d^S = \frac{\langle f_z \rangle_p}{\frac{1}{2}\rho_f A_p (w_s^S)^2}. \tag{7.8}$$

It can be shown that the fractional difference between the two definitions of the drag coefficients C_d and C_d^S can be expressed as:

$$\frac{|C_d^S| - |C_d|}{|C_d^S|} = 1 - \left(\frac{|w_s^S|}{|w_s|}\right)^2. \tag{7.9}$$

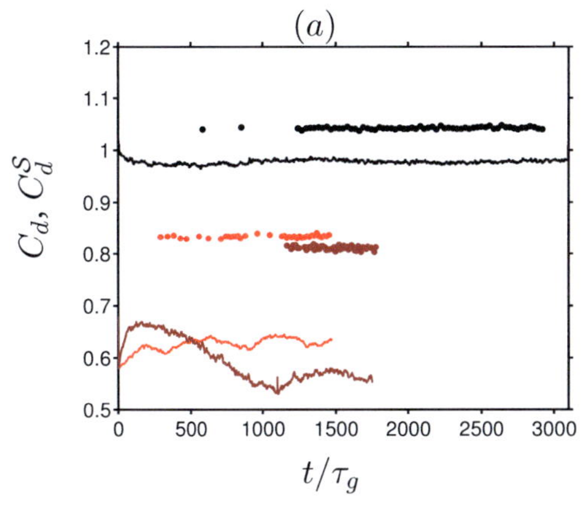

Figure 7.7: (a) Time evolution of the particle ensemble average of the vertical component of the hydrodynamic force, f_z, for case $M121$ (——), case $M178$ (——) and case $ML178$ (——). The force is normalized by the reference force defined in equation 3.7. The circular symbols depict the hydrodynamic force when normalized by the reference force based on the local fluid velocity $\boldsymbol{u}_f^S$ (•, •, •). (b) Mean particle drag coefficient for case $M121$, case $M178$ and case $ML178$ (color coding as in sub-figure (a)). Open symbols represent the time average of the of the values of the circular symbols in sub-figure (a). Filled symbols represent the time average of the hydrodynamic force (line plots) shown in sub-figure (a) at the stationary state. Black dashed line (- - - -) corresponds to the standard drag curve (Clift et al., 1978, table 5.2, s. 112) and the blue dashed line (- - - -) corresponds to the drag coefficient as defined by Schiller and Naumann (1933).

For case $M121$ this difference calculates to 6.2% and in the cases $M178$ and $ML178$ to 24.8% and 30.6%, respectively, indicating the large influence of the definition of the fluid velocity seen by the particles, especially in the clustering flow cases.

Figure 7.7a depicts the temporal evolution of the mean particle drag coefficient C_d and C_d^S for all three cases. It can be observed that the particle drag coefficient according to the definitions in equations 7.7 and 7.8 experience similar evolution with time as the corresponding mean particle settling rates w_s and w_s^S (cf. figure 7.1). In case $M121$ the curve of the drag coefficient remains essentially flat over the entire course of the simulation, whereas in case $M178$ and case $ML178$ the coefficient C_d experiences small temporal fluctuations. On the other hand, the values of C_d^S are approximately constant during the simulation. In figure 7.7b the time averaged of the coefficients C_d and C_d^S is shown. Please note that the average in all three cases is taken at times at which the flows are considered as statistically steady. For comparison the standard drag curves as defined in Clift et al. (1978, table 5.2, s. 112) and Schiller and Naumann (1933) are shown as well. Henceforth, the drag coefficients computed according to the correlations of Clift et al. (1978) and Schiller and Naumann (1933) will be denoted by C_d^{Clift} and C_d^{SN} respectively. As can be seen the values of C_d and C_d^S in case $M121$ are larger than the predictions of Clift et al. (1978) and Schiller and Naumann (1933), whereas in case $M178$ and $ML178$ the values of C_d (C_d^S) are over (under) predicted by C_d^{Clift} and C_d^{SN}. Here we would like to stress that the values of C_d^{Clift} and C_d^{SN} are obtained for a single fixed particle in uniform flow, which is in strong contrast to the flow conditions in the present study. For better comparison the exact values of the drag coefficients C_d, C_d^S, C_d^{Clift} and C_d^{SN} are shown in table 7.2.

In order to investigate the effects of mobility and clustering on the mean drag experienced by the particles, we have compared the values of the mean drag coefficient C_d obtained from

case	$Re_{p,rel}$ ($Re^{\mathcal{S}}_{p,rel}$)	I_{rV}	C_d^{Clift}	C_d^{SN}	$\langle C_d \rangle_t$	$\langle C_d^i \rangle_{p,t}$	$\langle C_d^{\mathcal{S}} \rangle_t$	$\langle C_d^{\mathcal{S}_i} \rangle_{p,t}$	$\langle C_d^- \rangle_t$	$\langle C_d^+ \rangle_t$
$M121$	141.1 (137.1)	0.203	0.914	0.93	0.98	1.058	1.042	1.068	1.338	0.754
$M178$	260.5 (225.7)	0.253	0.687	0.725	0.635	0.725	0.83	0.863	0.991	0.450
$ML178$	275.6 (230.4)	0.117	0.673	0.71	0.621	0.657	0.80	0.81	0.819	0.619

Table 7.2: Comparison of various definitions of the drag coefficient for the mean particle drag in all three cases $M121$, $M178$ and $ML178$: C_d^{Clift} according to the relation in Clift et al. (1978, table 5.2, s. 112); C_d^{SN} according to the relation in Schiller and Naumann (1933); C_d, $\langle C_d^i \rangle_{p,t}$, $C_d^{\mathcal{S}}$, $\langle C_d^{\mathcal{S}_i} \rangle_{p,t}$ as in equations 7.7 - 7.11; C_d^+ (C_d^-) indicate the drag coefficient based on the mean drag force plus (minus) the r.m.s. particle force in the streamwise direction, normalized by the reference force based upon the mean slip velocity w_s plus (minus) the r.m.s. streamwise fluid velocity.

the simulations with the fixed particles ($F121$ and $F178$) and the mobile particles ($M121$ and $M178$). In the case with the smaller Reynolds number ($Re_{p,rel} = 141$) the observed values of C_d in case $M121$ are slightly smaller than in case $F121$, with the difference measuring approximately 3.9%. The difference becomes even smaller, approximately 2.16%, when instead the value of $C_d^{\mathcal{S}}$ is used in case $M121$ for comparison. This indicates that the effect of mobility in case $M121$, under this configurations (particle-to-fluid density ratio and solid volume fraction) has only minor effect on the mean drag experienced by the particles. On the other hand the difference between the values of C_d in case $F178$ and $M178$ was measured to be as high as 20.6% with significantly smaller values in the case of mobile particles $M178$. Indicating that the inhomogeneous particle distribution in case $M178$ leads to a significant reduction of the mean drag coefficient of the particles (based on the global average values). However, a significantly different picture is observed, when the value of the drag coefficient based on the local fluid velocity $C_d^{\mathcal{S}}$ is considered in case $M178$. The difference than reduces to 3.7%, indicating that locally the particles in case $F178$ and $M178$ are exposed to flow conditions causing similar drag values.

Please recall that the values of the mean drag force in the fixed particle simulations $F121$ and $F178$ resembled closely the values of the drag force of the single particles in case $S121$ and $S178$. Thus, the difference between the drag of the single settling particle and the mean drag C_d ($C_d^{\mathcal{S}}$) of the multiple settling particles results in the case with $Re_{p,rel} = 141$ to 2.2% (4%) and in the case with $Re_{p,rel} = 245$ to 18.6% (6.4%), respectively. It appears that the higher mean settling rate of the particles w_s in case $M178$ (when compared to the single particle) resulting, most probably, from the agglomeration of the particles into clusters correlates with the significantly smaller values of the drag coefficient. Therefore, it is tempting to relate the smaller drag observed in case $M178$ to the particles located inside clusters.

In order to investigate the effect of the clusters on the particle drag more qualitatively, we have computed, the joint distribution of the Voronoï volume fluctuations and the instantaneous fluctuations of the normalized drag force experienced by each individual particle C_d^i, where C_d^i is defined as:

$$C_d^i(t) = \frac{f_z(i,t)}{\frac{1}{2}\rho_f A_p w_{p,rel}^2(i,t)}, \qquad \forall i = 1, \ldots, N_p. \tag{7.10}$$

The corresponding joint distribution can be found in figure 7.8 for all three cases. It can be observed that similar to the joint p.d.f. of the Voronoï cell volume and the particle settling velocity, the two quantities in case $M121$ correlates weekly with a correlation coefficient of 0.002. On the other hand, the graphs for case $M178$ and case $ML178$ evidence a positive

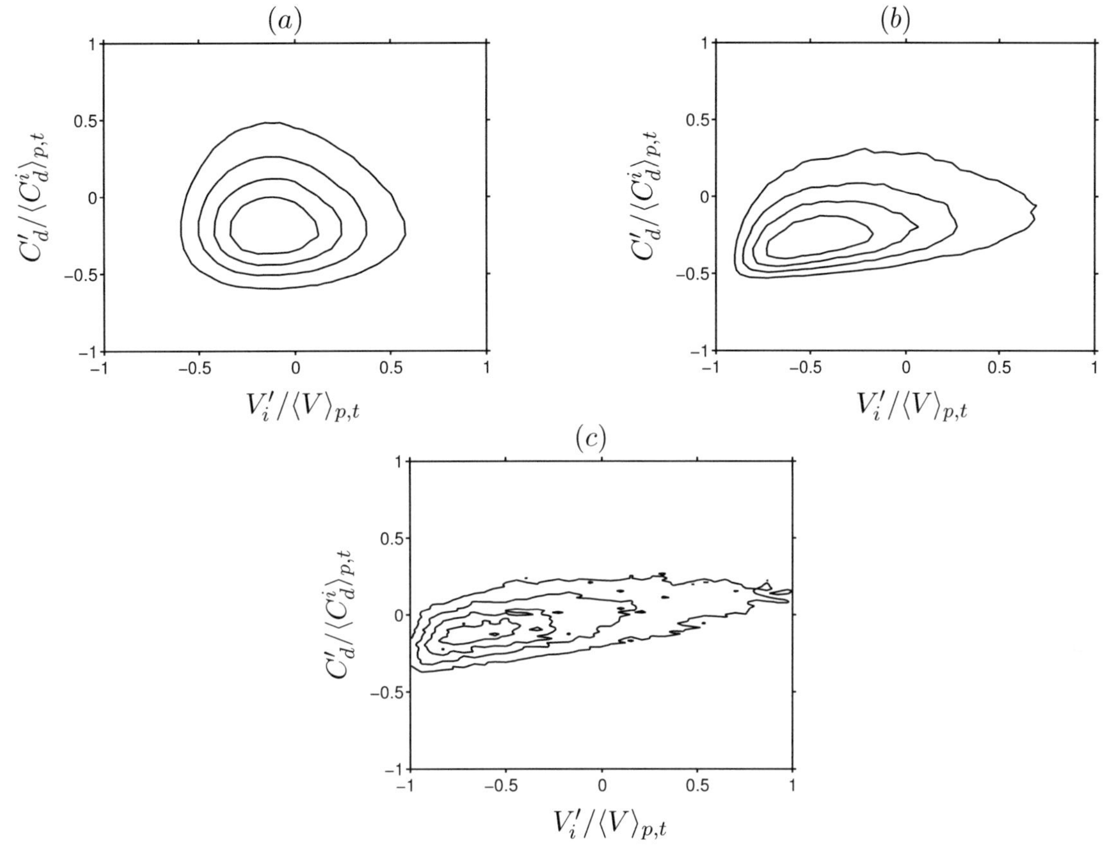

Figure 7.8: Joint p.d.f.s of the drag coefficient fluctuations $C'_d = C^i_d - \langle C^i_d\rangle_{p,t}$ and the fluctuations of the Voronoï cell volumes $V'_i = V_i - \langle V\rangle_{p,t}$. (a) Case $M121$, (b) case $M178$ and (c) case $ML178$. In all three plots the lines indicate contours at $0.2, 0.4, 0.6, 0.8$ times the maximum number of occurrences.

correlation between the Voronoï volumes and the particle drag coefficient. Particles associated with small Voronoï cells (i.e. particles located inside clusters) tend to have smaller than the average drag coefficient and vice versa for voids. Similar as in figure 7.2 the correlation in case $ML178$ appears to be more pronounced than in case $M178$, resulting in larger values of the correlation coefficient which was calculated to 0.315 in case $ML178$ and to 0.471 in case $ML178$. This implies that the higher clustering levels in case $ML178$ correspond to smaller values of the drag coefficient than in case $M178$ (cf. table 7.2). We have further calculated the drag coefficient in the clustering cases $M178$ and $ML178$ conditioned on the location of the particles, i.e. we calculated the drag coefficient for all particles located inside clusters and voids respectively. In case $M178$ it was observed that the drag of the particles located inside clusters (voids) was 2.3% (2.5%) smaller (larger) than the average. In case $ML178$, however the difference between the drag of the particles associated with clusters (voids) from the average was found to be substantially larger, viz. 12.1% (21.6%) smaller (larger) than the average drag inside clusters (voids).

The smaller deviation of the mean drag coefficient C^S_d based on the local fluid velocity w^S_s from the values of the drag coefficient in of the single particle simulations $S121$, $S178$, respectively the fixed particles $F121$, $F178$ indicates that the local instantaneous drag experienced by the particles is less dependent on the spatial distribution of the particles. An evidence of this can be found in the corresponding joint p.d.f. (plot omitted) of the Voronoï cell volume fluctuations and the instantaneous drag coefficient based on the local fluid velocity $C_d^{S_i}$, which is defined as:

$$C_d^{S_i}(i,t) = \frac{f_z(i,t)}{\frac{1}{2}\rho_f A_p (w_{p,rel}^S(i,t))^2}, \qquad \forall i = 1,\ldots,N_p. \tag{7.11}$$

The corresponding correlation coefficients was measured in case $M121$ to 0.007, whereas the values for case $M178$ and case $ML178$ were evaluated to much smaller values of 0.021 and 0.019, respectively.

For the particle Reynolds numbers as in the present flow cases, it is known that the drag coefficient depends in a non-linear manner on the particle Reynolds number (Clift et al., 1978). As discussed above, the mean drag is however dependent on the definition of the mean fluid velocity seen by the particle. The non-linear dependence of the mean drag can be represented by the difference of drag coefficient defined in equation 7.7 (equation 7.8) and the particle ensemble average of the coefficient defined in equation 7.10 (equation 7.11) (Bagchi and Balachandar, 2003), i.e. $\Delta C_d(t) = C_d(t) - \langle C_d^i(t)\rangle_p$ ($\Delta C_d^S(t) = C_d^S(t) - \langle C_d^{S_i}(t)\rangle_p$). Table 7.3 depicts the time averaged values of ΔC_d and ΔC_d^S for all three cases. Concerning the difference of ΔC_d, the largest difference is observed for case $M178$ and measures approximately 14%, while in case $M121$ and $ML178$ similar values can be observed, viz. $\langle \Delta C_d\rangle_t = 7.3\%$ and $\langle \Delta C_d\rangle_t = 5.8\%$ respectively. It is interesting to note that the difference of the drag coefficient based on the local fluid velocity in vicinity of the particles $\langle \Delta C_d^S\rangle_t$ measure in all three cases significantly smaller values than $\langle \Delta C_d\rangle_t$ with the largest value again observed in case $ML178$. This results indicate that the non-linear drag effect is largest in case $M178$, where the largest values of the relative turbulence intensity I_{r_V} were measured. As noted above, high values of the relative turbulent intensity would lead to large variations of the particle relative velocity, hence large variation in the drag coefficient of the individual particles (e.g. C_d^i). Consequently the particle ensemble average of the drag coefficient C_d^i can deviate from the mean drag based on the average particle relative velocity C_d. Therefore, non-linear drag dependence is expected to be significant for large levels of the relative turbulence intensity. The substantial difference $\langle \Delta C_d\rangle_t$ measured for case $M178$ also suggest the importance of the particle clustering or potential preferential particle trajectory on the mean drag. However, as evidenced in table 7.3, the difference $\langle \Delta C_d^S\rangle_t$ was measured to be similar for all values of the relative turbulence intensity showing only weak dependence wit I_{r_V}. This indicates that the definition based on the local relative velocity is somehow more robust with respect to the values of the relative turbulence intensity in the present flow configurations.

The non-linear drag effect can be further investigated by considering the mean drag force plus (minus) the r.m.s. drag force normalized by the reference force based upon the mean slip

case	$\langle \Delta C_d\rangle_t$	$\langle \Delta C_d^S\rangle_t$
$M121$	7.3%	2.5%
$M178$	14%	3.9%
$ML178$	5.8%	1.25%

Table 7.3: Difference between the time average of the drag coefficients defined in equations 7.7 and 7.10 (equations 7.8 and 7.11), i.e.: $\langle \Delta C_d\rangle_t = \langle C_d - \langle C_d^i\rangle_p\rangle_t$ ($\langle \Delta C_d^S\rangle_t = \langle C_d^S - \langle C_d^{S_i}\rangle_p\rangle_t$).

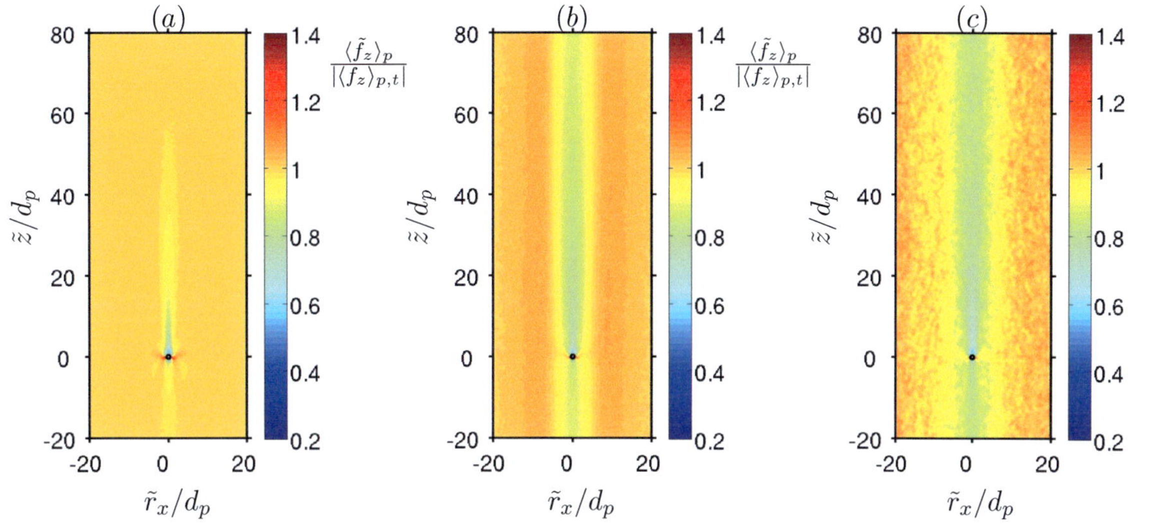

Figure 7.9: Spatial map of the particle-conditioned average value of the vertical hydrodynamic force, $\langle \tilde{f}_z \rangle_p$ for (a) case $M121$, (b) case $M178$ and (c) case $ML178$. The velocity $\langle \tilde{f}_z \rangle_p$ is plotted as function of the position $(\tilde{r}_x, \tilde{z})$ with respect to a test particle (vertical plane passing through the center of the reference particle).

velocity w_s plus (minus) the r.m.s. of the vertical component of the fluid velocity, viz.:

$$C_d^{\pm}(t) = \frac{\langle f_z \rangle_p \pm \sigma(f_z)}{\frac{1}{2}\rho_f A_p (w_s \pm \sigma(w_{p,rel}))^2}. \tag{7.12}$$

The non-linear drag effect manifests itself in a larger increase of the drag coefficient C_d^- than a decrease of C_d^+, i.e. in the non-linear drag regime the drag of the particles decrease at a slower rate with the particle Reynolds number than the rate at which the Reynolds number increases. The time averaged values for all three cases are depicted in table 7.2. As can be observed, the values of both $\langle C_d^- \rangle_t$ and $\langle C_d^+ \rangle_t$ deviate significantly from the mean drag coefficient $\langle C_d \rangle_t$ with bias towards $\langle C_d^- \rangle_t$ leading to a net increase of the drag coefficient. In particular the value of the drag coefficient $\langle C_d^+ \rangle_t$ in case $M121$ was measured to be 22% smaller than the average value $\langle C_d \rangle_t$, whereas the value of $\langle C_d^- \rangle_t$ was 26% larger than $\langle C_d \rangle_t$. In case $M178$, where the relative turbulence intensity is the largest, the drag coefficients $\langle C_d^+ \rangle_t$ and $\langle C_d^- \rangle_t$ deviate from the averaged value by approximately 29% and 35% respectively. The corresponding values in case $ML178$ were calculated to 21% and 24% for $\langle C_d^+ \rangle_t$ and $\langle C_d^- \rangle_t$. Therefore, even for particles with a normal distributed velocity field, the distribution of the normalized drag force will eventually deviate from the normal distribution. We have checked this by performing this experiment of Monte-Carlo type (plots omitted), that indeed this leads to a net increase of the drag coefficient. More qualitative analysis for the present flow cases can be performed by studying the probability distribution of the hydrodynamic force, which will be conducted in the next section.

In order to investigate the microscopic properties of the hydrodynamic force we have performed particle-conditioned averaging of the vertical component of the hydrodynamic force exerted by the fluid on the particle. The averaging is performed in the same way as in section §4.3.3. Figure 7.9 shows the two-dimensional map of the vertical component of the particle averaged force, $\langle \tilde{\boldsymbol{f}}(\tilde{x}, \tilde{z}) \rangle_{p,t}$. The force has been normalized with the absolute value of the unconditioned long-time average of the vertical component of the hydrodynamic force,

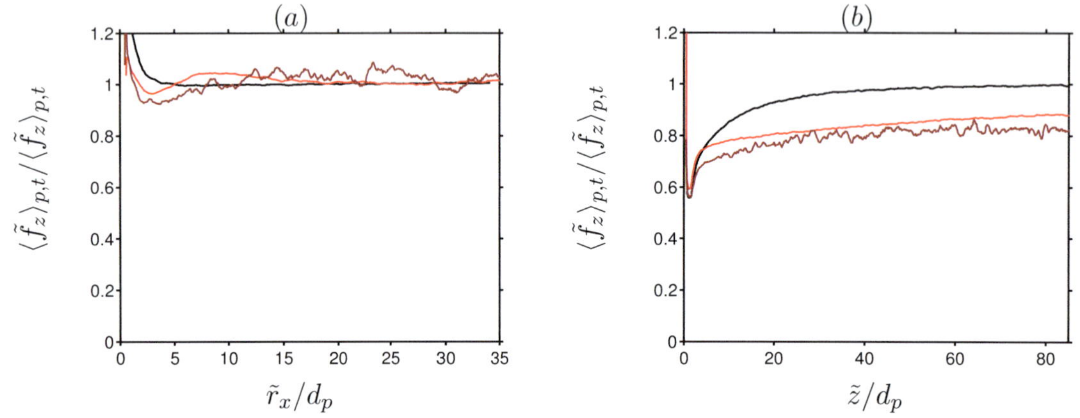

Figure 7.10: Variation of the particle conditioned average value of $\langle \tilde{f}_z \rangle_p$ from figure 7.9 along the two axis passing through the center of the reference particle, i.e. $(\tilde{x}, \tilde{z}) = (0, 0)$. (a) in the horizontal direction $(\tilde{x}, \tilde{z}) = (\tilde{r}_x, 0)$. (b) in the vertical direction $(\tilde{x}, \tilde{z}) = (0, \tilde{z})$. $\tilde{r}_x$ denotes the radial distance from the particle surface in the horizontal direction. Case $M121$ (——). Case $M178$ (——). Case $ML178$ (——).

i.e. $|\langle f_z \rangle_{p,t}|$. As can be seen the particle conditioned vertical component of the hydrodynamic force experience the same patterning as the particle conditioned settling velocity in figure 7.4. In all three cases the spatial distribution of $\langle \tilde{f}_z \rangle_{p,t}$ in the vertical plane is highly anisotropic oriented preferentially in the vertical direction. In all three cases the vertical component of the hydrodynamic force downstream from the reference particle is significantly smaller than the average value. In case $M121$ the averaged value is recovered approximately $60d_p$ downstream from the particle. In case $M178$ and $ML178$ on the other hand, the pattern extends throughout the entire vertical length of the computational domain. In case $M178$ the vertical stripe with values of the drag force smaller than the mean extends in the horizontal direction approximately $5d_p$ on both sides of the reference particle (almost independent from the vertical coordinate $\tilde{z}$). However, the domain of influence of the reference particle in case $ML178$ is observed to be much wider in the horizontal direction than in case $M178$, extending almost $40d_p$, which is in line with the observation in figure 7.4. Comparison with figure 7.4 clearly shows the correlation of the particle settling velocity with the drag force, i.e. the regions with higher than the average particle settling velocity corresponds to the regions with smaller than the average drag force.

The profiles of the particle-averaged drag force on the horizontal and vertical axis through the center of the test particle are shown in figure 7.10. In the horizontal direction it can be observed that in all three cases close to the test particle the force $\langle \tilde{f}_z \rangle_{p,t}$ is larger than the average force. In case $M121$ the force decreases with the horizontal distance and converges for $\tilde{r}_x > 5d_p$ towards the average value. In the two clustering cases the force decreases and becomes smaller than the average for distances approximately up to $5d_p$ in case $M178$ and $10d_p$ in case $ML178$. Please note that this extensions corresponds roughly to the horizontal extensions of the particle clusters. Beyond this distances the force becomes larger than the average force and converges slowly towards unity.

Regarding the vertical direction downstream from the reference particle, the drag force decreases immediately behind the test particle after which it increases slowly towards its asymptotic value. In all three cases the minimum of the particle-conditioned drag force appear to be at approximately the same distance of $1.3d_p$ behind the reference particle. As can be seen the force drops nearly 45% in all three flow cases. While in case $M121$ the normalized force reaches unity, in the clustering cases $M178$ and $ML178$ the force do not converge to unity in the present configurations.

In this section we studied the dominant component of the hydrodynamic force. The results showed that the drag coefficient (normalized drag force), based on the unconditioned particle settling velocity w_s in the clustering cases is significantly smaller than the standard drag and the hydrodynamic force of the corresponding simulations with single settling particle. This was largely due to the strong clustering observed in both cases $M178$ and $ML178$, as evidenced by the joint distribution of the Voronoï cell volume and the fluctuations of the drag coefficient. In case $M121$ however, the difference was measured to be much smaller. Further, we showed that the deviation of the drag force was the largest in the case with the largest relative turbulence intensity and vice versa the smallest deviation was observed for the smallest value of the relative turbulence intensity. When the particle relative velocity based on the local fluid velocity, w_s^S, was used for the normalization of the drag force, the drag coefficient in all three cases was found to be close to that of a single particle. Moreover, the deviation from the average was substantially smaller than when w_s was used. As expected, the effect of the non-linear drag was found to be more prominent in the case with the largest relative turbulence intensity, leading to a net increase of the drag. Finally, particle-conditioned average of the drag force showed an anisotropic spatial distribution of the hydrodynamic force and a strong correlation with the particle settling velocity. The results for the particle hydrodynamic force and the particle settling velocity appear to support the above suggested wake attraction mechanism responsible for the clustering of the particles, i.e. particles in the wake of a leading particle tend to experience smaller drag and have larger settling velocity than the leading one.

7.2 Fluctuating quantities

In this section the influence of the self-induced pseudo-turbulent flow field upon the motion of the particles and the resulting statistics, such as the fluctuations and the distribution of (i) the particle translational and angular velocity components and (ii) the force and torque of the particles, is going to be discussed. The Lagrangian particle statistics are presented and discussed as well.

Besides the mean particle settling velocity, the fluctuations of the particle velocity are of particular interest, especially since the surprising theoretical divergence of the particle velocity fluctuations for particles settling in Stokes flows showed by Caflisch and Luke (1985). Since then, a significant amount of theoretical as well as experimental and numerical works were accumulated. However, most of the studies considered low Reynolds number flows, e.g. Koch and Shaqfeh (1991); Koch (1993); Nicolai and Guazzelli (1995). Recent studies of Bunner and Tryggvason (1999); Cartellier and Riviére (2001); Climent and Maxey (2003) and Yin and Koch (2008) considered flows with finite but small particle Reynolds number of the order of $\mathcal{O}(10)$. A recent comprehensive review on the topic can be found in Guazzelli and Hinch (2011). Surprisingly, our review of the literature revealed, that little is known about the particle velocity fluctuations of particles settling with particle Reynolds number of the order of $\mathcal{O}(100)$. As soon as the sedimentation of many particles is considered, the motion of the particles is influenced by the neighboring particles as result from the large scale hydrodynamic interactions between the particles. Thus, in most of the mentioned works, the particle wakes were inferred to be the key factor responsible for the particle velocity fluctuations. Furthermore, the hydrodynamic interaction between the particles could lead to change of the micro-structure of the dispersed phase without even changing the macroscopic structure of the suspension (as in case $M121$, cf. section 6.3). Therefore, the micro-structure of the dispersed phase was identified also as one of the key parameters determining the particle velocity fluctuations. Consequently, even very dilute mono-disperse suspensions with small Reynolds number and no evidence of particle agglomeration will encounter deviations of the particle velocity from its mean.

Let us return to the fluctuations of the particle translational velocity. In section §5.2 the initial transient and the time evolution of the particle velocity fluctuations was already briefly discussed (figure 5.3b). As aforementioned, the velocity of the particles in all three cases experienced significant fluctuations where the vertical component can reach a maximum value of up to 25% (case $M178$) of the terminal settling velocity. Moreover, the fluctuations of the particle velocity evidenced strong anisotropy, the fluctuations of the vertical velocity component in all three cases are approximately 2.2 times the fluctuations of the horizontal velocity components. This suggests, that significant amount of momentum is transferred from the vertical to the horizontal direction in all three cases. Please recall that asymptotically an isolated particle in case $S121$ settles with a constant velocity along a straight axi-symmetric trajectory, while an isolated particle in case $S178$ settles steadily along an oblique path in a random horizontal direction. Therefore, ensemble average of many realizations of isolated particles as in case $S121$ would lead to zero r.m.s. values of the particle velocity. However, the same ensemble averaging for many particles as in case $S178$ would lead to non-zero value of the r.m.s. of the horizontal component of the particle velocity, due to the randomness of the horizontal orientation of the velocity vector. In particular, it can be shown that the normalized r.m.s. value of the horizontal velocity component (e.g. considering the x-direction) results in:

$$\frac{\langle u'_p u'_p \rangle_P^{1/2}}{w_s} = \frac{u_{pH}}{u_{pV}} \frac{1}{\sqrt{2}}, \tag{7.13}$$

where u_{pH} and u_{pV} are the absolute values of the particle velocity in the horizontal plane and in the vertical direction, respectively, obtained from the simulation of an isolated particle. The quantity in equation 7.13 was evaluated in cases $M178$ to 0.11 and in case $ML178$ to 0.09. This values proved to be larger than the value obtained for an ensemble of cases $S178$, i.e. $\langle u'_p u'_p \rangle_P^{1/2} / w_s = 0.065$ (Uhlmann and Dušek, 2014). This results show that the simple superposition of isolated particles settling in an uniform undisturbed flow cannot approximate satisfactory the velocity fluctuation amplitude of the particles in case $M178$ and $ML178$. Moreover, such a superposition yields a zero fluctuation amplitude of the vertical particle velocity component, which is clearly not true in the present flow configurations. However, there is a clear dependence with the solid volume fraction. As one would expect, in the limit of vanishing solid volume fraction ($\Phi_s \to 0$) a simple superposition will eventually reproduce the velocity fluctuation amplitude. Unfortunately, due to high computational cost we were not able to determine the critical value of the solid volume fraction and this needs to be addressed in future works.

The strong anisotropy of the particle velocity fluctuations is not unusual for sedimenting suspensions and was observed in the majority of the studies considering the sedimentation or the rise of particles. The work of Parthasarathy and Faeth (1990b) is one of the most relevant studies to the present work. They performed experiments of particles settling under the influence of gravity in an nearly stagnant water bath with particle Reynolds number larger than 150. The authors observed significant fluctuations of the cross stream velocity particularly for particles with particle Reynolds number larger than 150 and concluded that one of the factors responsible for this behavior are the unsteady lateral forces on the particles due to eddy shedding at higher terminal particle Reynolds number. Recently, Zenit et al. (2001) performed experiments of mono-disperse bubbles rising in a vertical channel with high Reynolds number. They reported values of the bubble velocity variance in the order of $\mathcal{O}(0.1)$ and the variance ratio of the vertical to the horizontal velocity components was measured to be 2.01. Surprisingly, similar values of the vertical to horizontal particle velocity ratio were observed in the studies with small, but finite Reynolds numbers, e.g. Bunner and Tryggvason (1999) and Climent and Maxey (2003) reported a ratio of three and Bunner and Tryggvason (2002) observed a value of approximately two for the lowest volume fraction considered in their studies.

As next we will decompose the particle velocity in order to investigate the different contributions to the fluctuations of the particle velocity, especially those resulting from the motion of the fluid surrounding the particles. Using the definition of the fluid velocity seen by the particle, $\boldsymbol{u}_f^S$ (cf. section §7.1.1), it can be shown that the variance of the particle velocity components can be decomposed as follows (Uhlmann and Doychev, 2014), e.g. for the vertical velocity component:

$$\langle w_p' w_p' \rangle_p = \langle w_f^{S\prime} w_f^{S\prime} \rangle_p + \langle w_{p,rel}^{S\prime} w_{p,rel}^{S\prime} \rangle_p + 2 \langle w_{p,rel}^{S\prime} w_f^{S\prime} \rangle_p, \quad (7.14)$$

where all velocity fluctuations are defined with respect to the instantaneous particle ensemble average:

$$\begin{aligned} w_p'(t) &= w_p(t) - \langle w_p(t) \rangle_p, & (7.15)\\ w_{p,rel}^{S\prime}(t) &= w_{p,rel}^S(t) - \langle w_{p,rel}^S(t) \rangle_p, & (7.16)\\ w_f^{S\prime}(t) &= w_f^S(t) - \langle w_f^S(t) \rangle_p. & (7.17) \end{aligned}$$

The relation in equation 7.14 defines implicitly the contributions to the particle velocity variance stemming from the (r.h.s. of equation 7.14 from left to right) (i) fluctuations of the fluid velocity in the vicinity of the particles, (ii) particles' relative motion with respect to the local fluid velocity and (iii) cross correlation between the former two velocity fluctuations.

The variance of the vertical particle velocity component and the above introduced three contributions to it are depicted in figure 7.11 for all three cases (note that this is the square of the quantity shown in figure 5.3b). It can be observed that in all three cases the contribution from the cross-correlation $\langle w_{p,rel}^{S\prime} w_f^{S\prime} \rangle_p$ is substantially smaller than the remaining two contributions, indicating that the fluctuations of the local fluid velocity and the particle velocity are only weakly coupled. This confirms the observations made in the previous section §7.1.3 that the drag coefficient based on the local relative velocity w_s^S correlates weakly with the location of the particles in the flow. In all three cases the dominant contribution stems from the fluid velocity in vicinity of the particles $\langle w_f^{S\prime} w_f^{S\prime} \rangle_p$ and it accounts for approximately 61.5% in case $M121$ and for 82% and 59.3% in the cases $M178$ and $ML178$, respectively. However, the values in the corresponding flow cases differ significantly. In case $M178$ the time average value of $\langle w_f^{S\prime} w_f^{S\prime} \rangle_p / w_{ref}^2$ is approximately three times larger than in case $M121$ and 2.33 times larger than in case $ML178$. Moreover, its temporal evolution exhibits a similar shape as the total particle velocity fluctuation energy. On the other hand, the energy of the fluctuations of the particle velocity relative to the surrounding fluid, $\langle w_{p,rel}^{S\prime} w_{p,rel}^{S\prime} \rangle_p / w_{ref}^2$, is in all three cases

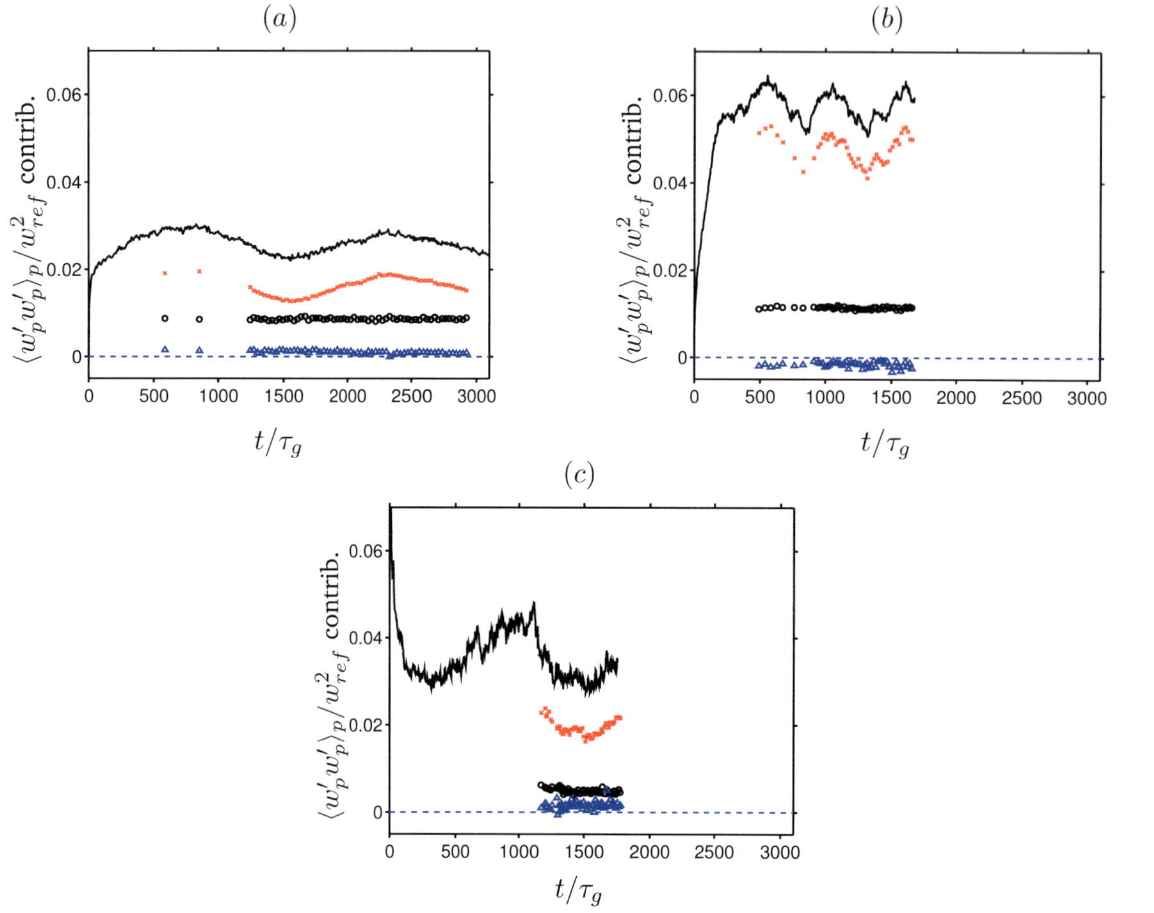

Figure 7.11: Variance of the particle vertical velocity component, $\langle w'_p w'_p\rangle_p$ (——), and its three contributions (cf. equation 7.13): $\langle w_f^{S\prime} w_f^{S\prime}\rangle_p$ (×), $\langle w^{S\prime}_{p,rel} w^{S\prime}_{p,rel}\rangle_p$ (○), $2\langle w^{S\prime}_{p,rel} w_f^{S\prime}\rangle_p$ (△). (a) Case $M121$. (b) Case $M178$. (c) Case $ML178$.

approximately constant in time and have nearly the same contribution to the total fluctuation of the particle velocity $\langle w'_p w'_p\rangle_p / w^2_{ref}$. Comparison between the three flow cases revealed that the actual difference between the observed values is rather small, when compared to the difference in the local fluid velocity contributions. The time average in case $M178$ was evaluated to be approximately 23.7% larger than in case $M121$ and nearly 57% larger than in case $ML178$. The rather flat curves of $\langle w'_p w'_p\rangle_p / w^2_{ref}$ in figure 7.11 suggest that the temporal behavior of the energy of the local fluid velocity fluctuations is responsible for the shape of the total particle velocity fluctuation energy.

The above results imply that the larger overall fluctuating energy in case $M178$ is mostly due to an increase of the fluctuating energy of the fluid velocity seen by the particles. When comparing the clustering and the non-clustering flow cases, it is tempting to relate the higher fluctuating energy of the particle velocity in case $M178$ to the occurrence of particle clustering and therewith resulting increase of the energy of the surrounding fluid velocity which the particles experience, $w_f^{S\prime}$. However in case $ML178$, at the same Galileo number and smaller solid volume fraction, the fluctuating energy is observed to be smaller than in case $M178$ (recall, that the clustering in case $ML178$ is stronger than in case $M178$). The variation of the fluctuating energy of the particle velocity, respectively, their contributions between all three cases is, however, consistent with the values of the relative turbulence intensity observed in the corresponding simulations. The largest fluctuating energy is measured for the highest values of the relative turbulence intensity. Therefore, the relative turbulence intensity could be used as a indirect measure of the variance of the particle velocity.

Concerning the horizontal component of the particle velocity the above analysis obtains qualitatively the same results (plots omitted), i.e. the fluctuations of the local fluid velocity $\mathbf{u}_f^{S\prime}$ are the dominant contribution to the fluctuation of the particle velocity and responsible for their temporal evolution.

The fluctuations of the horizontal particle velocity (cf. figure 5.3b) indicate that the flow field induces motion of the particles in the horizontal direction. Recall that an isolated particle in case $M121$ settles steadily on a vertical path, while an isolated particle in case $S178$ settles steadily along an oblique path in a random horizontal direction (cf. chapter §3). Thus, the question arise to what extent the motion of the single particle is relevant for the motion of the particle suspension. For this reason we have investigated the distribution of the instantaneous angle α of a particle's trajectory relative to the average fluid velocity with respect to the vertical axis z. This study will allow us to investigate the influence of the presence of multiple particles on the settling regime of the individual particles in the particle suspension. The angle α can be simply calculated by the following relation:

$$\alpha(i,t) = \tan^{-1}\left(\frac{\sqrt{u_{p,rel}^2(i,t) + v_{p,rel}^2(t)}}{|w_{p,rel}(i,t)|}\right). \tag{7.18}$$

Figure 7.12a depicts the probability density function of the angle α as computed in equation 7.18. As can be seen the most probable angles in the flow cases $M178$ and $ML178$ is close to that of the single particle from case $S178$ (depicted by the dashed vertical line). This indicates that, at the considered solid volume fractions and despite the strong agglomerations found in $M178$ and $ML178$, the motion of the particles is on average close to that of a corresponding isolated particle. The relative difference to the angle of the single particle was calculated in case $M178$ to 7.4% and, surprisingly, in case $ML178$ to only 2.1%. The standard deviation in case $M178$ (i.e. $\sigma(\alpha) = 4.3$ degrees) was found to be larger than in case $ML178$ (i.e. $\sigma(\alpha) = 3.1$ degrees). On the other hand, in case $M121$ the most probable angle deviates substantially from that of an isolated particle (which is equal to zero). The peak observed in the p.d.f. was found at $\alpha = 3.2$ degrees. The standard deviation was measured at slightly smaller (larger) value $\sigma(\alpha) = 3.7$ than in cases $M178$ ($ML178$). Again, the deviation from the mean value correlates with the values of the relative turbulence intensity. The non-zero value of the most probable angle in case $M121$ indicates that the hydrodynamic interactions with the particle-induced pseudo-turbulent flow field have large effect on the instantaneous motion of the individual particles.

The decay of the fluctuations of the angle α from the mean can be better discussed in a semi-log scale, as shown in figure 7.12b. As can be seen, the right tails of the p.d.f.s for sufficiently large α exhibit purely exponential decay, i.e. $p.d.f.(\alpha) \sim \exp(-a\alpha)$ with positive values for a. In case $M178$ the exponential behavior is clearly visible for $\alpha \gtrsim 15$ degrees, whereas in case $M121$ the exponential decay (at the same rate) matches a small range of angles around $\alpha \approx 30$ degrees. Concerning case $ML178$, the decay is at faster rate than in the flow cases with the higher value of the solid volume fraction.

The fluctuations of the particle velocity in the horizontal direction as well as the broad probability distribution of the angle α indicate that the particles are most probably subjected

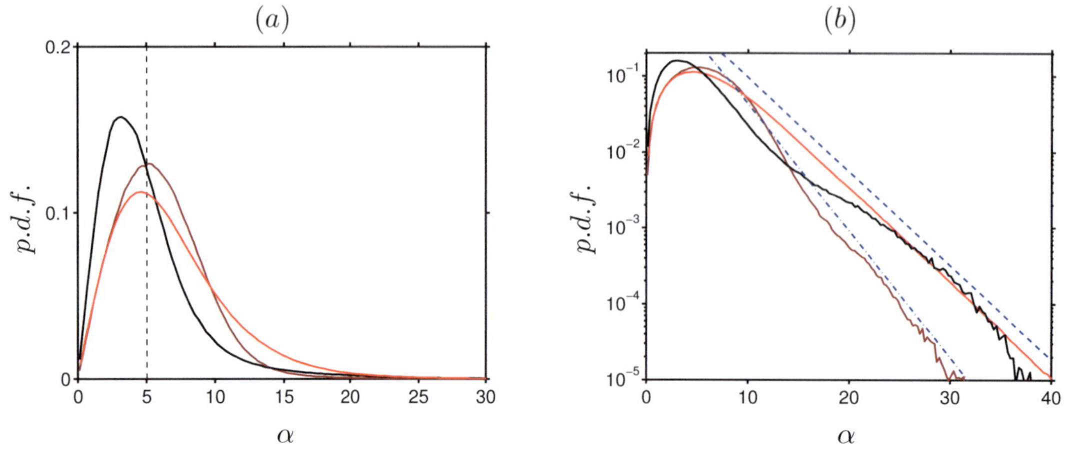

Figure 7.12: Probability density function of the angle α between the relative particle velocity vector and the vertical axis, defined as $\tan(\alpha) = (u^2_{p,rel} + v^2_{p,rel})^{1/2}/|w_{p,rel}|$, measured in degrees, for case $M121$ (——), case $M178$ (——) and case $ML178$ (——). The vertical dashed line (- - - -) denotes the value obtained in the single particle simulation $S178$. (b) shows the same data in a semi-log scale. The blue dashed line (- - - -) is proportional to $\exp(-a\alpha)$ with $a = 0.285$, whereas the blue chain-dotted line (—·—) corresponds to decay rate of $a = 0.385$.

to rotation around their own axis, especially in cases $M178$ and $ML178$. In the following the fluctuations of the particle angular velocity will be discussed. Figure 7.13 depicts the temporal evolution of the fluctuations of the particle angular velocity for all three cases. It can be observed that in all three cases the amplitude of the fluctuations of the particle angular velocity components is significantly smaller than of the particle translational velocity (cf. figure 5.3b). In all three flow cases the angular velocity fluctuations exhibit anisotropic behavior (horizontal component larger than the vertical component), due to the predominant vertical motion of the particles and the fluid, respectively. Please note, that fluctuations of the vertical fluid velocity are mainly responsible for the rotation around the horizontal axis, while the rotation around the vertical axis is generated by fluctuations of the horizontal fluid velocity component. In case $M121$ the fluctuations for both angular velocity components reach quickly their asymptotic value and the curves remain rather flat over the entire course of the simulation. On the other hand, in case $M178$ the fluctuating energy increase more slowly with time and reach their asymptotic value after approximately $200\tau_g$. Please recall that the particles statistics in case $ML178$, at the time the simulation is started, are already at steady state, thus the large initial values for the fluctuations in case $ML178$.

The current results appear to be consistent with the observations in Kajishima (2004a) with values of the horizontal angular velocity fluctuations being of the same order of magnitude as as in the present work. In their work the authors considered similar flow configuration as in the present study and investigated the influence of the particle rotation on the particle clusters. Further they argued that due to a Magnus force generated by the rotation of the particles tend to destabilize the particle agglomerations leading to periodical break-up of the particle clusters. For the clustering cases in the present work we could not find statistical evidence of this statement. In fact, as we saw in figure 6.6 the variance of the normalized Voronoï volumes fluctuates in time, but the fluctuations remain rather small. However, the Voronoï analysis provides a global view of the clustering, thus a further detailed analysis of the particles inside and on the periphery of clusters is required.

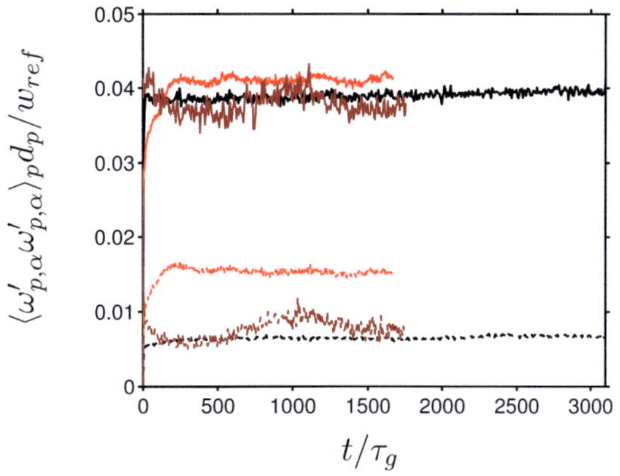

Figure 7.13: Temporal evolution of the fluctuating energy of the particle angular velocity components for case $M121$ (——,- - - -), case $M178$ (——,- - - -) and case $ML178$ (——,- - - -). Solid lines correspond to horizontal velocity components ($\alpha = 1, 2$), dashed lines show the vertical velocity component ($\alpha = 3$). The reference velocity w_{ref} denotes the absolute value of the terminal settling velocity $|w_\infty|$ obtained from the single particle simulations $S121$ and $S178$.

7.2.1 Probability distribution functions

As next the probability distribution of particle related quantities such as the particle translational and angular velocities, particle torque and force acting upon the particles are going to be discussed. The number of samples used for the calculation of the p.d.f.s in this section was in all three cases of the order of $\mathcal{O}(10^8)$ - $\mathcal{O}(10^9)$, assuring statistical convergence.

Particle velocity:
Figure 7.14 depicts the probability distribution function (p.d.f.) of the particle relative velocity components with respect to the global averaged fluid velocity (i.e. $w_{p,rel}$) for all three cases. As can be seen the particle velocity p.d.f.s in all flow cases exhibit distribution different from the Gaussian distribution. Moreover the fluctuations of the horizontal velocity components exhibit strong super-Gaussian distribution, where the vertical velocity component experience distribution close to the Gaussian. The non-Gaussian shape of the p.d.f.s is further quantified by the calculation of the higher moments of the distribution, in particular we have calculated the third and the fourth moments of the p.d.f.s in figure 7.14, known as the skewness and the kurtosis (flatness) of the probability density distribution. The exact values of the skewness and the flatness of the corresponding p.d.f.s in all three cases can be found in tables 7.4 - 7.5. As expected, the probability density function of the horizontal velocity component is in all three cases symmetric, which is confirmed by the zero values of the skewness. The distribution of the vertical velocity component, on the other hand, is in all flow cases negatively skewed, indicating the higher probability of finding particles with smaller than the mean velocity (note that larger negative values imply higher settling velocities). The largest value of the skewness is observed in case $ML178$ where also the highest clustering levels are observed. As can be seen, the vertical component of the particle relative velocity exhibits similar intermittency as the Gaussian distribution and indeed the values of the fourth moment, which measures the intermittency of the distribution, of the vertical velocity component measures values close to three. Regarding the horizontal velocity component, it is interesting to note that the highest value of the kurtosis is measured in the non-clustering case $M121$, experiencing value approx-

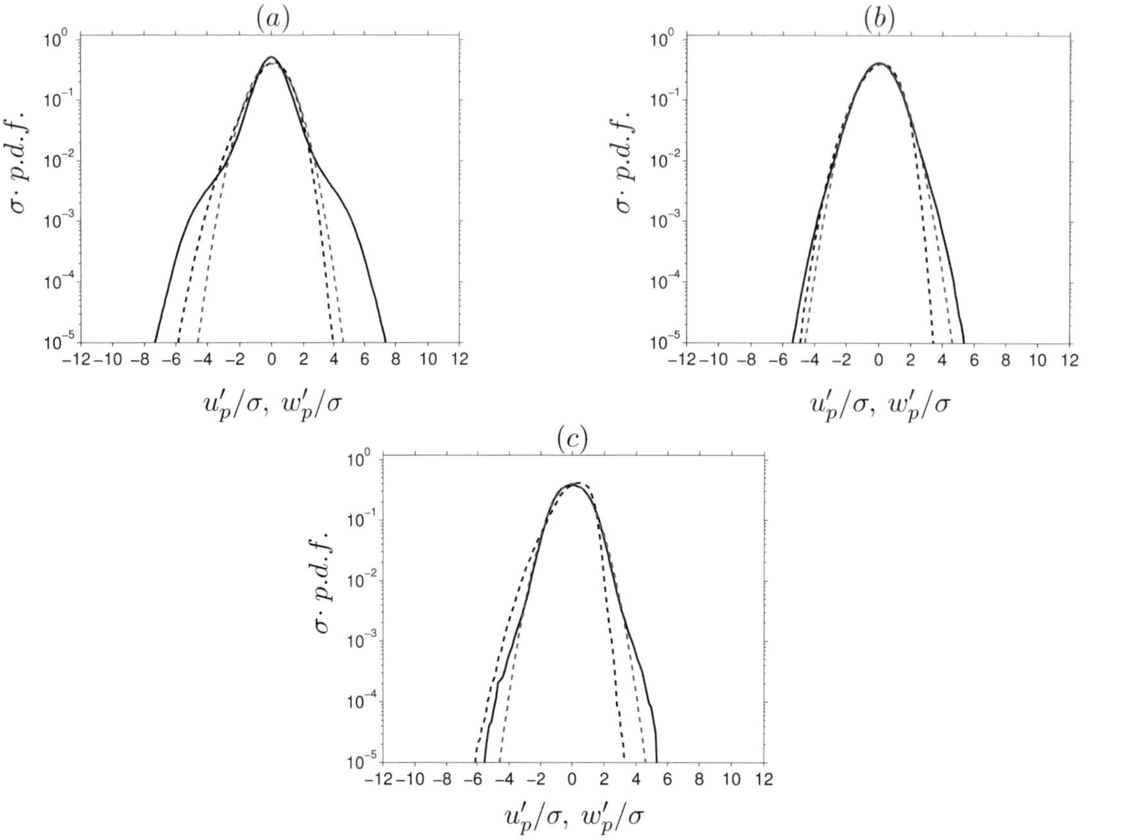

Figure 7.14: Probability density functions of the particle velocity components for (a) case $M121$, (b) case $M178$ and (c) case $ML178$. Horizontal components (——), vertical component (- - - -). Gaussian distribution (- - - -).

imately two times the values in the other two flow cases. It can be observed that the p.d.f. of the horizontal velocity component in case $M121$ exhibits increased probability of finding "extreme" events with values of approximately ± 6 standard deviations. Similar shape of the velocity p.d.f. with similar kurtosis values was observed in the work of Martínez-Mercado et al. (2010), where the clustering of bubbles rising in a pseudo-turbulent flow was experimentally studied. Comparable values for the fourth moment of the particle velocity were reported by Voth et al. (2002); Volk et al. (2011), where the dynamics of neutrally buoyant particles in a turbulent flow was studied. The authors reported values of the kurtosis in the range $2.4 - 3.2$. More recently Martínez-Mercado et al. (2012) measured the kurtosis of the velocity of light particles in a homogeneous isotropic turbulence to be close to that of a Gaussian distribution.

Consistent across all three cases the particle relative velocity experience distribution different than Gaussian. Thus, it appears that the pseudo-turbulent flow field (wake-induced flow field) in the present cases have a systematic effect on the particle relative velocity different from that induced by a classic homogeneous isotropic turbulence (Aliseda et al., 2002; Bunner and Tryggvason, 2002; Volk et al., 2011; Bellani and Variano, 2012b; Martínez-Mercado et al., 2012), where the particle velocities were observed to have essentially a Gaussian distribution.

Particle angular velocity:
Now let us turn to the distribution of the particle angular velocity components depicted in

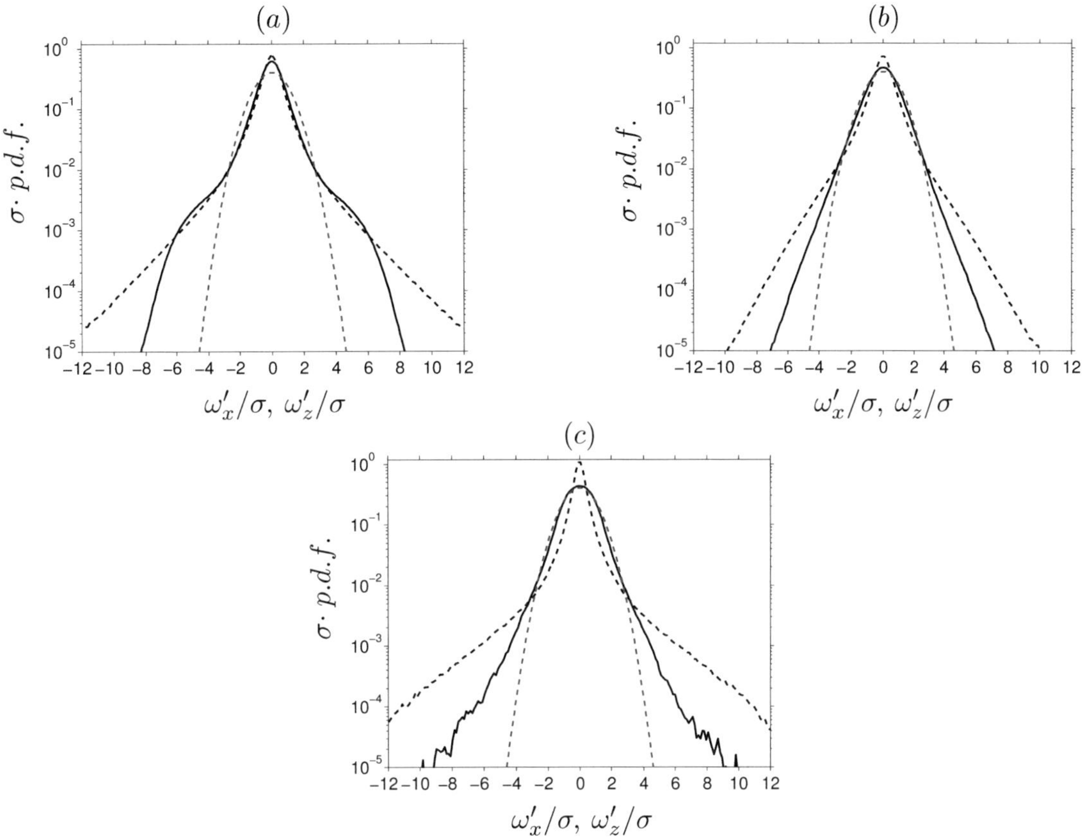

Figure 7.15: Probability density functions of the particle angular velocity components for (a) case $M121$, (b) case $M178$ and (c) case $ML178$. Color coding as in figure 7.14.

figure 7.15. It can be observed that the angular velocity in all flow cases have strong intermittent character. The horizontal components of the angular velocity experience in all flow cases similar skewness values as the translational particle velocity (i.e. symmetric distributions), except in case $ML178$, where a slightly negative value of the skewness is measured. However, the flatness values are larger in all flow cases than for the linear particle velocity, indicating that the distribution of the horizontal angular velocity component is more intermittent. The largest value of the p.d.f.'s flatness was again observed in case $M121$. Concerning the vertical component of the particle angular velocity, the probability density functions are in contrast to that of the linear particle velocity symmetric with nearly zero skewness. This is mainly due to the invariance of the flow field in the horizontal direction (recall, that the relative motion of the fluid in the horizontal directions generates rotation around the vertical axis). The intermittency of the angular velocity fluctuations in the vertical direction is however significant in comparison to that of the horizontal components. In case $ML178$ the fourth moment of the p.d.f. was evaluated to 29.54 being almost three times larger than in case $M178$. It is interesting to note that the flatness in the clustering case $M178$ is smaller than in the non-clustering case $M121$. The distribution of the angular velocity in case $M178$ appear to be similar to that obtained in the experimental work of Zimmermann et al. (2011), where the rotation intermittency experienced by large neutrally buoyant particles in a turbulent flow was investigated. The authors reported values of the flatness for the horizontal component of approximately four, where in case $M178$ we measured 4.58. Comparing the flatness values

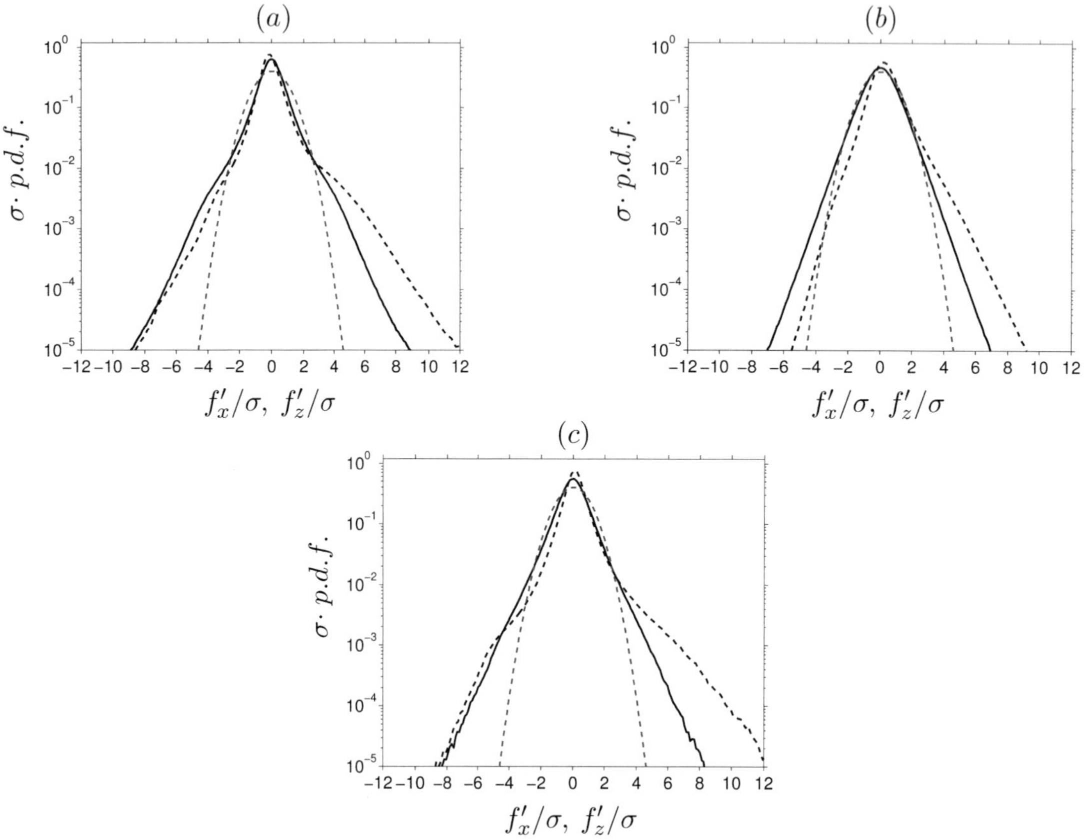

Figure 7.16: Probability density functions of the particle force components for (a) case $M121$, (b) case $M178$ and (c) case $ML178$. Color coding as in figure 7.14.

of the vertical angular velocity component for all flow cases, it appears that the values are inversely correlated to the relative turbulence intensity, i.e. for the smallest values of I_{r_V} the largest values of the flatness and vice versa.

Hydrodynamic force:

Let us now turn to the distributions of the acceleration of the particles. The particle acceleration is directly linked by Newton's law to the hydrodynamic force acting upon the particles, thus studying the hydrodynamic force of the particles is equivalent to studying the particle acceleration. The distribution of the particle acceleration has received recently a lot of attention from the scientific community. A recent review of the state of the art on the topic can be found in a review paper by Toschi and Bodenschatz (2009). The study of the particle acceleration is of particular interest, since the acceleration statistics reflects directly the action of the hydrodynamic forces upon the particles. Moreover, the segregation and the clustering of the particles were linked to the particle acceleration (Bec et al., 2006; Toschi and Bodenschatz, 2009). The probability distribution function of the particle acceleration for the present cases is depicted in figure 7.16. Here we would like to highlight that we have filtered all particle records corresponding to particle collisions in order to avoid the contribution of the artificial repulsion force used in the modeling of the solid-solid contacts. A simple visual examination of figure 7.16 reveals that the p.d.f.s of the particle acceleration are highly non-Gaussian with wide tails indicating the high probability for the occurrence of extreme events. This is in line with the observations made in the literature that the particle acceleration statistics, even for heavy finite-size particles, are highly intermittent (Bec et al., 2006; Ayyalasomayajula et al., 2006; Volk et al., 2008; Qureshi et al., 2008; Calzavarini et al., 2009a; Prakash et al., 2012). Most of this studies considered, however, the interaction of particles or bubbles with a turbulent carrier flow, while in the present flow configurations the interaction of the particles with a self-induced pseudo-turbulent flow is studied. More insight on the distribution of the particle acceleration statistics can be gained by analyzing the higher moments of the particle acceleration distribution. Considering the horizontal component, in all three flow cases the distribution is observed to be symmetric with zero skewness, confirming the isotropy of the flow in the horizontal direction. The vertical component of the hydrodynamic force, on the other hand, is positively skewed in all three flow cases with skewness values of 1.76 in case $M121$ and 1.68 and 1.71 in the clustering cases $M178$ and $ML178$, respectively. This indicates that the probability of finding larger than the average values of the drag force is higher than finding smaller values. This is in line with the results obtained in section §7.1.3 for the hydrodynamic force of the particles and supports the aforementioned non-linear drag effect, i.e. in the present flow conditions, the drag coefficient decays more slowly with increasing Reynolds number than $Re_{p,rel}^{-1}$. This would lead to a positively skewed distribution of the drag force even in case of symmetric distribution of the velocity fluctuations as noted by García-Villalba et al. (2012), where the sedimentation of finite-size particle in a plane vertical channel under turbulent conditions was studied. For negatively skewed distribution of the velocity fluctuations, as in the present flow cases, this effect is expected to be even more prominent. Regarding case $M121$ and $M178$, our results appear to support this argument by measuring the larger value of the skewness in case $M121$, where the larger (negative) value of the skewness of the particle velocity p.d.f. is observed. On the other, hand the results in case $M178$ appear to be in contradiction with this argument, thus requiring further investigation.

Considering the fourth moment of the acceleration distribution, reveals that in all three flow cases the flatness values of the vertical component of the particle acceleration is more intermittent than the horizontal component. Regarding the effect of the Galileo number, the flatness values of the particle acceleration p.d.f. for both components are in the clustering case $M178$ smaller than in the non-clustering case $M121$. For the horizontal components the flatness value in case $M121$ is nearly 40% larger than in case $M178$, while for the vertical component the flatness in case $M121$ is approximately 50% larger than in case $M178$. The smaller values of the particle acceleration flatness in case $M178$ indicates that the probability of finding extreme events in the particle acceleration are on average smaller than in case $M121$. Extreme events in the particle acceleration are often related to sudden short time changes of the particle velocity as result of the hydrodynamic interaction with the surrounding flow field (here we exclude contributions related to particle-particle collisions). Thus, the intermittency of the particle acceleration is generally believed to be a consequence of the fluid motion's small-scale intermittency (Qureshi et al., 2007). So, the question arise of the levels of intermittency of the local flow field surrounding the particles in the present configurations. In order to address this question, we have calculated the probability density function of the fluid velocity in vicinity of the particles (cf. section §8, figure 8.5). It turned out that indeed the local flow field in case $M178$ is less intermittent than in case $M121$, experiencing smaller flatness values, which supports the above considerations. The exact values of the corresponding p.d.f.s' higher moments can be found in tables 8.1 - 8.2. This indicates that the flow field induced by the particles and the particle cluster in case $M178$, although with higher intensity, leads to attenuation of the probability for the particles' hydrodynamic force to experience extreme

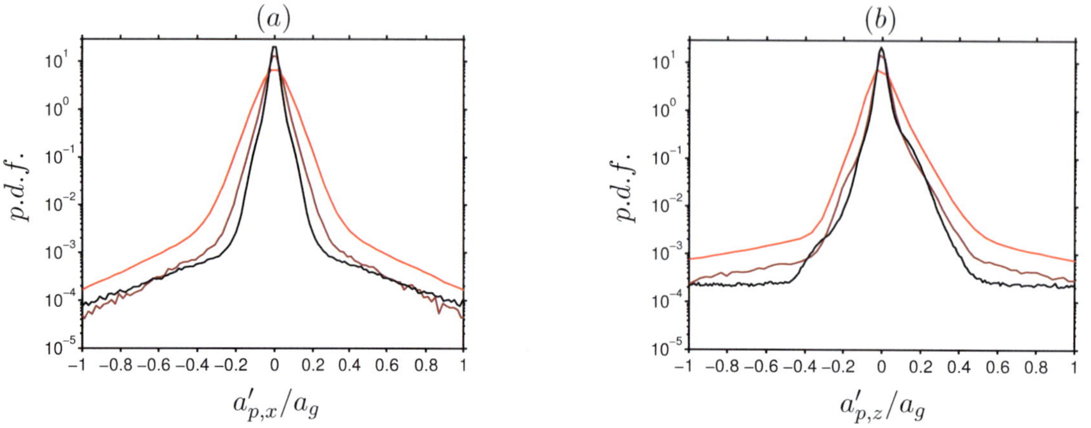

Figure 7.17: Probability density functions of the particle acceleration $\boldsymbol{a}_p$ components for case $M121$ (——), case $M178$ (——) and case $ML178$ (——). (a) horizontal component and (b) vertical component. The particle acceleration has been normalized by the characteristic acceleration $a_g = u_g^2/d_p$ with u_g defined as in equation 2.11.

events (in terms of the standard deviation of the hydrodynamic force). With respect to the solid volume fraction, i.e. case $ML178$, the results for the intermittency of the particle acceleration appear to support the above considerations. The flatness values in case $ML178$ are larger than in case $M178$, stemming (as conjectured above) most probably from the larger intermittency of the local fluid velocity $\boldsymbol{u}_f^S$ in case $ML178$.

However, the higher intermittency of the hydrodynamic force in case $M121$ than in case $M178$ does not imply a higher intensity of the force fluctuations. Please note that the force in figure 7.16 is normalized by its standard deviation, which is in case $M178$ larger than in case $M121$. The standard deviation for the horizontal (vertical) force components in case $M178$ is approximately 57% (6.7%) larger than in case $M121$. In fact, as we saw in section §6.2 the particles in case $M121$ are more homogeneously distributed than in case $M178$ (even than a random distributed particles), additionally the relative intensity of the flow is significantly smaller than in case $M178$. As result, the occurrence of extreme events due to hydrodynamic interactions of the particles in case $M121$ is expected to be less frequent than in case $M178$. Therefore, any occurrence of such an event would have a larger contribution (weight) to the intermittency (flatness value) of the particle acceleration, which would result in higher intermittency. Similarly, due to the significantly smaller solid volume fraction in case $ML178$ than in case $M178$, the probability for the particles' hydrodynamic force to experience extreme events would decrease, thus increasing the relative contribution of such an event to the intermittency of the hydrodynamic force.

As a matter of fact, the probability to find large deviations from the average value of the hydrodynamic force is higher in case $M178$ than in case $M121$ or case $ML178$. This can be best illustrated by considering the non-normalized p.d.f.s (no scaling by the standard deviation) of the particle acceleration components, which are depicted in figure 7.17. The particle acceleration, $\boldsymbol{a}_p$, have been normalized by the characteristic acceleration a_g, viz. $\boldsymbol{a}_p/a_g$, where $a_g = u_g^2/d_p$ with u_g defined as in equation 2.11. It can be observed that, indeed, the intermittency is highest in case $M178$. The distribution of the particle acceleration components in case $M121$ and case $ML178$ are not as broad as in case $M178$, indicating the smaller standard deviation in the corresponding flow cases. We have evaluated the flatness values

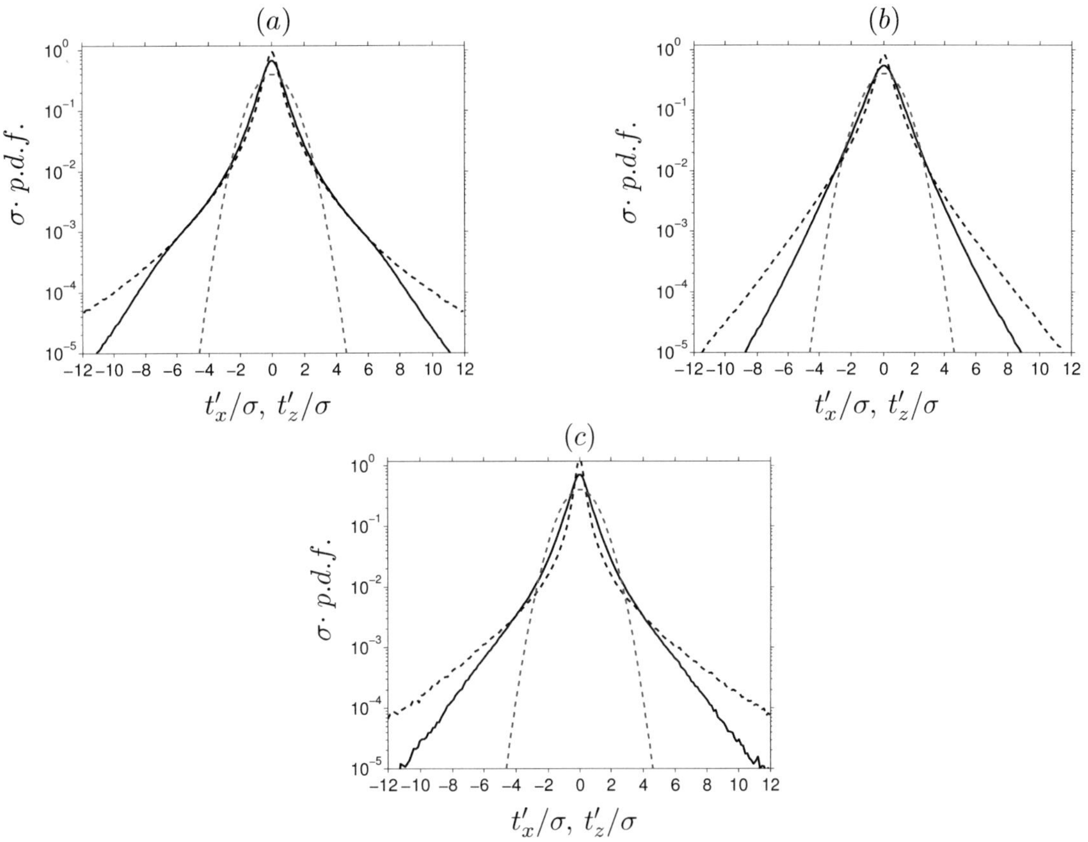

Figure 7.18: Probability density functions of the particle torque components for (a) case $M121$, (b) case $M178$ and (c) case $ML178$. Color coding as in figure 7.14.

in case $M178$ for the horizontal (vertical) component to 8.24 (10.26), whereas in case $M121$ a flatness value of 6.94 (2.59) was observed. The exact values for the third and the fourth moments of the p.d.f.s from figure 7.17 can be found in table 7.6.

As aforementioned, the particle acceleration is related to the instantaneous changes of the particle velocity. Therefore, the particle acceleration is often investigated by studying the particle velocity increments for very small time increments, i.e.:

$$\delta_{u_{p,i}}(\tau_{sep}) = u'_{p,i}(t + \tau_{sep}) - u'_{p,i}(t), \tag{7.19}$$

where τ_{sep} is the time increment. We checked (plots omitted) that indeed the distribution of the velocity increment $\delta_{u_{p,i}}(\tau_{sep})$ at very small values of τ_{sep} (note, that smallest value for τ_{sep} is limited by the time step of the simulation) resembles closely the distribution of the particle acceleration. On the other hand, for sufficiently long time increments at which the velocity signals are completely decorrelated, García-Villalba et al. (2012) showed that the kurtosis values of the increment will have an asymptotic value of $K(\delta_{u_{p,i}}) = K(u'_{p,i})/2 + 3/2$. Indeed, in all flow cases this values were obtained, except for case $M178$, where for the largest time increment the asymptotic value was observed to be slightly larger than the predicted one. This is most probably due to the long correlation times of the particle velocity in case $M178$ resulting from the particle clustering (as we will see in the next section §7.2.2).

In the end, let us mention that the clustering of the particles appear to have similar effect as the so called *sampling* effect (particles tend to sample regions of the flow with low vorticity)

case	$S(u_p)$	$S(w_p)$	$S(f_{p,h})$	$S(f_{p,z})$	$S(\omega_{p,x})$	$S(\omega_{p,z})$	$S(t_{p,x})$	$S(t_{p,z})$
$M121$	0.01	−0.42	0.0	1.76	0.01	0.01	−0.01	0.0
$M178$	0.0	−0.34	0.0	1.68	0.0	0.01	0.0	0.0
$ML178$	0.0	−0.73	0.0	1.71	−0.02	−0.13	0.01	0.05

Table 7.4: Skewness of the probability density functions of the particle velocity, force, angular velocity and torque components for all three cases $M121$, $M178$, $ML178$. Data corresponds to the p.d.f.s in figures 7.14 - 7.18.

case	$K(u_p)$	$K(w_p)$	$K(f_{p,h})$	$K(f_{p,z})$	$K(\omega_{p,x})$	$K(\omega_{p,z})$	$K(t_{p,x})$	$K(t_{p,z})$
$M121$	7.86	3.53	7.51	15.04	10.66	17.05	13.23	19.21
$M178$	3.46	2.90	4.51	7.08	4.58	10.45	6.87	13.47
$ML178$	3.13	3.62	6.26	15.73	5.53	29.54	12.30	28.15

Table 7.5: Kurtosis of the probability density functions of the particle velocity, force, angular velocity and torque components for all three cases $M121$, $M178$, $ML178$. Data corresponds to the p.d.f.s in figures 7.14 - 7.18.

case	$\sigma(a_{p,x})$	$\sigma(a_{p,z})$	$S(a_{p,x})$	$S(a_{p,z})$	$K(a_{p,x})$	$K(a_{p,z})$
$M121$	0.03	0.06	0.0	0.13	6.94	2.59
$M178$	0.07	0.09	0.01	0.34	8.24	10.26
$ML178$	0.04	0.05	0.01	0.43	8.28	7.13

Table 7.6: Standard deviation, skewness and kurtosis of the probability density functions of the particle acceleration for all three cases $M121$, $M178$, $ML178$. Data corresponds to the p.d.f.s in figure 7.17.

in turbulent flows (Fessler et al., 1994). We would like to stress that the particle clutering in the present case has completely different origin than that observed for turbulent flows as in Fessler et al. (1994). For more detailed discussion of the phenomenology of the *sampling* effect for finite-size particles we refer the interested reader to the discussion in Qureshi et al. (2008).

Particle torque:
The distribution of the angular acceleration, i.e. particle torque, is depicted in figure 7.18. It can be observed that the distribution of the particle torque in all three flow cases is highly non-Gaussian with high probable occurrence of extreme events. The third moment of the p.d.f.s reveal that the distribution in all flow cases is symmetric with zero skewness. Considering the flatness values, it appears that the particle torque is more intermittent than the particle acceleration with higher flatness values for both components of the particle torque. It is interesting to note that the flatness values observed in the present simulations are considerably higher than the observed in Zimmermann et al. (2011), where they report values of $K = 7 \pm 1$, whereas the flatness in our case can reach values as high as $K \approx 28$ (case $ML178$). However, the flow considered in Zimmermann et al. (2011) is highly turbulent with $Re_\lambda \sim 300$ and the particle considered in their experiments was neutrally buoyant with size of the order of the integral length scale $D/L_{int} \sim 0.6$. For the present numerical experiments the pressure force does not contribute to the torque applied at the particle center. Thus, the angular acceleration

has only a single contribution from the viscous hydrodynamic stresses:

$$\mathbf{I}_p \dot{\boldsymbol{\omega}}_p = \mathbf{T}_H, \tag{7.20}$$

where $\mathbf{T}_H = \int_{\mathcal{S}} \mathbf{r}_c \times (\boldsymbol{\tau} \cdot \mathbf{n}) d\sigma$ with $\mathbf{r}_c$ being the distance vector from the particle center, $\boldsymbol{\tau} = \rho_f \nu (\partial u_{f_i}/\partial x_j + \partial u_{f_j}/\partial x_i)$ the viscous stress tensor, $\mathbf{n}$ the outward pointing normal vector at the particle surface and $\mathcal{S}$ the surface of the sphere. In the above equation $\mathbf{I}_p = \rho_p D^5 \pi/60$ denotes the moment of inertia in case of a homogeneous mass density. Thus, it is tempting to relate the local flow field experienced by the particles to the high values of the particle torque intermittency and merits further investigation in future works.

7.2.2 Lagrangian statistics

Let us now turn to the analysis of the correlation times of the Lagrangian statistics. The study of the Lagrangian statistics, such as the Lagrangian autocorrelations will provide further insight into the multiscale interaction between the dispersed phase and the carrier phase. Moreover, the study of the Lagrangian statistics will help to reveal the complex particle response to the flow field as the particles settle throughout the computational domain. In the following we will discuss the Lagrangian autocorrelation of various particle quantities, such as the particle linear and angular velocities and the hydrodynamic force and torque exerted upon the particles by the fluid.

The Lagrangian autocorrelation function of the fluctuations of a particle related quantity $\psi_i'(t) = \psi_i(t) - \langle \psi(t) \rangle_p$ is defined as follows:

$$R^p_{\psi_i \psi_i}(t, \tau_{sep}) = \frac{1}{\int_{T_0}^{T_n} \psi_i'(t) \psi_i'(t) dt} \int_{T_0}^{T_n} \psi_i'(t) \psi_i'(t + \tau_{sep}) dt, \tag{7.21}$$

where τ_{sep} denotes the time increment and the interval $[T_0, T_n]$ defines some time interval of interest. Please note that in the above equation no summation is implied over the repeated indices. The Lagrangian autocorrelation function is normalized by the autocovariance and gives a measure of how long a particle quantity is correlated with its value at some earlier time. For statistically stationary flows, the autocorrelation function is independent of the initial time t and depends only on the separation time τ_{sep}.

First let us discuss the Lagrangian autocorrelation of the particle velocity components, which are depicted in figure 7.19. The Lagrangian particle velocity autocorrelation could be interpreted as a measure of the particle dispersion and is related to it by the well known Taylors's theorem (Taylor, 1921):

$$\langle Y^2(T) \rangle - 2 \langle u' u' \rangle \int_{T_0}^{T_n} \int_0^t R_{uu}(\tau) d\tau dt, \tag{7.22}$$

where $R_{uu}(\tau)$ is the Lagrangian autocorrelation function of the fluctuating velocity following a fluid point $u'(t)$. The equations are valid for the dispersion of heavy particles, by substituting the particle velocity for the velocity u'. Hence, the study of the Lagrangian particle velocity autocorrelation has been of great interest in the past. Yudine (1959) studied the process of dispersion of heavy particles with large terminal velocities, later Csanady (1963) investigated the turbulent diffusion of heavy particles in the atmosphere. They developed the physical ideas behind the so called *crossing trajectories* and the associated *continuity effect* which are related

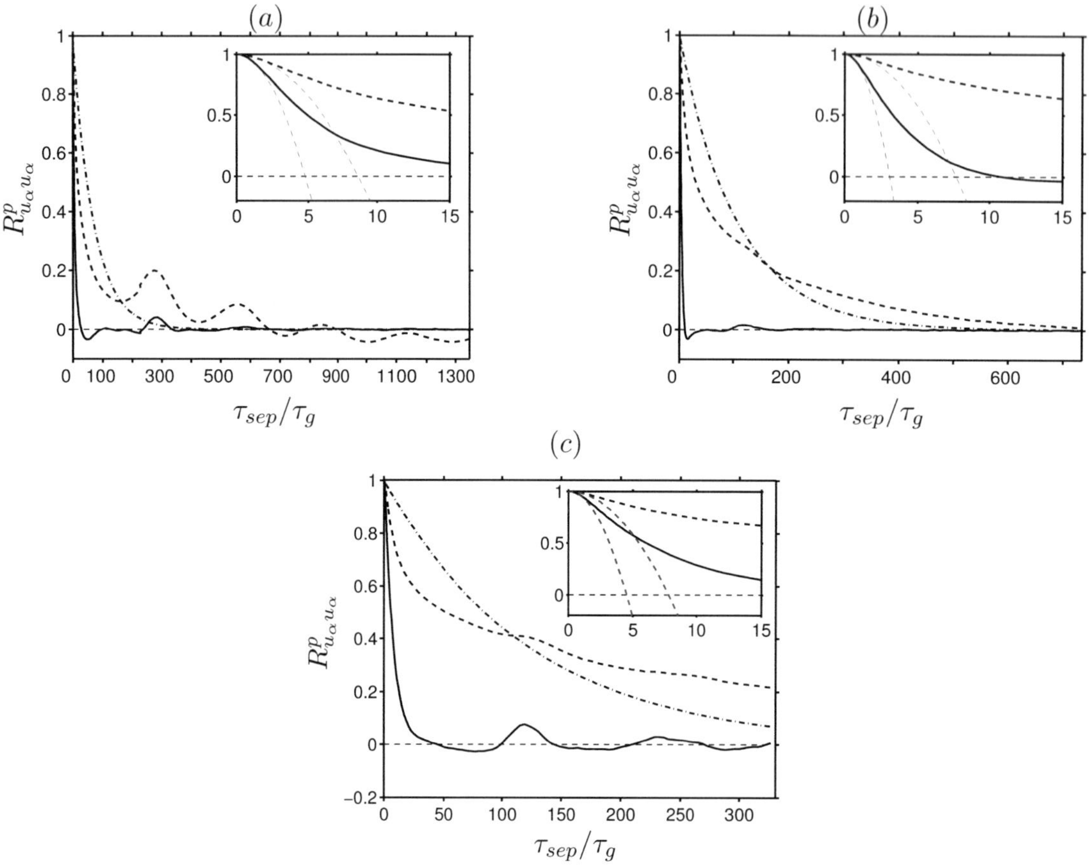

Figure 7.19: Lagrangian autocorrelation of the particle velocity components as function of time lag τ_{sep} for (a) case $M121$, (b) case $M178$ and (c) case $ML178$. Line coding: horizontal component (——), vertical components (- - - -), Langevin approximation $R^L_{u_\alpha u_\alpha} \approx \exp(-\tau_{sep}/T_{int})$, (—·—). The insets show a closeup of the same data for small separation times τ_{sep}. The dashed blue lines - - - - represents the osculating parabola at $\tau = 0$. The values of the integral time scale and the Taylor micro-scale are reported in tables 7.7 - 7.8.

to the evolution of the particle velocity autocorrelation. Later Snyder and Lumely (1971) performed measurements of the Lagrangian particle velocity autocorrelation in a turbulent flow. Squires and Eaton (1991) investigated the effect of the particle drift and particle inertia of heavy particles in a isotropic (decaying) turbulence on the particle Lagrangian velocity autocorrelation. They found out that particles with high inertia exhibit higher correlation coefficients and the Lagrangian correlation of the particles in direction of the particle drift velocity was more persistent than the autocorrelation in the directions perpendicular to the particle drift velocity. More recently Ahmed and Elghobashi (2001) investigated the particle dispersion in a homogeneous turbulent shear flow. They performed a DNS of small point particles with one-way coupling and observed the crossing trajectory effect and continuity effect on the Lagrangian particle velocity autocorrelation. Bunner and Tryggvason (2002) investigated the dispersion of bubbles by a pseudo turbulence generated by the motion of the bubbles. Similar to the work of Squires and Eaton (1991) they stated that the vertical (streamwise) fluctuating velocity component remains correlated for longer times than the horizontal components.

Particle velocity:
The Lagrangian autocorrelation functions of the particle velocity components for all three flow cases are depicted in figure 7.19. It can be observed that the anisotropic character of the flow is also reflected in the particle velocity Lagrangian autocorrelation function. In all cases the vertical (streamwise) component of the particle velocity is correlated over long time period, whereas the horizontal components decay more rapidly and actually become negative for small time separations before converging to its asymptotic value (i.e. zero) for long separation time.

While all flow cases experience negative loops (negative correlation) for small time separations of the horizontal velocity components, the rate at which the correlation decays towards zero and the time at which the minimum values are observed vary between the cases. The fastest decay is observed in case $M178$ with reaching the first minimum value of R_{uu} at $\tau_{sep} = 16.3\tau_g$. The corresponding values for case $M121$ and case $M178$ are measured to $\tau_{sep} = 46.6\tau_g$ and $\tau_{sep} = 84.9\tau_g$, respectively. It appears that the time at which the minimum is observed correlates with the relative turbulence intensity of the flow. The occurrence of the negative loops for small time separations are often related to the continuity effect mentioned above. The continuity effect, as defined in Csanady (1963), implies the existence of a *backflow* (directed upward with respect to the mean flow) necessary to satisfy continuity in the continuous phase. As can be seen the autocorrelation of the horizontal velocity component after reaching its initial minimum appear to exhibit some small oscillation with prominent positive peaks that decay quickly with increase of the separation time. The frequency of this oscillations appear to correlate with the frequency of similar oscillations of the correlation function of the vertical velocity component (best seen in case $M121$). The intervals seem to correspond to an average return time defined as the ratio of the vertical extension of the computational domain L_z and the apparent velocity lag $w_{p,lag}$. The average return time in case $M121$ was evaluated to $290\tau_g$ gravitational time units, whereas in case $M178$ and $ML178$ the average return time was calculated to $117.1\tau_g$ and $110.6\tau_g$, respectively. As can be taken from the figure this values are in good agreement with the time intervals at which the peaks in the Lagrangian autocorrelation occur.

Let us first consider case $M121$, where the oscillations are most prominent. A possible reason for this oscillations could be the existence of coherent flow structures in conjunction with the periodic boundary conditions, causing the particles to encounter a similar flow field after an average return time. This implies, that this flow structures are temporally correlated over long time periods. For the time period simulated in the present work we observed up to four such prominent peaks of the Lagrangian autocorrelation function in case $M121$. Thus, a lower bound of the time scale of this structures can be estimated to be at least four turnover periods.

Concerning case $M178$, no such prominent peaks in the Lagrangian autocorrelation function of the vertical velocity component can be observed, despite the smaller extension of the computational domain in the vertical direction, when compared to case $M121$. As discussed in the previous chapter §6 the flow structures in case $M178$ are extending throughout the entire computational domain, i.e. for the present configuration the structures appear to have an infinite length. Thereby, the particles as they settle through the computational domain, tend to experience on average the similar flow environment (infinite flow structures) leading to the absence of a prominent peaks in the autocorrelation in case $M178$. On the other hand in case $ML178$, the smaller value of the solid volume fraction appears to reduce this effect, leading to the two peaks in the Lagrangian autocorrelation function in figure 7.19c.

Despite the statistical bias of the particle velocity autocorrelation due to the finite streamwise length of the computational domain we calculated the integral time scale by integrating the autocorrelation functions:

$$T_{int} = \int_0^\infty R^p_{u_\alpha u_\alpha}(\tau)d\tau \tag{7.23}$$

The values for both cases and the different velocity components are presented in table 7.7. The integral time scale provides an estimation of the time period over which the particle velocity components remain correlated. Concerning the horizontal velocity component, the integral time scale in case $M121$ was measured to $T_{int} = 6.98\tau_g$, whereas in he clustering cases significantly larger values were obtained, viz. $T_{int} = 14.38\tau_g$ and $T_{int} = 8.78\tau_g$ in case $M178$ and $ML178$, respectively. As can be taken from the figure, the vertical velocity component remains correlated over substantially longer time intervals. The integral time scales were evaluated in case $M121$ to $T_{int} = 68.8\tau_g$ and in case $M178$ ($ML178$) to $T_{int} = 106.6\tau_g$ ($T_{int} = 122.9\tau_g$), respectively. The large integral time scales of the vertical velocity component are mostly due to the strong particle clustering and the resulting coherent flow structures, which are evidenced in case $M178$ and case $ML178$. This kind of vertically elongated flow structure facilitate the existence of the large integral time scales of vertical velocity component in cases $M178$ and $ML178$. Comparison of the integral time scale of the horizontal and the vertical velocity components evidences once again the strong anisotropy of the dispersed phase.

As aforementioned, the Lagrangian autocorrelation function is related to the dispersion of the particles (Taylor, 1921). In the past, several studies have been devoted to the modeling of the particle dispersion, viz. Pozorski and Minier (1998); Pai and Subramaniam (2012); Iliopoulos et al. (2003). A large number of the studies related to the particle dispersion are often based on a Langevin type of equation for the dispersion of the particles. This models assume that the autocorrelation function of the particle velocity follows an exponential decay and the Lagrangian autocorrelation function can be approximated as $R^L_{u_\alpha u_\alpha} = \exp(-\tau/T_{int})$. Due to the exponential nature of the Langevin model, such models are not capable of capturing the continuity effect (negative loops in the velocity autocorrelation) described above. Nevertheless, we have applied the above approximation of the particle velocity autocorrelation for the present simulations. The function $R^L_{u_\alpha u_\alpha}$ is superposed in figure 7.19 for comparison. The integral time scale determined from the Lagrangian autocorrelation function was used in the determination of $R^L_{u_\alpha u_\alpha}$. Computing the integral time scale by integration of the Langevin function results in a very good agreement with the integral time scales obtained form the simulations, indicating the exponential decay of the Lagrangian autocorrelation function. The corresponding values are presented for completeness in table 7.7.

Now let us turn to the short time behavior of the Lagrangian autocorrelation function. As aforementioned, the initial decay of the particle velocity autocorrelation experience different temporal behavior in the three flow cases. The initial decay can be characterized through the Taylor micro-scale, which measures for the horizontal velocity components 4.78, 3.08 and 4.05 gravitational time units. This is in line with the observations made above that autocorrelation in case $M178$ decay at the fastest rate. For the vertical component of the particle velocity similar trend is observed, the smallest value of the Taylor micro-scale is measured in case $M178$, i.e. $\tau_\lambda(w_p) = 7.67\tau_g$. Comparison with the value for case $ML178$ ($\tau_\lambda(w_p) = 7.74\tau_g$) reveals that the solid volume fraction has only a minor effect on the short time behavior of the vertical component of the particle velocity, with slightly larger (0.9%) Taylor micros-scale in case $ML178$. The largest value is, consequently, observed in case $M121$ and measures $\tau_\lambda(w_p) = 8.65\tau_g$. The Taylor micro-scale for all velocity components for both cases can be

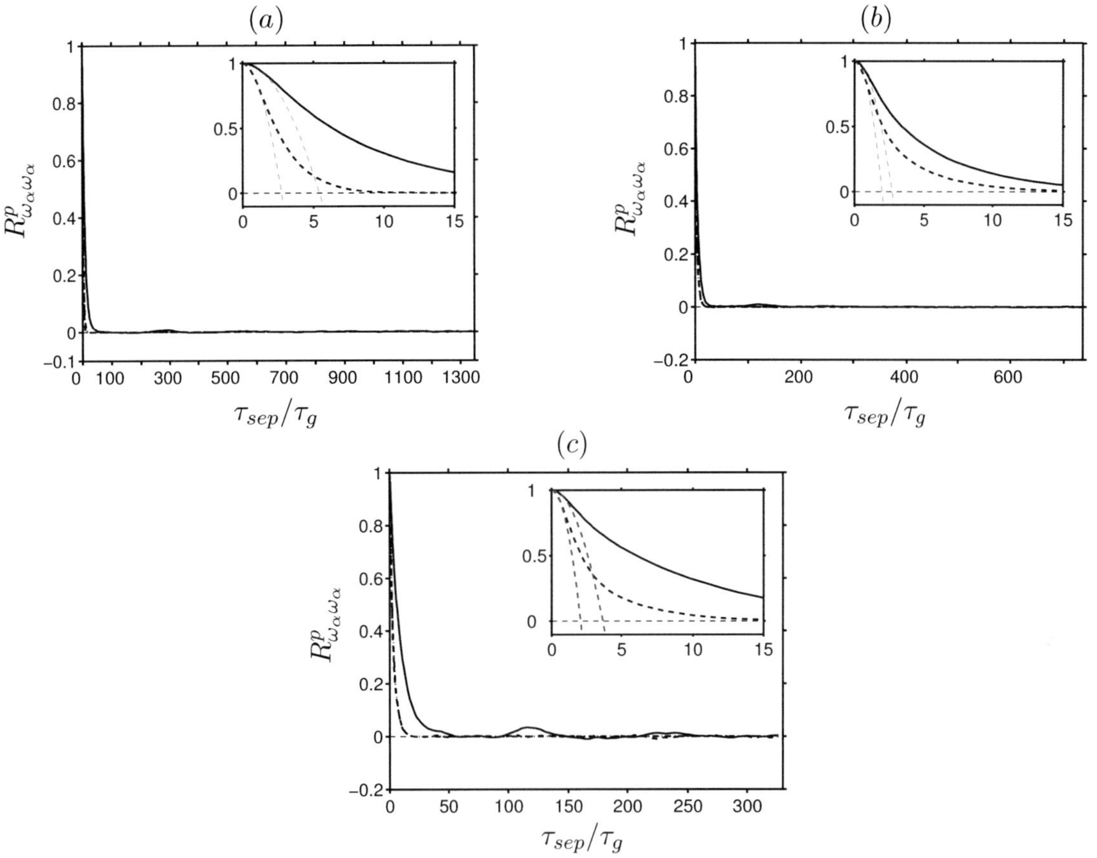

Figure 7.20: Same as in figure 7.19, but for the particle angular velocity components.

found in table 7.7. It is interesting to note that despite the faster initial decay of the particle velocity autocorrelation in the clustering cases $M178$ and $ML178$, the velocity is correlated over longer time intervals than in case $M121$.

The overall evolution of the Lagrangian autocorrelation function is in line with the observations made in the above mentioned references (Squires and Eaton, 1991; Bunner and Tryggvason, 2002), evidencing longer correlation times of the vertical velocity component. Please note that the particle respectively the bubble Reynolds number in this works are an order of magnitude smaller than the Reynolds number of the particles considered in the present study. Moreover, the analysis of the Lagrangian particle velocity autocorrelation function showed that the initial decay of the correlation function in case $M178$ is fastest for all velocity components and slowest in case $M121$. Regarding the long time behavior, the same trend was observed for the horizontal velocity components, while the opposite was observed for the vertical velocity components, which was accounted as direct result from the particle clustering in case $M178$. Moreover, the strong anisotropy of the flow caused a anisotropy of the integral time scale as well as of the Taylor micro-scale.

Particle angular velocity:
Figure 7.20 depicts the autocorrelation function of the particle angular velocity components for all three flow cases. As can be seen the correlation functions in all flow cases behave very similar, the autocorrelation for both components decays quickly without experiencing negative

values. The autocorrelation function for the horizontal component decays, however, more slowly than for the vertical component and the difference is largest in case $ML178$, while in case $M178$ the smallest difference between the decay rates of the autocorrelation of both velocity components is observed. This can be best seen by comparing the Taylor micro-scales of the horizontal and vertical components. For this purposes we calculated the ratio of the corresponding Taylor micro-scales, viz. $\tau_\lambda^{xz} = \tau_\lambda^x/\tau_\lambda^z$. In case $M121$ this ratio was evaluated to $\tau_\lambda^{xz} = 1.98$, whereas in case $M178$ and case $ML178$ we obtained $\tau_\lambda^{xz} = 1.4$ and $\tau_\lambda^{xz} = 1.8$. Comparison of the Taylor micro-scale values for the three flow cases reveals that the initial decay in the clustering cases is faster than in the non-clustering case $M121$. Similar as for the particle velocity, the smaller value of the solid volume fraction lead to a small increase of the initial decay rate, which is evident by the larger values of the Taylor micro-scale in case $ML178$ than in case $M178$.

Regarding the integral time scale, the horizontal components in all flow cases are correlated over longer time intervals than the vertical components (recall that vertical fluid velocity generates rotation of the particles around the horizontal axis and vice versa). The relatively similar values between the clustering and the non-clustering cases implies that the particle clustering does not significantly affect the long time behavior of the particle angular velocity autocorrelation. However, the effect the solid volume fraction appear to affect the long time behavior of the horizontal angular velocity component with approximately 50% larger value of the integral time scale in case $ML178$ than in case $M178$.

Here again the approximation of the decay rate through an exponential function of a Langevin type agrees very well with the rsults obtained from the simulations. The exact values of the integral time scale and the Taylor micro-scale are depicted in table 7.7. Unfortunately, we could not find any reference data relevant to our simulations in the literature to compare with and hope that this study will arouse the interest of the scientific community.

Particle hydrodynamic force:
We now present the autocorrelation function of the hydrodynamic forces, which are exerted upon the particles by the flow field. As mentioned above the hydrodynamic force of the particles is directly related to the particle acceleration and both will be used in the following interchangeably. Figure 7.21 depicts the autocorrelation of the particle acceleration for both cases. Here we would like to note, that in the process of calculation of the autocorrelation function, the influence of particle collisions was excluded by neglecting all particles which are colliding in a particular time interval $[t,\ t + \tau_{sep}]$. As a consequence, the largest time interval is limited to the largest collision free interval, leading to the shorter time intervals over which the correlation functions were computed in comparison to the particle velocity and angular velocity.

As can be seen the autocorrelation functions in all flow cases exhibits very similar behavior. Both components of the particle hydrodynamic force decay quickly and experience negative loops. Moreover, the correlation of the horizontal component of the hydrodynamic force decays more quickly than the correlation of the vertical component. The autocorrelation function of the horizontal (vertical) component first cross the zero axis after approximately $2.33\tau_g$ $(4.6\tau_g)$ in case $M121$, whereas in case $M178$ this time is evaluated to $1.69\tau_g$ $(2.81\tau_g)$ and to $2.25\tau_g$ $(3.24\tau_g)$ in case $ML178$. This values appear to be in good agreement with the values reported by Yeung and Pope (1989), where the authors studied numerically the Lagrangian statistics of homogeneous isotropic turbulence. The authors identified that the characteristic time at

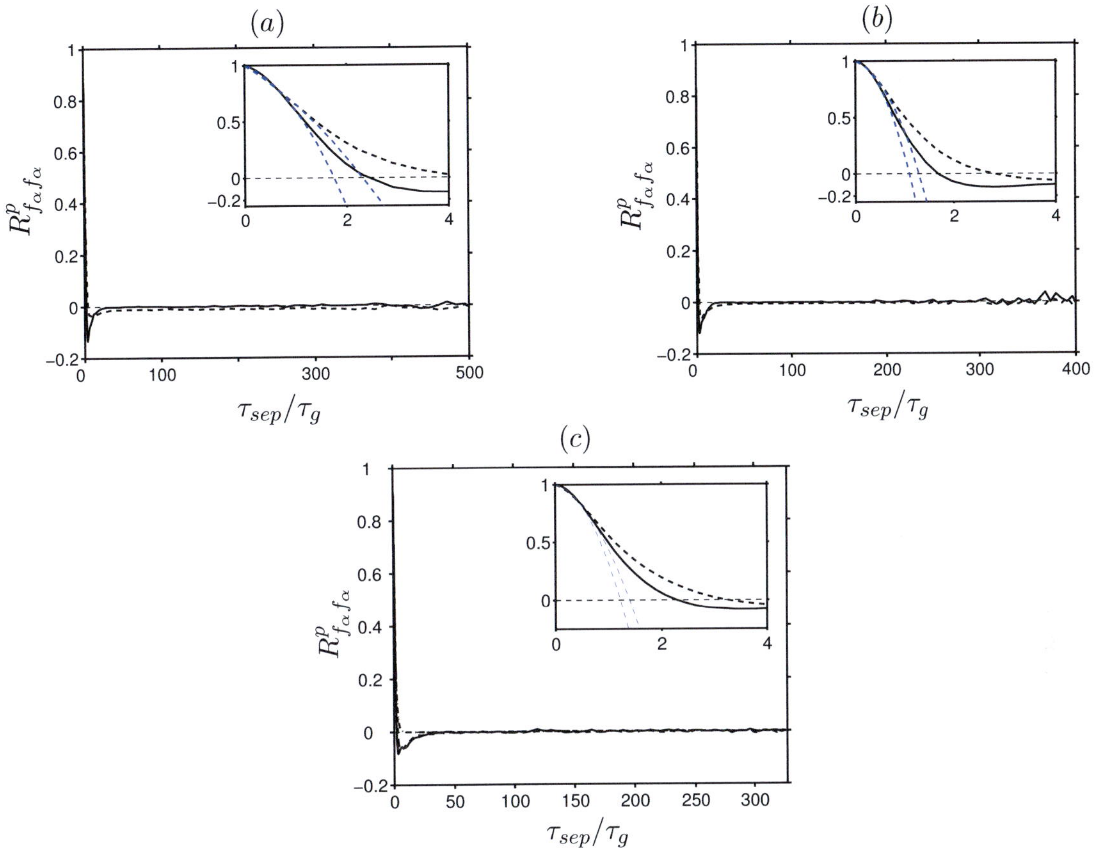

Figure 7.21: Same as in figure 7.19, but for the components of the particle hydrodynamic force.

which the autocorrelation of the vertical component first crosses the zero line is approximately $2.2\tau_\eta$ (τ_η denotes the Kolmogorov time scale). This was shown to apply also to finite size particles with size larger than the Kolmogorov length scale, e.g. Mordant et al. (2004); Xu and Bodenschatz (2008). The mentioned works considered however turbulent flows with high turbulent intensities. Mordant et al. (2004) also calculated the autocorrelation of the acceleration magnitude and showed that it decays much slower than the correlation of the components, they attributed this to the intermittency of the turbulent flow. In contrast to Mordant et al. (2004) we did not observe such a behavior, the correlation function of the acceleration magnitude (not shown here) followed the evolution of the correlation function of the vertical component, indicating the dominance of the vertical component of the particle acceleration in the present flow configurations. Assuming that the gravitational time scale is the relevant time scale of the small scales, the present results appear to be in a reasonably good agreement with the value observed in Mordant et al. (2004); Xu and Bodenschatz (2008). Please note that the value of $2.2\tau_\eta$ is observed for a highly turbulent flows with wide range of Reynolds numbers. Nevertheless, the value in case $M178$ and case $ML178$ shows better agreement with the value observed in the literature than that obtained in case $M121$.

Similar to the correlation functions of the particle velocity and angular velocity, the initial decay of the correlation functions for both components in case $M178$ is faster than in case $M121$, as evidenced by the larger values of the Taylor micro-scale in case $M121$ than in case $M178$ (exact values of the correlation function for both cases can be found in table 7.8). The same tendency was also observed in the long time behavior of the autocorrelation function, as the integral time scales in case $M121$ shows higher values than in case $M178$. As for the particle angular velocity, the particle clustering in case $M178$ appears to have only a minor influence on the temporal dynamics of the hydrodynamic forces, since they are mostly influenced by the flow field in vicinity of the particles.

Regarding case $ML178$, the results are consistent with that obtained for the particle translational and angular velocities, i.e. the values of the Taylor micro-scale and the integral time scale are larger than in the case with the higher solid volume fraction $ML178$, indicating faster initial decay and longer long time correlation of the hydrodynamic force.

A characteristic time for the evolution of a particle response to flow changes can be obtained from the autocorrelation function of the particle acceleration by integrating the correlation function over its positive part:

$$T_I = \int_0^{\tau_0} R^p_{f_\alpha f_\alpha}(\tau) d\tau, \qquad \tau_0 = \min_\tau \{R^p_{f_\alpha f_\alpha}(\tau) = 0\}, \tag{7.24}$$

where τ_0 is the first zero crossing time (Calzavarini et al., 2009b). For the horizontal (vertical) component in case $M121$ we obtained $T_I = 1.2\tau_g$ ($T_I = 1.58\tau_g$), whereas in case $M178$ the values are evaluated to $T_I = 0.83\tau_g$ ($T_I = 1.09\tau_g$) and in case $ML178$ to $T_I = 1.063\tau_g$ ($T_I = 1.29\tau_g$). This implies that the characteristic time of the forcing experienced by the particles is longer in case $M121$ and case $M178$ than in case $M178$. This is in line with the observations made in section §7.1.3 and §7.2.1, that due to the higher relative turbulence intensity the particle force in case $M178$ is more frequently subjected to extreme events, which would lead eventually to faster decorrelation of the hydrodynamic force than in case $M121$ and $ML178$.

Finally, we would like to note that the short-living character of the particle acceleration and the overall evolution of the correlation function appears to be in line with previous studies of the particle acceleration, e.g. (Voth et al., 1998, 2002; Mordant et al., 2004; Volk et al., 2008; Xu and Bodenschatz, 2008; Homann and Bec, 2009; Calzavarini et al., 2009b; Volk et al., 2011). However, most of this studies considered highly turbulent flows, where a pseudo turbulent flow is considered in the present study.

Particle torque:

Finally, let us briefly discuss the autocorrelation function of the particle torque, which is depicted for all three flow cases in figure 7.22. As the autocorrelation of the particle acceleration, the autocorrelation of the particle torque was calculated by discarding all particle time records that are incorporated in particle-particle collisions, thus the maximum length of the time interval considered for the calculation of the correlation function is shorter, as in the case of the autocorrelation of the particle acceleration. It can be observed that the correlation function of the particle torque components experience similar evolution with time as the particle force. The correlation function for both components decays quickly and negative loops at small separation times with small negative values can be observed. Beyond this times, the correlation function converges rapidly towards its asymptotic value, indicating the complete decorrelation of the particle torque. As can be observed in the close-ups (insets of figure 7.22), the vertical component of the particle torque decorrelates faster than the horizontal component and the decay rate in all three cases appear to be very similar, showing small, but noticeable difference in the initial decay of the correlation functions. As aforementioned, the initial decay of the autocorrelation function is best represented by the Taylor micro-scale, which

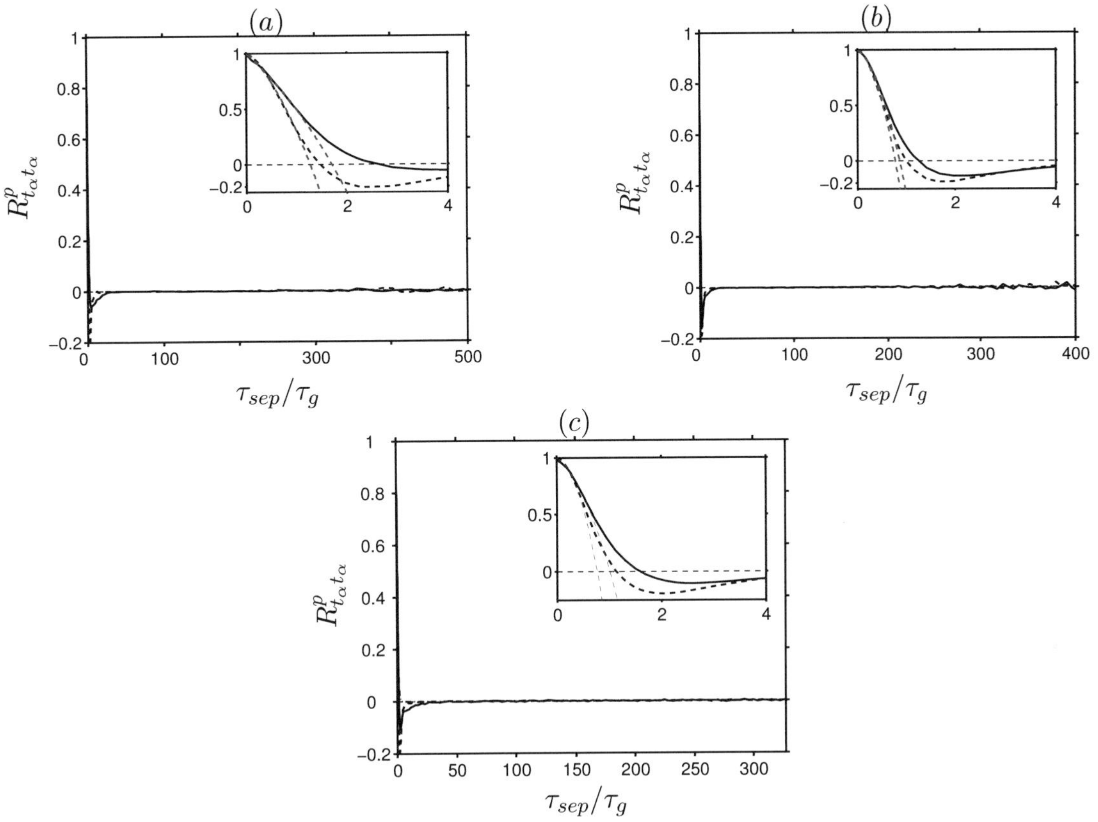

Figure 7.22: Same as in figure 7.19, but for the components of the particle torque.

is shown in table 7.7 for both components. Please consult table 7.7 for exact values of the Taylor micro-scale. Detailed study of the Taylor micro-scale values of both components in all three flow cases reveals that the difference between the initial decay of the horizontal and the vertical components is the smallest in case $M178$, where as in case $M121$ and $ML178$ the difference is very similar. As for the particle angular velocity, the decay of the autocorrelation function is well approximated by an exponential function of Langevin type. The same trend is observed in the long time behavior of the autocorrelation function. The integral time scale of the torque components in case $M121$ are larger than on case $M178$ and case $ML178$. As result the correlation function of the particle torque shows virtually an isotropic character. Please note that as in the case of the angular velocity, the dominant torque component is in the horizontal direction. Finally, we would like to note that our research of the literature did not show any information about the temporal dynamics of the particle torque.

7.3 Conclusion

In the present chapter we have investigated the dynamics of heavy finite-size particles settling in an initially quiescent fluid. We considered the three flow cases $M121$, $M178$ and $ML178$ introduced in chapter §5.

A comprehensive study of the particle settling velocity has been performed. When computed with respect to the volume averaged fluid velocity, it was observed that the particles in the

case	$\tau_\lambda(u_p)$	$\tau_\lambda(w_p)$	$\tau_\lambda(\omega_{p,x})$	$\tau_\lambda(\omega_{p,z})$	$\tau_\lambda(f_{p,h})$	$\tau_\lambda(f_{p,z})$	$\tau_\lambda(t_{p,x})$	$\tau_\lambda(t_{p,z})$
$M121$	4.78	8.65	5.41	2.73	1.7	2.3	1.7	1.3
$M178$	3.08	7.67	2.66	1.95	1.05	1.25	0.84	0.75
$ML178$	4.5	7.74	3.69	2.08	1.23	1.40	1.02	0.77

Table 7.7: Taylor micro-scale, τ_λ calculated from the autocorrelation functions of the particle linear and angular velocity components as well as for the particle hydrodynamic force and torque components for all three flow cases $M121$, $M178$ and $ML178$. Data corresponds to the autocorrelation functions from figures 7.19 - 7.22. All values have been scaled with the gravitational time scale τ_g.

case	$T_{int}(u_p)$	$T_{int}(w_p)$	$T_{int}(\omega_{p,x})$	$T_{int}(\omega_{p,z})$	$T_{int}(f_{p,h})$	$T_{int}(f_{p,z})$	$T_{int}(t_{p,x})$	$T_{int}(t_{p,z})$
$M121$	6.98	68.77	9.13	2.93	1.21	1.58	1.06	0.77
	(7.00)	(68.83)	(9.15)	(2.94)	(1.21)	(1.59)	(1.06)	(0.78)
$M178$	4.38	106.6	6.07	3.04	0.83	1.09	0.59	0.51
	(4.39)	(106.5)	(6.08)	(3.04)	(0.83)	(1.1)	(0.59)	(0.52)
$ML178$	8.78	122.9	10.24	3.13	1.059	1.27	0.72	0.597
	(8.79)	(114.3)	(10.25)	(3.13)	(1.06)	(1.27)	(0.73)	(0.598)

Table 7.8: Integral time scale, T_{int}, for the autocorrelation functions of the particle linear and angular velocity components as well as for the particle hydrodynamic force and torque components for all three flow cases $M121$, $M178$ and $ML178$. Data corresponds to the autocorrelation functions from figures 7.19 - 7.22. The integral time scale was calculated as the integral of the Lagrangian autocorrelation function, $T_{int} = \int_0^\infty R_{\alpha\alpha}(\tau)d\tau$. Additionally the integral time scale (the value in the brackets) as estimated by a Langevin model is given as well, $T_L = \int_0^T exp(-\tau_{sep}/T_{int})$. All values have been scaled with the gravitational time scale τ_g.

flow cases with the higher Galileo number, where strong agglomerations of the particles were observed, settle on average faster than an isolated particle with the same parameters (case $S121$ and case $S178$). However, when the particle relative velocity is computed with respect to the local fluid velocity, using our definition of the fluid velocity seen by the particles (equation 7.1), we determined the average particle relative velocity to be similar to the one of a corresponding isolated particle at both investigated Galileo numbers. As a consequence, the particles in the clustering case were found to be preferentially located in regions with downward fluid motion (when compared to the box averaged fluid velocity). Moreover, it was found that the settling velocity of the particles correlate with the particle concentration in case $M178$ and case $ML178$, with the particles settling on average faster than the average in regions with high particle concentration. Particle conditioned averaging revealed that the particle velocity is spatially distributed in an anisotropic manner, i.e. in vertically oriented structures extending in the clustering cases throughout the entire vertical length of the computational domain. In all flow cases it was shown that the particles located downstream of a test particle settle, on average, at higher rates than the global average. The length scales extracted from the particle conditioned analysis corresponded closely to that obtained in section §6.

Concerning the fluctuations of the particle motion, it is found that a simple superposition of isolated particle data cannot account for the observed amplitudes of the particle suspension. However, the levels of anisotropy of the particle velocity fluctuations proved to be in good agreement with similar works in the literature (e.g. Parthasarathy and Faeth (1990b)). As expected, the highest intensity was observed in the clustering case $M178$. A decomposition of the particle velocity variance was performed, which separates the contribution due to the particle motion with respect to the surrounding fluid from the contribution due to the fluctuations of the fluid velocity in vicinity of the particles. The decomposition revealed that the larger fluctuations of the particle vertical velocity component in the clustering case is due to an increase of the fluctuating energy of the fluid velocity seen by the particles, which also dominate the temporal evolution (shape of the curve) of the total particle velocity fluctuation energy. The results revealed that the observed increase in the particle velocity fluctuations amplitude in case $M178$ is due to a substantial amount of particles being located in clusters or void regions, which constitute large scale fluid velocity fluctuations with respect to the volume averaged value. With regard to the effect of the solid volume fraction, the results indicated a correlation with the relative turbulence intensity. The largest fluctuating energy is measured for the highest values of the relative turbulence intensity.

We have further performed a detailed study of the vertical component of the particle hydrodynamic force. The results evidenced that in the clustering cases the particle drag coefficient (normalized drag force with respect to the unconditioned mean particle settling velocity w_s) is significantly smaller than the drag of the corresponding single particle simulation $S178$. This was attributed to the strong clustering of the particles in case $M178$ and $ML178$. Further, we have showed that the deviation from the average value of the drag correlates with the relative turbulence intensity. The largest deviation was observed in the case with the largest value of the relative turbulence intensity and vice versa. The mean drag based on the average of the instantaneous drag experienced by the individual particles with respect to the local fluid velocity w_s^S was found to be close to the drag of the single particle in case $S121$ and case $S178$. The effect of the non-linear drag proved to be more prominent in the case with the largest relative turbulence intensity, leading to a net increase of the drag coefficient. Particle conditioned average of the particle hydrodynamic force showed, as the particle velocity, an anisotropic spatial distribution and a strong correlation with the particle settling velocity.

The results for the particle hydrodynamic force and the particle relative velocity appear to support the so called wake attraction mechanism responsible for the phenomenon of drafting-kissing-tumbling (Hu et al., 1993) (particles in the wake of a leading particle tend to experience smaller drag and have larger settling velocity than the leading one), thus leading most probably to the clustering of the particles in case $M178$ and $ML178$.

Study of the probability density function of the particle velocity components revealed that the particle velocity in all three case experienced distribution different from the Gaussian distribution, indicating that the pseudo-turbulent flow field in the present simulations have a systematic effect on the particle relative velocity, different from that induced by a classic homogeneous turbulence. The largest deviation of the particle velocity p.d.f.s was observed in case $M121$. In fact, the distribution of all particle quantities (particle translational and vertical velocity, particle hydrodynamic force and torque) was more intermittent in the non-clustering case than in the clustering ones. The probability density function of the particle hydrodynamic force evidenced for all cases positive skewness, which was attributed to the non-linear drag effect (García-Villalba et al., 2012). The intermittency of the particle acceleration was related to the intermittency of the fluid small-scale motion in vicinity of the particles.

Finally we have performed a Lagrangian analysis of particle related quantities. It was found out that the vertical velocity component remained correlated over longer time intervals than the horizontal velocity component. The particle agglomeration in case $M178$ and case $ML178$ caused a substantial increase of the correlation time, when compared to the non-clustering case. Moreover, the finite size of the computational domain was evidenced to have an effect upon the decay of the autocorrelation function, implying that an even longer computational domains are required. Study of the correlation function for the particle acceleration showed that the characteristic time of the forcing experienced by the particles was very similar in all three cases with small bias towards sorter times in case $M178$. Comparison of the results in case $M178$ and case $ML178$ revealed that the smaller value of the solid volume fraction leads to an increase of the correlation times for all particle related quantities.

A more qualitative results for the effect of the particle clustering can be obtained by performing a conditioning of the above statistics on the position of the particles by using the intrinsic cluster definition (cf. section §6), e.g. calculating of the particle velocity and hydrodynamic force distribution for particles located in clusters or void regions. As evidenced, the local flow field in vicinity of the particles has a significant effect upon the particle statistics, especially on the particle velocity and hydrodynamic force, thus a more detailed study of the consequences for a specific non-linear drag model need to be further investigated in future works.

Chapter 8

Flow field induced by the particles

One common feature of dispersed two-phase flows is that the relative motion of the dispersed phase with respect to the continuous phase will result in generation of velocity fluctuations in the continuous phase. Even the motion of a single particle in an ambient fluid at very low Reynolds numbers will generate disturbances in the liquid. The agitation of the flow field will become even more complicated in the case of multiple freely moving particles. The turbulence generated solely by the settling (rising) of heavy (light) particles in a fluid otherwise at rest has been shown to have different characteristics from the conventional turbulence in the absence of the dispersed phase, which is mainly caused by velocity gradients in the liquid. To emphasis this difference, the particle-induced turbulence is often referred to as *pseudo-turbulence*. In the past, the interaction of the classical shear-induced turbulence with the dispersed phase has been subjected to numerous studies in the scientific community and is still considered as not completely understood and was identified in section §1.2.2 as one of the (main) open questions with regard to particulate flows. A study of the literature revealed, that surprisingly little is known about the flow field generated by settling particles. For example, the decay rate of the energy spectra of the flow appears to experience different behavior than the classic turbulence and until now there is still no definitive explanation for this difference. Motivated by these observations, the objective of the following sections is to determine the intensity of the self-induced fluid motion. This analysis is different, but related to the study of the modulation of existing (background) turbulence due to the addition of particles. The following analysis aims at the characterizing the *pseudo-turbulence* of the flow field and will involve two parts. In the first part the integral statistics of the flow field, such as standard deviation, probability distribution of the fluid velocity field, turbulent kinetic energy budget and spectra (section §8.2) are discussed, then the turbulent kinetic budget and energy spectra and the corresponding two-point correlation are analyzed (sections §8.3 and §8.4). In the second part the local properties of the flow field in the vicinity of the particles are investigated. This involves the averaging of the flow field and fluctuations conditioned on the particle positions, as well as the mean particle wake statistics (section §8.5). Finally, the chapter is closed in section §8.6 by brief summary and recommendations for future works.

8.1 Qualitative flow description

Before we proceed with the results, let us first give some visual impressions of the flow field. Figures 8.1 and 8.2 depict iso-surfaces of the negative velocity fluctuations of the vertical fluid

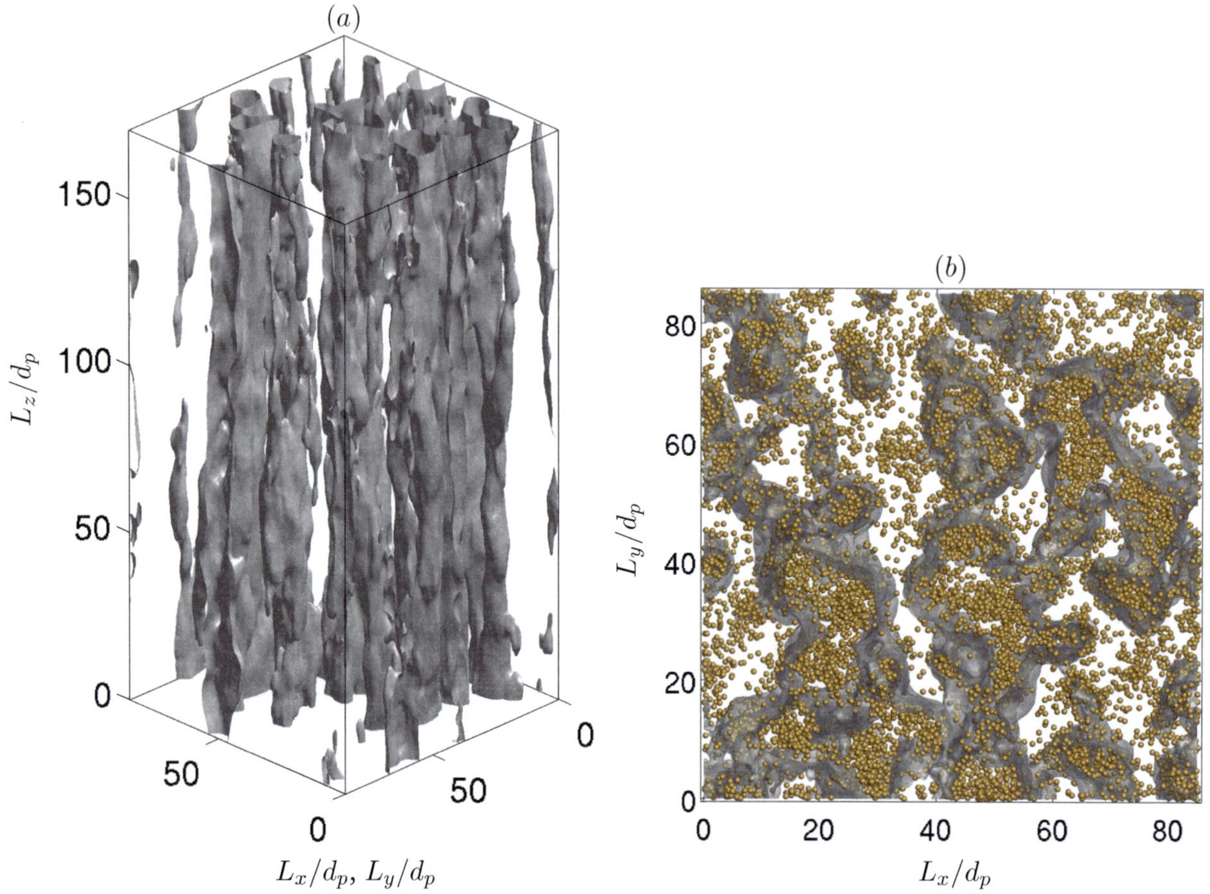

Figure 8.1: (a) Iso-surfaces of the vertical fluid velocity fluctuations w'_f (with respect to the box-averaged value) in case $M178$ at one instant in the statistically steady regime. The velocity has been box-averaged with a filter of $5d_p$. The graph shows a value of $-0.25u_g$. (b) The same graph along with the particle positions, but viewed from the top.

velocity component (with respect to the box-averaged value) for both clustering cases, $M178$ and $ML178$. The iso-surfaces corresponds to $w'_f = -0.25u_g$. It can be immediately observed that the flow field in both cases exhibits columnar like vertical flow structures extending throughout the entire vertical length of the computational domain. Please note that similar structures for positive fluid fluctuations ($w'_f = 0.25u_g$) can be observed located in the interspace between the structures observed in figures 8.1 and 8.2 (plot omitted). The graphs in sub-figures (b) show the fluid fluctuation of w'_f along with the instantaneous particle positions viewed from the top. Clearly the vertical structures observed in sub-figures (a) correlate with the particle position and with the local agglomeration of particles. The negative fluid velocity "streamers" encompass regions with high particle number density. This is best observed in figure 8.2b for case $ML178$, where large number of particles are located in the cores of such flow structures and large void regions with no evidence of such structures are observed. This is in line with the observations made in figure 7.3, where the negative fluctuations of the vertically averaged fluid velocity were shown to correlate with the location of the particle clusters and the increase of the particle settling velocity. Further, figures 8.1 and 8.2 confirm that the particles are preferentially located in regions with downward fluid motion. In figure 8.3 the vortices in the

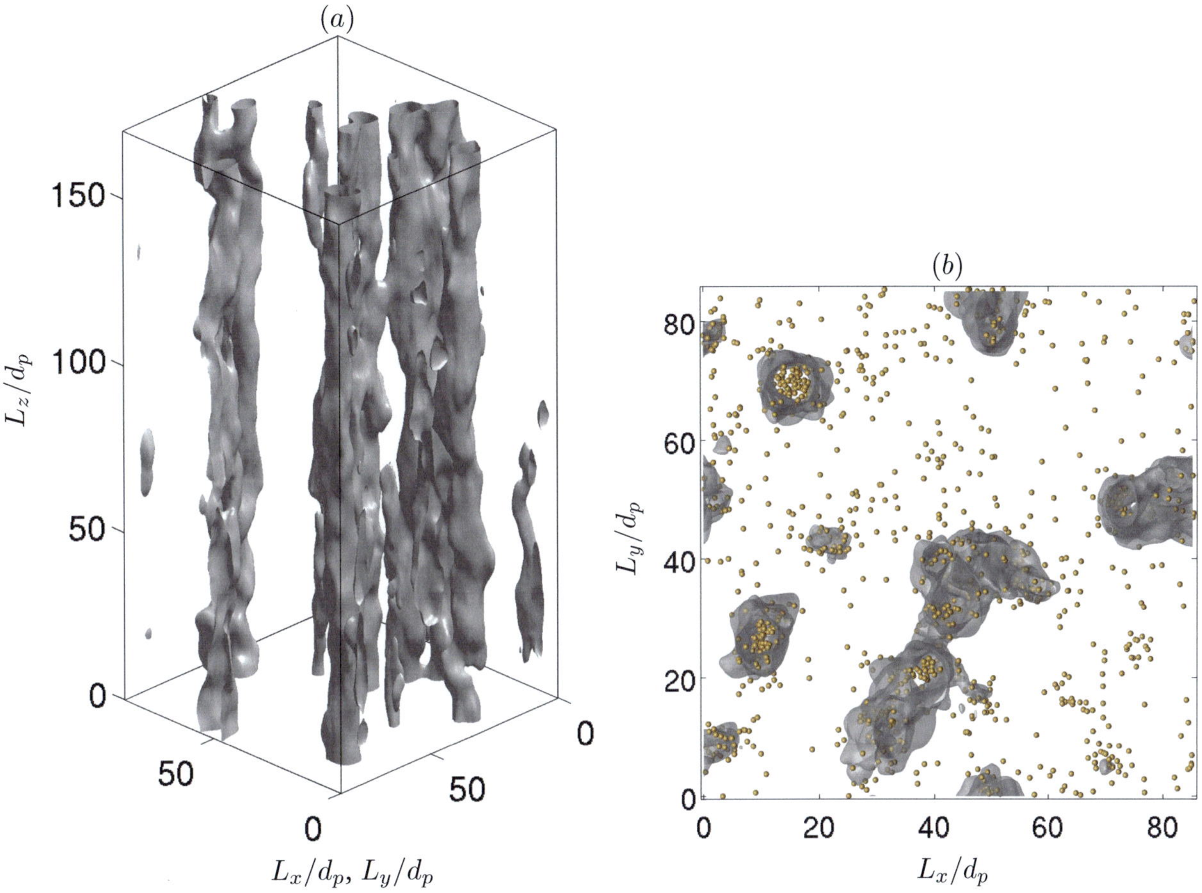

Figure 8.2: The same as in figure 8.1, but for case $ML178$.

particle wakes are visualized by the vortex eduction criterion proposed by Jeong and Hussain (1995). The graphs in the figures show selected sub-domains of the computational domain containing a part of one strong downward current. It can be observed that in both cases large number of particles are located inside this currents, where complex interaction between the wakes of the particles takes place. On the other hand, in the void regions surrounding the downward currents the wakes of the particles are observed to be close to the double-threaded wake of an isolated particle at the same Galileo number (cf. case $S178$, figure 3.7).

In Case $M121$, where the amplitude of the fluctuations is much smaller and the particles are more evenly distributed, such large flow structures are not encountered (plot omitted) and the wakes of the individual particles consist of attached toroidal vortex ring, similar to those found in the corresponding single particle simulation (cf. case $S121$, figure 3.6).

8.2 Statistics of the fluctuating field

The correct prediction of the flow induced by the movement of particles have been always of great interest to the scientific community. As aforementioned, the particles in the present flow cases were placed into an initially quiescent fluid. Therefore, any fluid motion results from the motion of the particles, which in turn is only due to the gravitational potential. In

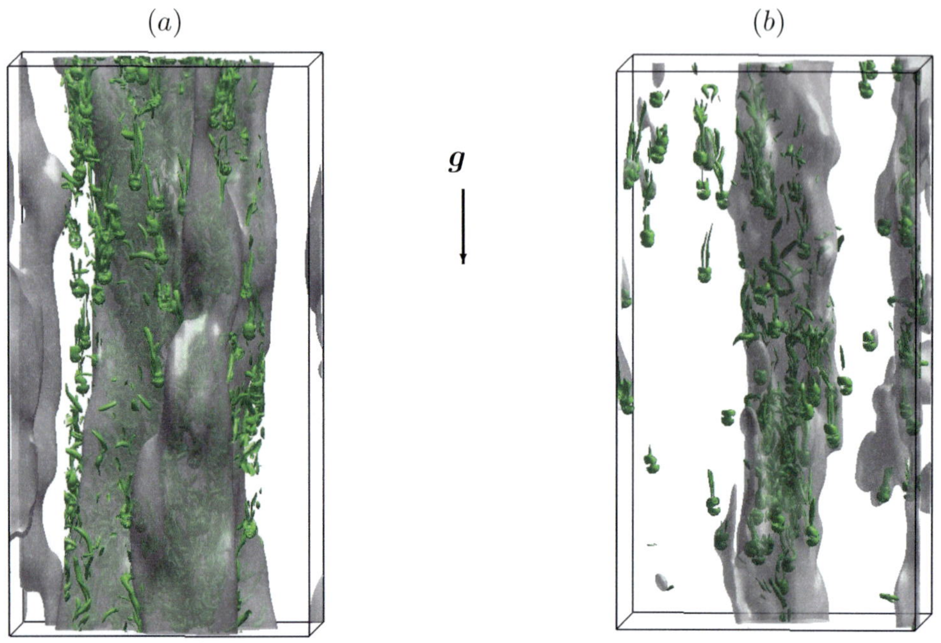

Figure 8.3: Iso-surfaces of the vortex eduction criterion proposed by Jeong and Hussain (1995) at a value of $\lambda_2 = -0.27u_g^2/d_p^2$ (green color) showing a sub-volume of size $32d_p \times 32d_p \times 58d_p$ at the same instant as in figures 8.1 - 8.2. (a) case $M178$ and (b) case $ML178$. The iso-surface of the vertical fluid velocity component is at the same iso-value as in figures 8.1 - 8.2.

the present cases, due to the relatively high relative velocity between the two phases, the flow field generated by the settling particles is expected to be dominated by the particle wakes. In such flows, the decay of the velocity disturbance, generated by a particle, with the distance to the particle play an important role. Before we proceed with the results let us give a short overview over what is known from the literature.

For dilute suspensions the principle of superposition of the disturbances caused by a single particle in the flow is widely used for the modeling of the velocity fluctuations in the fluid. Most of the models based on the superposition principle assume that the dispersed phase is homogeneously distributed and thereby the hydrodynamic interaction between the particles is relatively small. For such flows, the contribution of individual particles can be summed up. In the limit of vanishing solid volume fraction and very small Reynolds numbers ($Re_{p,rel} \ll 1$), the Stokes solution predicts that the flow perturbations behind the particles decay as r^{-1}, where r is the distance from the particle. Similar to the particle velocity fluctuations, Caflisch and Luke (1985) showed that due to the slow decay, the superposition of the flow induced by multiple settling particles will lead to the divergence of the fluid velocity variance with the size of the domain under consideration. At finite Reynolds numbers a wake develops behind the particles, wherein the velocity behind the particle decays as r^{-1} in the laminar case and $r^{-2/3}$ in the turbulent case (Pope, 2000), which again results in the divergence of the velocity variance (Parthasarathy and Faeth, 1990a; Mizukami et al., 1992). Parthasarathy and Faeth (1990a) studied the sedimentation of spherical particles in a nearly stagnant water bath in the very dilute regime with solid volume fraction $\Phi_s < 0.01\%$ for three different particle Reynolds numbers, 38, 156 and 545. They used the principle of superposition to predict the fluid velocity fluctuations. They authors overcome the divergence of the fluid velocity fluctuations, by assuming that the wake of a single particle vanishes for distances larger than 175 particle diameters. The attenuation of the particle wakes in a settling suspension of particles with moderate to large Reynolds numbers was identified as one of the main reasons for the convergence of the velocity variance. For moderate Reynolds numbers in the order of $\mathcal{O}(10)$, Koch (1993) proposed a screening mechanism of the velocity behind the particles, due to a deficit of a particle in the wake of the reference particle. Different mechanisms for the wake attenuation for dispersed two-phase flows with large Reynolds numbers can be identified and more detailed discussion can be found in Risso et al. (2008).

In a recent work Ellingsen and Risso (2001) studied the rise of a single bubble at high Reynolds number. They stated that the velocity fluctuations induced by the bubble had two contributions, one due to the potential flow and second due to the wake of the bubble. An analytical description of the contribution of the potential flow to the fluid velocity variance can be derived. In a pioneering work, Batchelor (1967) provided analytical description of the flow field induced by the moving of sphere in an unbounded irrotational flow. He stated that the fluid disturbance behind the sphere decays as r^{-3}, which is sufficient for the calculation of the fluid velocity variance in dispersed flows with multiple particles. Based on this, a prediction of the fluid velocity variance for dilute suspension can be derived by integrating the velocity induced by a single particle over the volume occupied by the fluid and multiplying by the solid volume fraction, viz.:

$$\langle u'_{f,i} u'_{f,i} \rangle = \begin{pmatrix} 3/20 & 0 & 0 \\ 0 & 3/20 & 0 \\ 0 & 0 & 4/20 \end{pmatrix} \Phi_s w_{p,rel}^2 \tag{8.1}$$

This model assumes an inviscid fluid and does not consider the influence of the particle wakes and the local distribution of the dispersed phase. As can be immediately seen, the approximation in equation 8.1 describes nearly isotropic fluid fluctuations. Thus, the application of the above approximation in case of *pseudo turbulence*, where the flow is dominated by the wakes of particles settling with high velocity and was proven to be highly anisotropic, is questionable. Nevertheless, this approximation for the velocity variance was applied to high Reynolds number bubbly flows: Biesheuvel and Wijngaarden (1984) studied the velocity fluctuations in a dilute bubbly flow with high Reynolds numbers. The bubbles were assumed to be spherical and the viscous forces were assumed to play only a minor role. The boundary layer and the velocity fluctuations were also neglected and thus the results should be treated with reservation. Later Lance and Bataille (1991) also adapted the potential approximation in their experimental study of the rise of large bubbles with high Reynolds number in grid turbulence. They reported that the velocity variance was mainly due to the potential flow and that the wake contribution was less than 20% from the total variance. Their results showed to be, nevertheless, well approximated by the potential flow approximation. Finally we want to remark, that the velocity variance based on the potential approximation still depends on the solid volume fraction and increases with increasing volume fraction. Zenit et al. (2001) performed experiments of bubble suspension with high Reynolds number and low Weber number (the Weber number measures the relative importance of the fluid's inertia compared to its surface tension). They compared the variance due to the potential approximation with their results and observed that the variance was strongly underestimated by the potential approximation. Bunner and Tryggvason (2002) studied the dynamics of homogeneous bubbly flow with Reynolds number in the order of $\mathcal{O}(10)$ by means of DNS. They reported that the estimated values were roughly 20% from the simulated values. This indicates that the contributions of the wakes to the total variance is substantial and is the dominating part to the fluid variance.

As mentioned above, the velocity variance in the fluid was observed to have contributions stemming from the potential flow and the particle wakes together. Therefore, models for the velocity variance should account for the velocity disturbances due to the particle wake. The wake by itself can be decomposed in three different regions, which would have different contributions to the velocity fluctuations in the liquid. Comprehensive studies on the particle (bubble) wakes were performed in the past with aim to identify the different regions of the wake. Chen and Faeth (2000) investigated the turbulent properties of the inter-wake region of homogeneous dilute particle laden flow. They considered similar experimental setup and parameters range as Parthasarathy and Faeth (1990a) and drawn analogies between the turbulent inter-wake region and the grid-generated turbulence and suggested that the wake-induced disturbances correspond to the developing region of grid-generated turbulence, whereas the inter-wake region corresponds either to the initial or final decay regions of grid-generated turbulence. Risso and Ellingsen (2002) and Risso et al. (2008) identified three different regions of the particle wake: (i) region close to the reference particle, where the flow is similar to that induced by a single particle, (ii) further behind the particle, an intermediate region, where the hydrodynamic interactions between the particles play a crucial role and the fluid velocity decays faster than behind a single particle and (iii) asymptotic region far behind the particle, where the fluid velocity reaches asymptotic value and is independent in the distance from the particle. Therefore, the different contributions and influence of this regions have to be considered for the particular case.

Now let us turn to the present results. The Reynolds stress tensor for the region occupied by the fluid is defined as:

$$\langle u'_{f,i}u'_{f,j}\rangle_{\Omega_f} = \frac{1}{\Omega_f}\int_{\Omega_f} u'_{f,i}u'_{f,j}dV. \tag{8.2}$$

In the above equation the velocity fluctuations are computed with respect to the box-averaged mean fluid velocity, i.e. $u'_{f,i} = u_{f,i} - \langle u_{f,i}\rangle_{\Omega_f}$. For exact definition of the averaging operator the reader is referred to appendix A. Due to symmetry in the horizontal direction and the absence of walls, the off-diagonal components of the tensor are identically zero and the velocity correlations in the horizontal directions are equal, i.e. $\langle u'_f u'_f\rangle_{\Omega_f} = \langle v'_f v'_f\rangle_{\Omega_f}$. Thus, in the following discussion only the horizontal component and the vertical component are going to be considered. Henceforth, the horizontal component of the fluid velocity fluctuations will be for convenience abbreviated as $\langle u'_f u'_f\rangle_{\Omega_f}$. Finally, the r.m.s. of the fluid velocity components can be calculated as the square root of the horizontal and vertical Reynolds stresses.

The temporal evolution of the fluid velocity fluctuations was already briefly discussed in section §5.2, figure 5.3a. The velocity field in case $M178$ experienced larger fluctuations than in case $M121$. In case $M121$ the r.m.s. values in lateral direction showed small increase over long time, while the fluctuations in vertical direction seem to have a long time oscillating behavior. The fluctuations for both velocity components in case $M121$ remained over the entire course of the simulation essentially flat, while the fluctuations in both clustering cases $M178$ and $ML178$ exhibit prominent temporal fluctuations, which were more distinctive for the vertical velocity component. Similar evolution of the fluid velocity fluctuations was observed in the numerical studies of Kajishima and Takiguchi (2002); Kajishima (2004a). The peaks in the fluctuations of the vertical fluid velocity component were related in their work to the life cycle of the particle clusters. As noted by the authors, the particles located in clusters tend to settle faster than the average and they generate a downward motion of the fluid surrounding the clusters, which results in the increasing of the amplitude of the fluid velocity fluctuations of the vertical velocity component (cf. section §7.1). The break-up of the clusters caused respectively attenuation of the velocity fluctuations. Following their reasoning we estimated the life time of the particle clusters in case $M178$ to be approximately $550\tau_g$. Unfortunately, we were not able to compute the autocorrelation of the fluid velocity, which would provide more qualitative analysis of the temporal evolution and behavior of the flow field. This needs to be addressed in future studies.

As a result of the wake-induced character of the fluid motion, in all three cases the fluid velocity fluctuations exhibit strong anisotropy with the vertical component being highly dominant. This high level of anisotropy suggest strong effects of the particle wakes, where the mean particle wake contributes to the fluctuating velocity field. The anisotropy of the flow field has been commonly observed in studies of *pseudo-turbulence*. Parthasarathy and Faeth (1990a) have made similar observations, where experiments of particles settling in an ambient fluid were performed. As in the present cases, they observed, that the fluctuations of the vertical fluid velocity component are significantly larger than the fluctuations of the horizontal velocity components, roughly by a factor of two. However, in their experiments the relative turbulence intensity is an order of magnitude smaller than in the present cases, except in case $ML178$ where similar values are observed (cf. figure 7.6). Bunner and Tryggvason (2002) also observed strongly anisotropic Reynolds stresses, in the lowest gas volume fraction. They noted the value of the ratio of the vertical to horizontal fluid velocity fluctuations to be 3.3. However, the considered volume fraction is four times larger than in the present study. More recently Kajishima (2004a) studied the collective behavior of solid particles in a particle induced turbulence by means of DNS and observed similar value of the ratio between the vertical and horizontal fluid velocity fluctuations as in the present study. In their work they particularly studied the influence of particle rotation and loading ratio on the interaction between particle clusters and particle-induced turbulence. The particle Reynolds number in Kajishima (2004a), however, was on average larger than 300, which is slightly larger than in the present flow cases $M178$ and $ML178$.

The high anisotropy of the flow field fluctuations suggest that the application of the potential flow model may not provide reasonable predictions for the fluid velocity variance. Nevertheless, we applied the potential flow prediction (equation 8.1) to the present data and calculated an approximation of the long time mean velocity fluctuations. The potential model under predicted the simulated values (e.g. $\langle w'_f w'_f \rangle_{\Omega_f}$) substantially by a factor of 6.6 in case $M121$, by a factor 7.6 and 12 in case $M178$ and case $ML178$, respectively. This result implies that the influence of the wakes and the hydrodynamic interaction in the inter-wake region play significant role and need to be considered in future models for the fluid velocity variance. Similar values of the ratio between the observed and predicted variance values were also observed in the literature, e.g. Zenit et al. (2001) and Bunner and Tryggvason (2002). Bunner and Tryggvason (2002) report ratio of five, which is reasonably close to the values observed here, despite the order of magnitude smaller Reynolds number they considered. In the limit of vanishing small particle Reynolds numbers ($Re_p \to 0$) the influence of the wake is expected to decrease and eventually for very small Reynolds numbers the difference to the potential model would continually decrease as well.

As aforementioned, the method of superposition is widely used in dilute suspensions. In order to assess the applicability of such model for the present flow conditions and investigate how well the fluctuations induced by single particle scale with that of multiple particles, we calculated the volume compensated ratio of the velocity fluctuations between the flow cases with single and multiple particles. For case $M121$ the difference between the multiple particles and the single particle is calculated to 8% and for cases $M178$ and $ML178$ to 30% and 35.8%, respectively. This implies that due to the clustering in case $M178$ and $ML178$, the fluid velocity variance cannot be simply calculated as superposition of the velocity variance induced by the single particle in case $S178$. This is in line with the observations made for the particle velocity fluctuations in section §7.2. Please recall that one of the constraints of the superposition principle was the homogeneous distribution of the dispersed phase, where in case $M178$ and case $ML178$ large inhomogeneities (strong clustering) of the particle distribution is observed. However, in case $M121$ the superposition principle appear to reasonably approximate the fluctuations induced by the particle suspension.

Parthasarathy and Faeth (1990a) proposed theoretical predictions for the velocity fluctuations in the liquid based on the linear superposition principle. They identified three different components, which contribute to the particle induced flow field: (i) modulation in the potential flow region, (ii) modulation due to the mean properties of the wake and (iii) modulation due to the turbulent properties of the wake. All three components can be evaluated separately and thus the relative level of influence of each one can be assessed for the particular flow problem. By using the asymptotic wake properties and as mentioned above terminating integration at $r/d_p = 175$, the following expressions for the fluid velocity fluctuations were obtained (Parthasarathy and Faeth, 1990a):

$$\langle u'_f u'_f \rangle_{\Omega_f}^{1/2} / w_{ref} = C_u [\varepsilon d_p (\theta/d_p)^{2/3} / w_{ref}^3]^{1/2} \tag{8.3}$$

$$\langle w'_f w'_f \rangle_{\Omega_f}^{1/2} / w_{ref} = C_w [\varepsilon d_p (\theta/d_p)^{2/3} / w_{ref}^3]^{1/2} \tag{8.4}$$

In the above equations the reference velocity was taken to be the terminal settling velocity of the isolated particles in the cases $S121$ and $S178$. The parameter $\varepsilon d_p(\theta/d_p)^{2/3}/w_{ref}^3$ on the right hand sides was denoted by the authors as the dissipation factor and is function of the dissipation rate ε, the particle diameter d_p, the reference velocity w_{ref} and the wake momentum diameter $\theta = (C_d d_p^2/8)^{1/2}$, where C_d is the particle drag coefficient. The constants C_u and C_w result from the time integration and depend on the length of the wake, which was taken to be 175 particle diameter. At this distance the mean velocity defect was observed in Parthasarathy and Faeth (1990a) to be comparable to the ambient velocity fluctuations and suggests that the wake properties were indistinguishable from the surrounding flow field. The exact values of the constants C_u and C_w are going to be discussed in the following. For more detailed derivation of the above equations we refer the reader to the aforementioned reference. Note that the above equations 8.3 and 8.4 already contain the contributions of the above mentioned three components. Finally, let us note that the constants C_u and C_w were calculated in Parthasarathy and Faeth (1990a) to $C_u = 4.63$ and $C_w = 6.84$.

Figure 8.4 depicts the values of the fluctuations for the horizontal and the vertical fluid velocity components for all three cases along the values approximated by the above described model of Parthasarathy and Faeth (1990a). Let us first discuss the results for case $M121$. Since no particle clustering was observed for this case and the particle Reynolds number is fairly close to the Reynolds number discussed in Parthasarathy and Faeth (1990a), we expect that the predictions for the fluid velocity variance should be reasonable. Since the vertical velocity component is the dominant one in the present simulations, the correct prediction of the vertical component velocity fluctuations is going to be critical for the particular model. Therefore, we first applied the above model for the vertical velocity component. First, the predicted velocity fluctuations over estimated the values observed in the simulation by as much as 27.5%. However, if we neglect the contribution due to the turbulent wake, i.e. only considering the mean properties of the wake and the potential flow region, the model yields very good approximation for the simulated fluid velocity fluctuations. The difference between the predicted and the simulated value was evaluated to 3.8%. The constant C_w without the contribution of the turbulent wake was calculated to $C_w = 4.72$. In order for the model to predict correctly the fluid velocity fluctuations for the horizontal component, the constant C_u needs to represent the level of anisotropy in the flow, since the fluctuations of both components are function of the same dissipation factor. Thus, we calculated the constant C_u to be equal to a fraction of the coefficient C_w corresponding to the level of anisotropy, i.e. $C_u = C_w/\alpha$, where α represents the observed anisotropy level. By doing this, we obtained $C_u = 1.1$, which provided very good agreement with the result from the simulation. Similar approach was undertaken in Mizukami et al. (1992), where they applied the above model for the prediction of the velocity fluctuations generated by the settling of spherical glass beads through stagnant air. They observed degree of anisotropy of roughly two and proposed to calculate the constant as $C_u = C_w/2$, which in their case agreed well with the measured velocity fluctuations. Here we would like to note that we used the drag coefficient from the simulation and not the standard drag approximation as in Clift et al. (1978). By resolving the equations 8.3 and 8.4 one obtains that the velocity fluctuations are proportional to the two-third power of the drag coefficient, i.e.:

$$\langle w'_f w'_f \rangle_{\Omega_f}^{1/2} / w_{ref} \sim \phi_s^{1/2} C_d^{2/3}. \tag{8.5}$$

Thus, the correct representation of the drag coefficient is crucial for the application of the present model and any deviation of the simulated drag coefficient from the standard drag coefficient will affect the prediction for the velocity fluctuations with the difference to the power of two-third.

Concerning case $M178$ and case $ML178$, we would like first to recall, that the above model assumes that the particle distribution is homogeneous, which is clearly not true in cases $M178$ and $ML178$, as shown in section §6. Nevertheless, we applied the above model in order to assess the applicability of the model for cases with non-homogeneous particle distribution. By using the constant $C_w = 6.84$ provided by Parthasarathy and Faeth (1990a), the model under estimates the velocity fluctuations for the vertical component obtained from the simulation $M178$ by as much as 20%, whereas in case $ML178$ the difference was calculated to 49.3%. As one would expect the larger difference is measured for the case with the higher clustering level. This clearly proves the limits of the above model for non-homogeneously distributed particles. The coefficient C_u was again obtained as a fraction of C_w proportional to the anisotropy levels observed in case $M178$ and case $ML178$, leading to the same deviations for the horizontal velocity component (note that in equations 8.3 - 8.4 the only difference is the value of the coefficients C_u and C_w). Neglecting of the turbulent part of the particle wake (as expected) lead to an increase of the deviation from the simulated values.

Now let us turn to the distribution of the fluid velocity. Figure 8.5 depicts the probability density function of the horizontal and vertical fluid velocity components for all three flow cases. As can be observed both velocity components exhibit super-Gaussian distribution with the horizontal velocity component being more intermittent than the vertical component, except for case $ML178$ where the opposite is observed. As for the fixed particle simulations $F121$

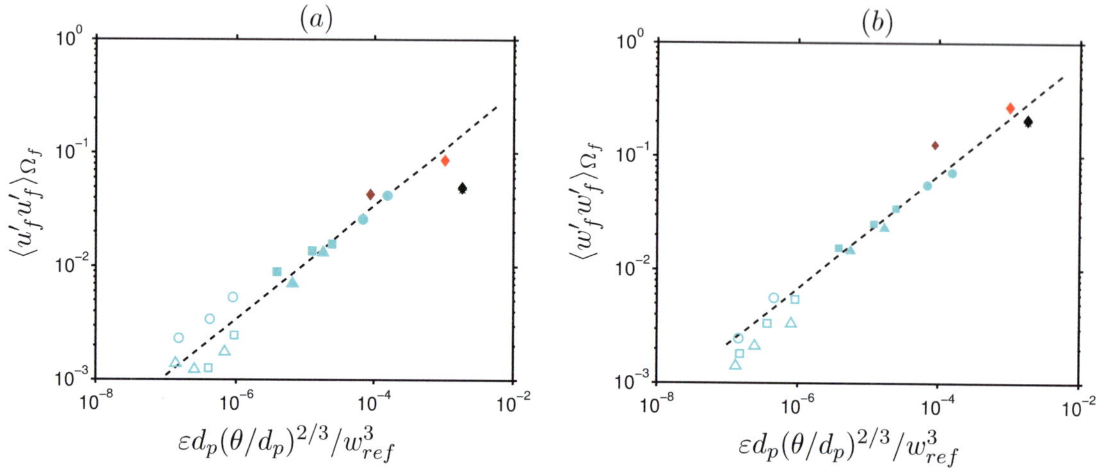

Figure 8.4: Fluid velocity fluctuations of the (a) horizontal and (b) vertical velocity components obtained from the simulations, case $M121$ (♦), case $M178$ (♦) and case $ML178$ (♦). The data is plotted along with the results from Parthasarathy and Faeth (1990a) and Mizukami et al. (1992), □, ■, △, ▲, ○, ●. The black star ($\star$) corresponds to the value predicted by the model in equations 8.3 - 8.4 without consideration for the turbulent part of the wake. The reference velocity w_{ref} was chosen to be the terminal settling velocity for the single particle from cases $S121$ and $S178$. d_p denotes the particle diameter, $\theta = (C_d d_p^2/8)^{1/2}$ is the wake momentum diameter and ε is the dissipation rate. The black dashed line corresponds to the constant $C_w = 6.84$.

and $F178$, the horizontal velocity component in case $M121$ and case $M178$ is approximately symmetrically distributed about its mean value, exhibiting only small skewness values of the order of $\mathcal{O}(10^{-2})$. The values of the fourth moment of the p.d.f.s reveal that the horizontal velocity component in the clustering case $M178$ is less intermittent than the non-clustering flow case $M121$. The values of the flatness in case $M121$ were measured to 10.4, whereas in case $M178$ nearly the half of this value was observed, i.e. $K(u_f) = 5.01$. Comparison with the flatness values for the fixed particle simulations shows that, for the present configurations, the mobility of the particles leads to decrease of the velocity intermittency, regardless if the particles agglomerate into clusters or not. In fact, the decrease of the flatness value of the horizontal velocity component appear to be in both cases by a similar factor, i.e. factor of 2.6 in the non-clustering case and 2.9 in the flow case with the higher particle Reynolds number.

Concerning the vertical velocity component, the velocity in case $M121$ and case $M178$ are negatively skewed, experiencing skewness values of $S(w_f) = -1.58$ in case $M121$ and $S(w_f) = -1.29$ in case $M178$. This indicates that intense velocity fluctuations occur more frequently in the downward direction than in the upward direction. As for the fixed particle simulations, this can be attributed to the wakes character of the flow, where fluid is entrained in the wakes of the particles. However, the here observed values are significantly smaller than in the case for fixed particles, suggesting that the mean wake in the present cases may be significantly attenuated compared to the mean wake of the fixed particle assemblies (as we will see in section §8.5, indeed, the mean wake of the particles in the present cases with mobile particles is attenuated, when compared to the wake of the fixed particle simulations). Figure 8.5 shows that the p.d.f. of the vertical fluid velocity component in case $M121$ exhibits similar prominent increase of the probability to find extreme values of the fluid velocity of the order of 6 standard deviations smaller than the mean value, which is manifested by a prominent peak in the p.d.f.. This feature can be accounted to the wake character of the flow. For values

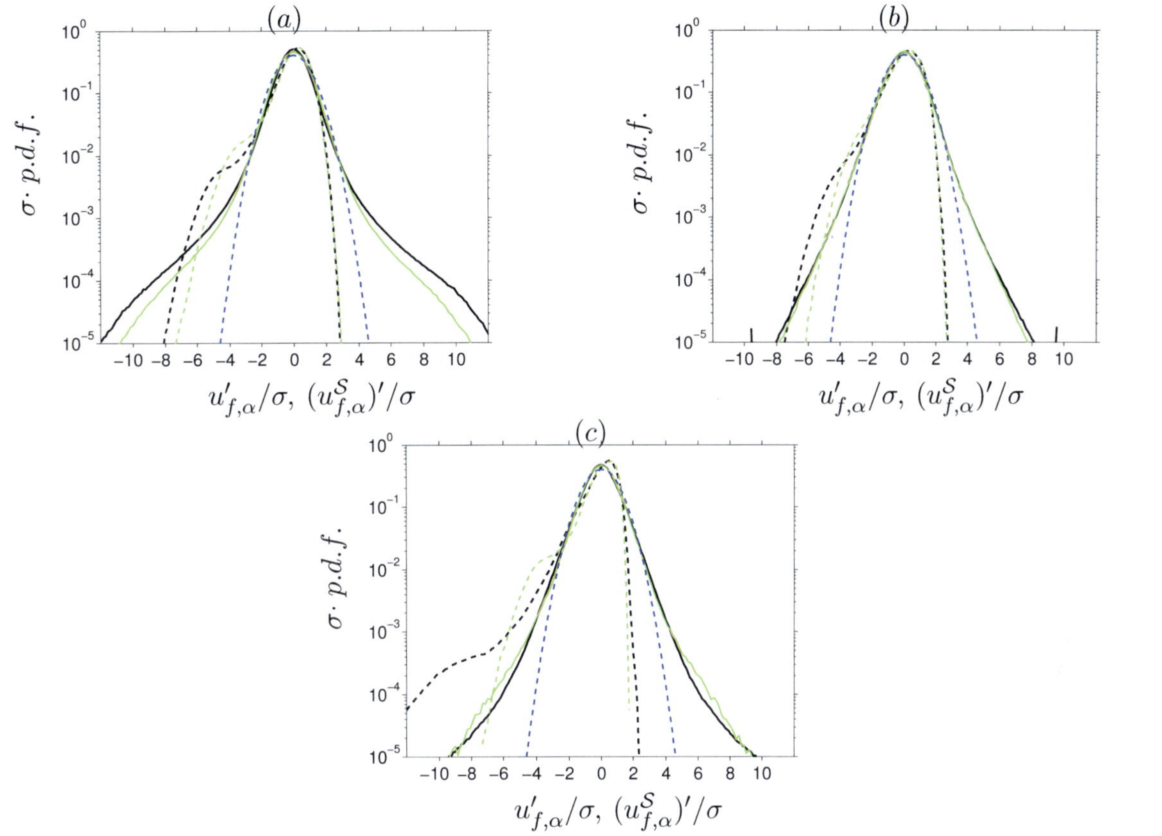

Figure 8.5: Probability density function of the horizontal ($\alpha = 1, 2$) and vertical ($\alpha = 3$) components of the flow velocities $\boldsymbol{u}_f^S$ (black color) and $\boldsymbol{u}_f$ (green color) (a) for case $M121$, (b) case $M178$ and (c) case $ML178$. Line coding: horizontal component (———, ———); vertical component (- - - -, - - - -); Gaussian distribution (- - - -).

of the solid volume fraction, as in the present study, the perturbed flow field by the particles contributes only to smaller number of velocity samples. Since this samples contribute to the most of the velocity fluctuations, the contribution of the flow field in vicinity of the particles to the total distribution consist mainly of an increase of the probability density function for large values of the velocity fluctuations. On the other hand in case $M178$, this feature appear to be attenuated by the motion of the particles, indicating that the fluctuations of the vertical fluid velocity component in case $M178$ are with less probability subjected to extreme events (large deviations of the mean value) than in case $M121$. This is evidenced also by the smaller flatness value of the vertical fluid velocity component in case $M178$ ($K(w_f) = 6.34$) than in case $M121$ ($K(w_f) = 8.27$). Again, comparison with the values obtained for the fixed particle simulations $F121$ and $F178$ reveals, that the motion of the particles leads to a decrease of both the intermittency and the skewness of the vertical fluid velocity component.

Regarding the effect of the solid volume fraction, the fluid velocity in case $ML178$ is more intermittent than in case $M178$, especially for the vertical velocity component, where the flatness value is approximately 2.5 times larger than in case $M178$. However, the difference for the horizontal velocity component is rather small and was measured to be less than 6%, indicating that the distribution of the horizontal fluid velocity is relatively independent of the solid volume fraction. The skewness of the fluid velocity increases as well for both velocity components in case $ML178$, which stems most probably from the stronger levels of clustering observed in case $ML178$. The occurrence of the above mentioned peak at large negative values of the fluid velocity fluctuations moved further away from the mean value and is observed to be at approximately -10 standard deviations, causing also an increase of the velocity intermittency. As mentioned above, this increase of the probability for the occurrence of negative fluctuations is related to the particle wakes. With decreasing of the sold volume fraction (i.e. number of particles) the number of wakes and thus the samples of the flow field located in the wake of a particle decrease accordingly. However, the less probable occurrence leads to higher values of the velocity flatness. The values of the skewness and the flatness for both fluid velocity components and all flow cases can be found in tables 8.1 - 8.2.

In order to better investigate the flow field in vicinity of the particles, we have calculated the probability density function of the particle conditioned fluid velocity $\boldsymbol{u}_f^S$, which is also depicted in figure 8.5. It can be observed that in case $M121$ and case $M178$ the fluid velocity is less intermittent than the unconditioned fluid velocity, evidenced by the smaller values of the fluid velocity flatness. The values of the skewness are reduced in all cases as well. Please refer to tables 8.1 and 8.2 for the exact values. This indicates that the fluid velocity in the vicinity of each individual particle is very similar and its p.d.f.s resemble the shape of the unconditioned p.d.f.. The overall distribution in case $ML178$ appears to follow the same trend as in case $M178$, except for the vertical velocity component, where the largest difference is observed. The distribution of the velocity w_f^S in case $ML178$ shows similar features as the one in case $M178$, indicating that the local flow field in both cases is very similar and nearly independent of the solid volume fraction (for the present flow conditions). The difference between the standard deviation of the vertical velocity component in case $M178$ and case $ML178$ measures only 3.1%. Similar observations were made for example by Risso and Ellingsen (2002), where the velocity fluctuations induced by a homogeneous dispersion of bubbles was investigated. The authors distinguished two regions in the flow: (i) in vicinity of the particles they found the flow to be similar to that generated by isolated bubble and as a consequence the fluid fluctuations herein do not depend on the solid volume fraction; (ii) in the second region (the complementary region) they stated that the velocity fluctuations are controlled by non-linear interactions between the wakes of all particles and evolve in a self-similar manner as the $\Phi_s^{0.4}$. The p.d.f. of the total fluctuations is then the sum of the p.d.f.s of the fluid velocity in this two regions weighted by the relative volume fraction they occupy. In order to address the scaling of the fluid velocity fluctuations with the solid volume fraction, we studied the standard deviation of the fluid velocity in case $M178$ and case $ML178$ (recall, the only difference between this two cases is the value of Φ_s, which is in case $ML178$ factor of ten smaller than in case $M178$). For the vertical component the standard deviation in case $M178$ was evaluated to $\sigma(w_f) = 0.23$, whereas in case $ML178$ significantly smaller value was measured, i.e. $\sigma(w_f) = 0.11$. As next we applied the scaling of the fluid velocity as proposed by Risso and Ellingsen (2002), viz. the

case	$S(u_p)$	$S(u_p^S)$	$S(w_p)$	$S(w_p^S)$
$M121$	0.09	0.01	-1.58	-1.58
$M178$	0.007	-0.01	-1.29	-1.0
$ML178$	-0.005	0	-2.2	-1.90

Table 8.1: Skewness of the probability density functions of the components of the fluid velocities $\boldsymbol{u}_f$ and $\boldsymbol{u}_f^S$ for all three cases $M121$, $M178$, $ML178$. Data corresponds to the p.d.f.s in figure 8.5.

case	$K(u_p)$	$K(u_p^S)$	$K(w_p)$	$K(w_p^S)$
$M121$	10.4	7.76	8.27	7.18
$M178$	5.01	4.68	6.34	4.56
$ML178$	5.30	5.68	15.23	8.27

Table 8.2: Kurtosis of the probability density functions of the components of the fluid velocities $\boldsymbol{u}_f$ and $\boldsymbol{u}_f^S$ for all three cases $M121$, $M178$, $ML178$. Data corresponds to the p.d.f.s in figure 8.5.

fluid velocity was normalized by $w_\infty(\Phi_s/\Phi_{ref})^{0.4}$, where w_∞ is the terminal settling velocity of an isolated particle (case $S178$), Φ_s the solid volume fraction and Φ_{ref} is some reference solid volume fraction and was chosen in the present work as the solid volume fraction in case $ML178$. With this scaling the standard deviation calculates to $\sigma(w_f) = 0.099$ and $\sigma(w_f) = 0.011$ in case $M178$ and case $ML178$, respectively. Thus, the difference between the standard deviation in both cases reduces to approximately 15%. It is interesting to note, that the scaling of the fluid velocity with the solid volume fraction is incompatible with the summation of the contributions of individual wakes, which would provide a p.d.f. with moments all proportional to the solid volume fraction (Risso et al., 2008). Moreover, the scaling of the fluid velocity depends on the velocity scale of a single particle, which is clearly independent of the volume fraction as noted by (Riboux et al., 2010). Similar observations for the fluid velocity distribution and the standard deviation were made recently by the numerical investigations of Risso et al. (2008); Riboux et al. (2010, 2013). The values of the normalized standard deviation were found to be in very good agreement with the values reported in Riboux et al. (2010).

8.3 Energy budget

In the following, the kinetic energy balance of the flow field for the present flow conditions is discussed. To this end, the different contributions to the turbulent kinetic energy (TKE) are identified and their relative importance to the total kinetic energy is studied. This study is of particular interest, since it allows us to investigate the energy production and dissipation of the dispersed phase. Please recall, that the fluctuations in the flow field are solely due to the relative motion of the particles with respect to the fluid. Therefore, this analysis is a good starting point for the study and modeling of turbulent particulate flows and especially the contributions of the dispersed phase to the kinetic energy of the fluid.

Before we proceed with the analysis, let us introduce the kinetic energy equation for the present flow conditions. As introduced in section §5, the only driving force in the present cases $M121$, $M178$ and $ML178$ is the gravitational acceleration. This allows us to make some simplifications of the kinetic energy equation, leading to the following form of the TKE equation:

$$\frac{\bar{D}k}{\bar{D}t} = -\varepsilon + \Psi_b + \Psi_p, \tag{8.6}$$

where $k = \langle \Phi_f u'_{f,i} u'_{f,i} \rangle_\Omega / 2$ is the turbulent kinetic energy (TKE) averaged over the entire domain, $\varepsilon = 2\nu \langle s_{ij} s_{ij} \rangle_\Omega$ is the average dissipation rate, where $s_{ij} = (u'_{i,j} + u'_{j,i})/2$ is the fluctuating rate of the strain tensor, Ψ_b is the buoyancy force, which represents the work done by the gravitational potential and Ψ_p is the correlation between the fluid velocity and the forcing term due to the presence of the particles. The fluid indicator function is denoted by Φ_f

(cf. appendix A) and the differential operator $\bar{D}/\bar{D}t$ is defined in equation D.72 in appendix D. The buoyancy term Ψ_b is defined as follows:

$$\Psi_b = w_s \Phi_s \left(\frac{\rho_p}{\rho_f} - 1 \right) g_z, \tag{8.7}$$

where w_s is the averaged relative velocity between the two phases (cf. equation 5.2) and Φ_s is the solid volume fraction. The contribution to the TKE stemming from the dispersed phase is given by:

$$\Psi_p = \langle \Phi_f u'_{f,i} f'_{p,i} \rangle_\Omega \tag{8.8}$$

where $f_{p,i}$ is the hydrodynamic force which imposes the rigid body motion of the particles on the fluid. This term is also known as the two-way coupling term and was studied recently for the case of finite-size particles in homogeneous isotropic turbulence in the absence of gravity by Lucci et al. (2010). For statistically steady state the term on the left hand side vanishes and the equation reduces to the following form:

$$\varepsilon = \Psi_b + \Psi_p \tag{8.9}$$

As can be seen the contributions to the TKE from the dispersed phase is compensated by the dissipation rate of the TKE. For heavy particles ($\rho_p/\rho_f > 1$) and gravity acting opposite to the vertical direction, the first term on the right hand side of equation 8.9 is always positive and acts as a source of TKE. Considering the term Ψ_p, no *a-priory* allegation of its sign can be made and further analysis is required to determine whether Ψ_p acts as sink or source of TKE. For detailed derivation of the turbulent kinetic energy equation and detailed definition of all terms the reader is referred to appendix D.

Figure 8.6 depicts the budget of the turbulent kinetic energy of the flow field (the contributions of all four terms in equation 8.6) as function of time for all three cases. As can be seen the TKE energy is dominated by the gravitational potential Ψ_b and the dissipation rate ε, whereas the two-way coupling term and the rate of change of TKE are in all cases negligibly small. As expected the potential energy act as a source of turbulent kinetic energy in all three cases and is compensated by the dissipation rate, which acts as a sink of TKE. Concerning the effect of the Galileo number, figure 8.6 reveals that the potential energy in case $M178$ is larger than in case $M121$ due to the higher relative particle velocity. The difference between both cases was evaluated to 19.9%. With respect to the effect of the solid volume fraction, as can be seen in equation 8.7 the potential energy is a linear function of the solid volume fraction and indeed the value of Ψ_b is in case $ML178$ approximately ten times smaller than in case $M178$. The rate of change of the TKE (term on the left hand side in equation 8.6) was measured in all three cases to be positive and of the order of $\mathcal{O}(10^{-6})$, indicating that the flow systems are in all three flow cases at statistically steady state. The two-way coupling term Ψ_p, although small in magnitude, turned out to be in all three cases negative, indicating that the fluctuating particle motion acts on average as a sink of kinetic energy in the present flow cases. It is interesting to note that in the study of Lucci et al. (2010) this term was found to have systematically positive values. Ferrante and Elghobashi (2003) found the term to have both negative and positive values, thus acting as a source or sink of TKE. This circumstance indicates that understanding of the exact physical mechanisms leading to the sign of the two-way coupling term is not complete and a more detailed study is required in future works, especially in the presence of background turbulence. Finally, let us mention that in number of experimental studies

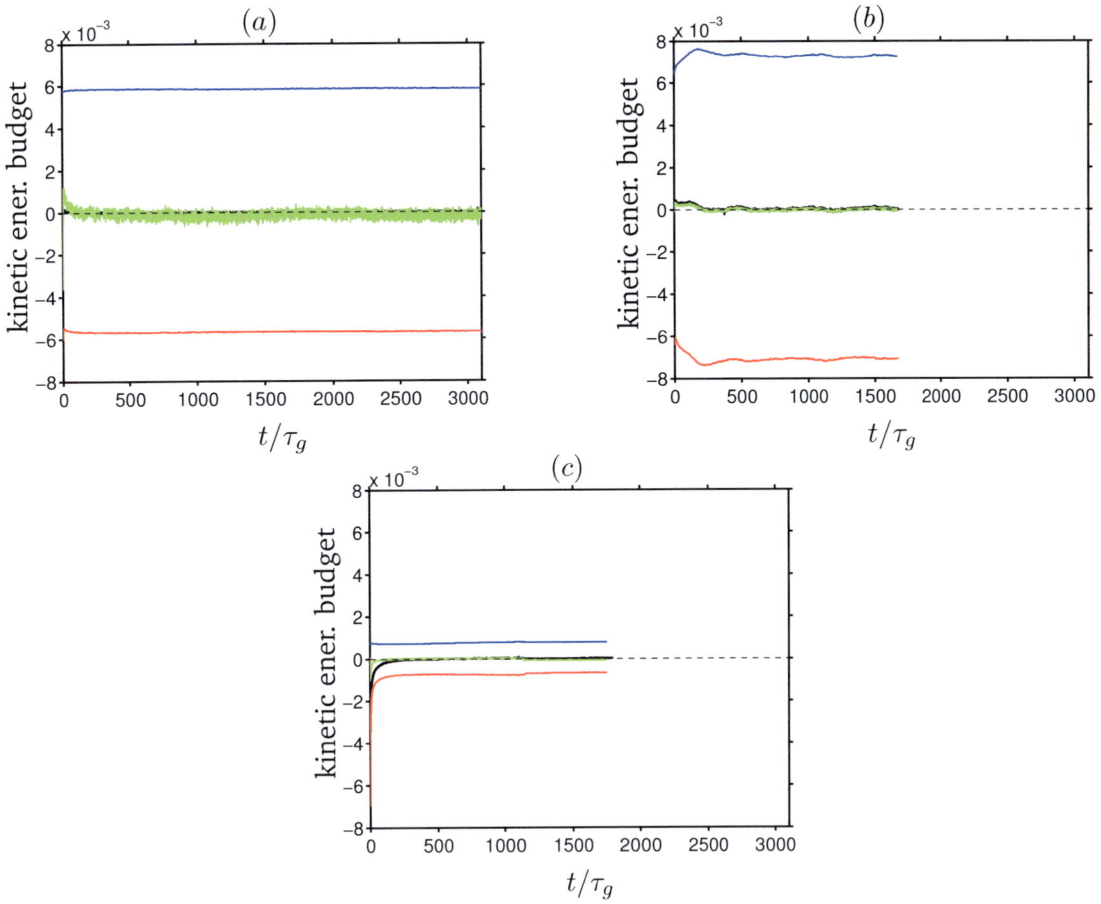

Figure 8.6: The four terms in the average kinetic energy equation 8.6: rate of change term $d_t E_k$ (——), rate of dissipation ε (——), average work due to potential energy $w_{p,rel}\phi_s(\rho_p/\rho_f - 1)g_z$ (——) and correlation between the fluctuating velocity ad particle force $\langle u'_{f,i} f_{p,i}\rangle_\Omega$ (——). All quantities are normalized with the particle scale u_g^3/d_p. (a) case $M121$, (b) case $M178$ and (c) case $ML178$.

Parthasarathy and Faeth (1990b); Mizukami et al. (1992); Lee et al. (2003) the kinetic energy due to the gravitational potential is used as a surrogate for the value of the dissipation rate, neglecting the two-way coupling and assuming a statistically steady state. As can be seen from figure 8.6 our data appear to support this hypothesis. We have checked that indeed the term Ψ_b is a good approximation of the total dissipation in the flow field by calculation the dissipation rate of the flow, i.e. $\varepsilon = \langle s_{ij}s_{ij}\rangle_\Omega$. The difference in all three cases was evaluated to be less than 7% (case $ML178$).

8.4 Energy spectrum and two-point correlations of the flow field

Figure 8.7 shows the one-dimensional energy spectrum of the flow field for all three flow cases, $M121$, $M178$ and $ML178$. The energy spectrum is calculated as in equation 4.6 and is normalized by the respective maximum value in each case. As introduced in section §1.2.2 the energy spectrum in dispersed bubble flows is observed to exhibit a decay -3 power law with the wave number κ. In the case of fixed particles, in section §4.3.2 was showed that the particle-induced flow field indeed followed a -3 power law. The objectives of the present section is to investigate whether the energy spectrum of the flow field in case of mobile particles under the present conditions exhibit a power law decay. First let us discuss the spectrum in the horizontal direction (cf. figure 8.7(a,c,e)). Please note that due to the invariance of the flow field in the horizontal direction, it is sufficient to consider the spectrum, for example only, in the x-direction. The spectrum in the y-direction can be easily obtained by substituting the v-component of the velocity for the u-component. It can be observed, that the energy spectrum in the horizontal direction in all three flow cases exhibits regions with decay rates proportional to κ_x^{-3}. However, the scaling behavior is maintained in a smaller range of wave numbers, compared to the fixed particle cases $F121$ and $F178$. The range in all three flow cases was measured to extend approximately between $\kappa_x d_p = [3, 10]$, which is nearly four times smaller than in the cases for the fixed particles. Beyond ten particle diameters, the energy spectrum decays at faster rate of approximately $\kappa_x^{-3.5}$. Comparing case $M178$ and case $ML178$ it is evident form the graph that the fluid velocity spectra decay significantly faster in case $ML178$ than in case $M178$, practically disappearing for values of the wave number larger than $30d_p$.

Regarding the spectrum in the vertical direction (figure 8.7(b,d,f)), the decay rate in case $M121$ deviates substantially from the -3 power law, whereas for fixed particles the opposite behavior was observed (cf. §4.3.2). This indicates that the motion of the particles and thus the dynamics of the hydrodynamic interaction between the particles play an important role for the scaling exponent of the energy decay. Note, that in case $M121$, except for the micro-structure in vicinity of the particles, the motion of the particles is relatively unaffected by the relative motion of the particles to each other. Nevertheless, the energy spectrum in case $M121$ deviates from that in the case of fixed particles.

On the other hand, in case $M178$ a -3 power law is observed for both velocity components for a wide range of the wave numbers, ranging from approximately $\kappa_x d_p = 2$ to $\kappa_x d_p = 68$. As a consequence, for the present flow conditions (particle-to-fluid density ratio, solid volume fraction, mobile particles), the decay power law appear to be a function of the Galileo number. Additionally, the comparison with the fixed particle case $F178$ suggest that the mobility of the particles affects the decay rate of the energy spectra. In case $ML178$ the overall decay of the energy spectrum for both velocity components appear to follow roughly the -3 power law, but not as distinct as in case $M178$, indicating that the solid volume fraction has an effect upon the decay power law as well. In a pioneering study Lance and Bataille (1991) showed that the classical $-5/3$ power law is substituted by an exponent of $-8/3$ for increasing values of the solid volume fraction. Thus, the present results appear to support the observations made in their work.

As in the case of the fixed particles, the energy spectra in both components clearly show a reverse cascade process and the effect of the particles can be observed in the energy enhancement at scales similar to the particle size.

Overall, the present results are in line with the observation from the literature. For example Martínez-Mercado et al. (2010) measured a -3.2 decay rate, which is maintained for nearly two decades of the wave number. They considered the agitations in the fluid induced by bubbles with high Reynolds number of the order of 1000. Bunner and Tryggvason (2002) observed a decay rate proportional to $\kappa_x^{-3.5}$ for bubble-induced flow field. A -3 scaling has been also given by Risso et al. (2008); Riboux et al. (2010). In the latter work the authors observed a -3 decay for wave numbers smaller than the particle wave number and a classical $-5/3$ scaling for larger wave numbers independent of the solid volume fraction. In a numerical

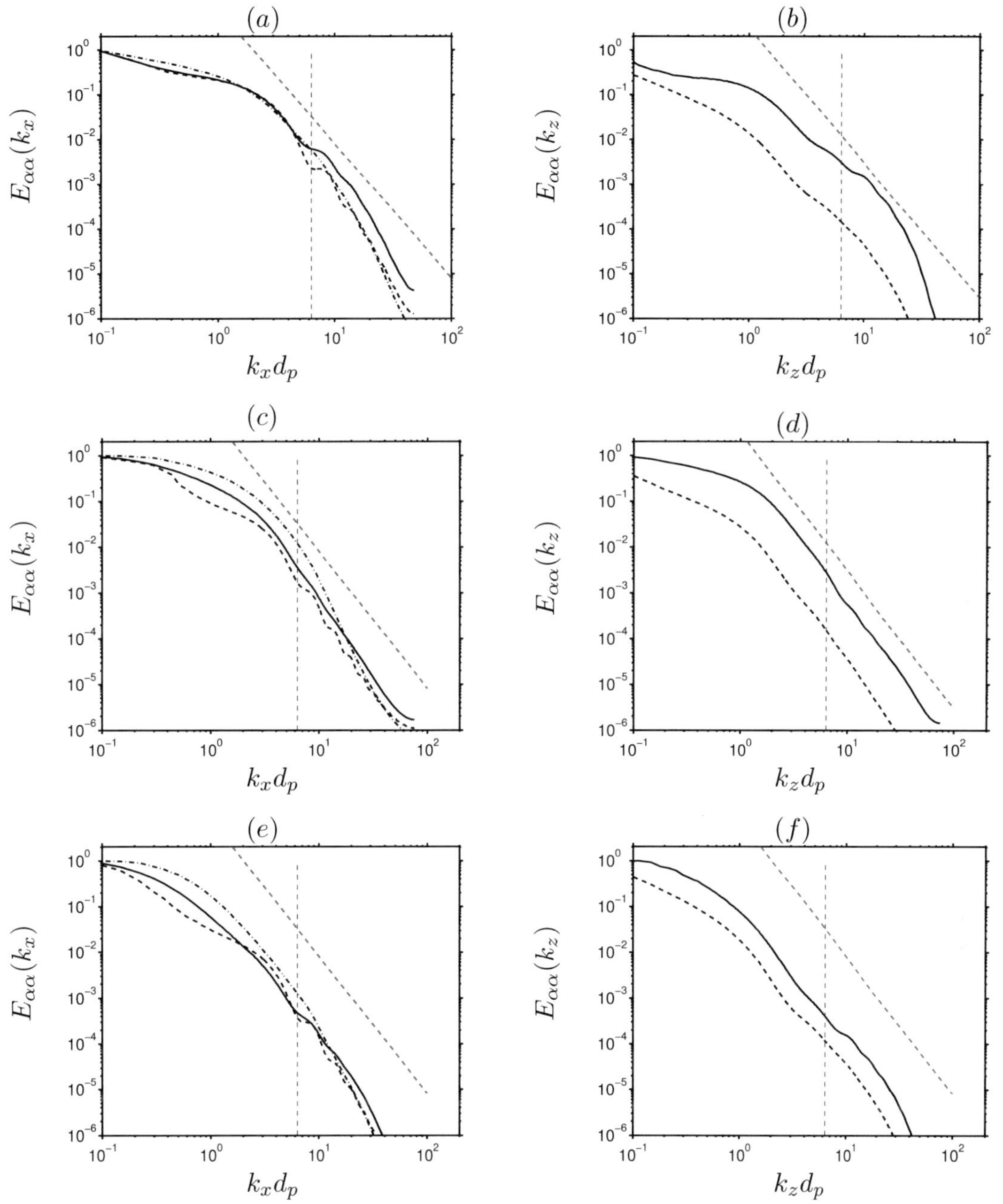

Figure 8.7: Energy spectra of fluid velocity fluctuations in case $M121$ (a,b), case $M178$ (c,d) and case $ML178$ (e,f). Horizontal direction (a,c,e) and in the vertical directions (b,d,f). Color coding of the velocity components: u: ——, v: —·— and w: ----. The dashed magenta lines (----) indicate decay rates proportional to k_α^{-3}.

study by Mazzitelli and Lohse (2009) of micro-bubbles a $-5/3$ decay power law was observed. In their study the authors however neglected the wakes of the bubbles. Thus, as suggested by Martínez-Mercado et al. (2010) the power law of the decay may be related to the wake structure of the particles.

A number of different explanation of the different energy scaling in pseudo-turbulence were proposed in the literature. Martínez-Mercado et al. (2010) argued that the hydrodynamic interactions, specifically the wakes of the particles are responsible for the -3 decay law. Similar argument was used in the study performed by Riboux et al. (2010). On the other hand Mendez-Diaz et al. (2013) argued that the specific details of the hydrodynamic interactions between the particles did not influenced the fluid fluctuations and stated that the disorder of the particle array is the reason fluid velocity fluctuations.

With respect to our results, the difference between case $M178$ and $ML178$ clearly indicates that the scaling exponent does not (or at least weakly) depend on the solid volume fraction, since in both cases very similar particle distribution is observed. Therefore, our results appear to support more the argument of Martínez-Mercado et al. (2010) that the hydrodynamic interactions among the particles, which depend on the solid volume fraction, are responsible for the decay rate of he energy spectra. However, this merits further investigation and needs to be addressed in future studies.

Finally, we would like to mention that all of the aforementioned works considered light particles, i.e. bubbles, where here heavy finite-size particles are considered.

Now let us turn to the two-point correlation of the fluid velocity field. Figure 8.8 shows the two-point correlation function for all three flow cases. It can be observed that the fluid velocity components in case $M121$ decorrelate in the horizontal direction in a very similar manner. The u and w velocity components first become negatively correlated at approximately $r_x/d_p = 20$ and beyond that distance remain negatively correlated at small values of -0.015. On the other hand in the clustering cases, the fluid velocity components exhibit different evolution with the radial distance in the horizontal direction. The vertical velocity component remains over longer distances correlated than the horizontal velocity components. While the horizontal velocity components decorrelate completely for large horizontal separations, the vertical component in both cases $M178$ and $ML178$ remains negatively correlated. However, in case $M178$ the value of the correlation function is significantly smaller than in case $ML178$. Closer look of the two-point correlation function of the vertical velocity component reveals small oscillations with the local maximum and minimum. The distance between this values represents approximately the average distance between the particle clusters in the horizontal direction (cf. figure 6.2). The large negative correlations of the vertical velocity component in case $ML178$ stems most probably form the large void regions in which the fluid velocity remains correlated over longer distances (cf. figure 6.13).

Regarding the two-point correlation in the vertical direction (figure 8.8 (b,d,f)), the horizontal fluid velocity component in all three cases decorrelates completely for large vertical separations. Thereby, the horizontal velocity component in case $M121$ and case $ML178$ evidences negative loops with small negative values. Concerning the vertical velocity component, the graphs in figure 8.8 show that the velocity remains correlated at separations of half the box vertical length. Comparison with the correlation function for the fixed particle simulations $F121$ and case $F178$, evidences that due to the clustering in cases $M178$ and $ML178$ the vertical velocity remains correlated in the vertical direction over the entire vertical length of the computational domain. This confirms the visual observations made in figures 8.1 and 8.2. Moreover, the correlation function clearly shows dependence with the solid volume fraction, experiencing substantially larger correlation values in case $ML178$ than in case $M178$. This is mainly due to the higher clustering levels in case $ML178$ (section §6.2).

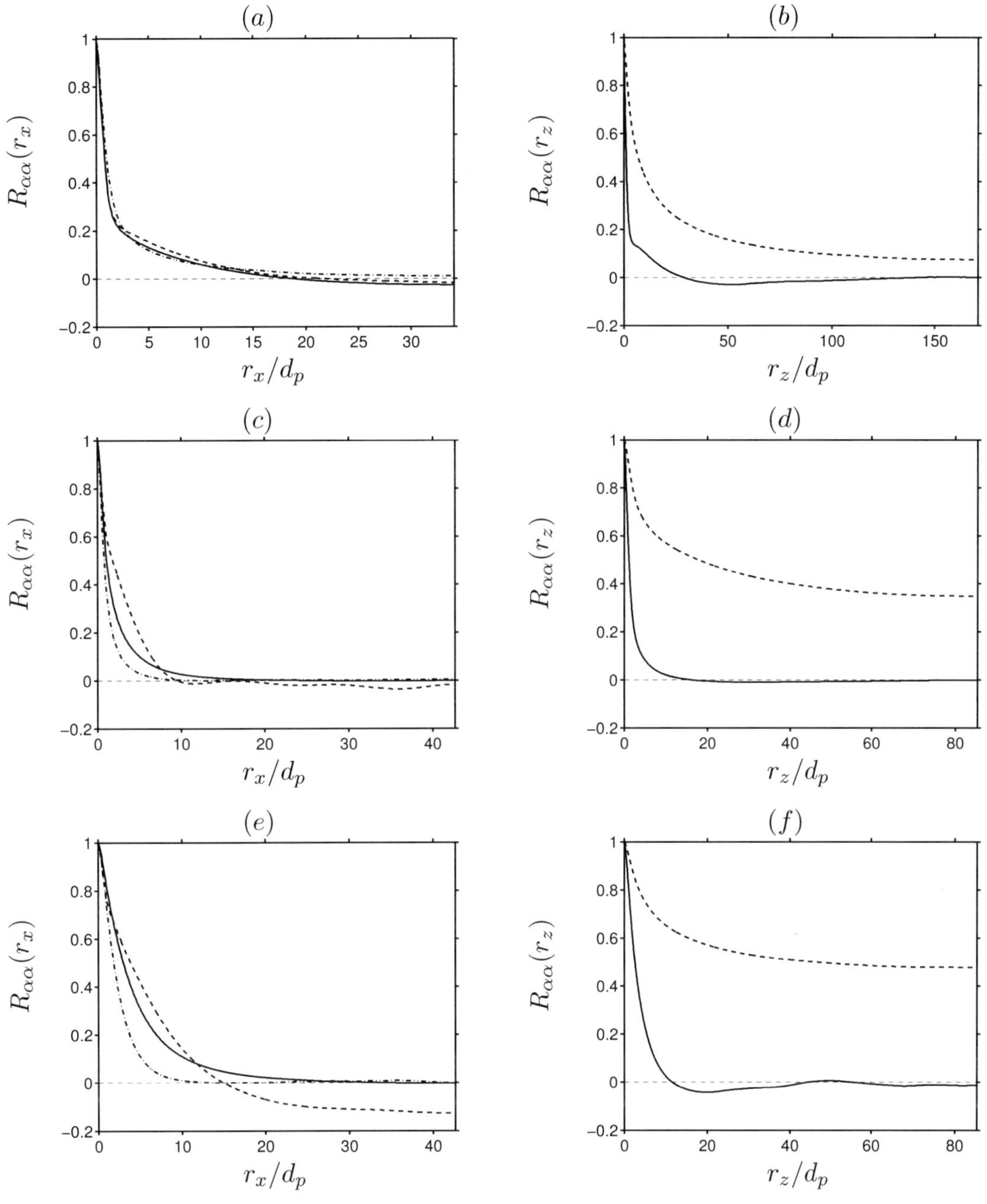

Figure 8.8: Two-point spatial autocorrelation function of fluid velocity fluctuations in case $M121$ (a,b), case $M178$ (c,d) and case $ML178$ (e,f) for separations in horizontal direction (a,c,e) and in the vertical directions (b,d,f). Line styles as in figure 8.7.

8.5 Particle wakes

In this section we will present the particle conditioned flow field in vicinity of the particles. To this end, the flow field in vicinity of the particles is conditionally averaged with respect to a reference particle. The averaging is performed as in section §4.3.4, where more detailed definition can be found. Figure 8.9 depicts contours of the vertical velocity component of the particle-averaged velocity $\tilde{w}_{f,rel}/u_g$ (equation 4.9), on a plane through the center of the

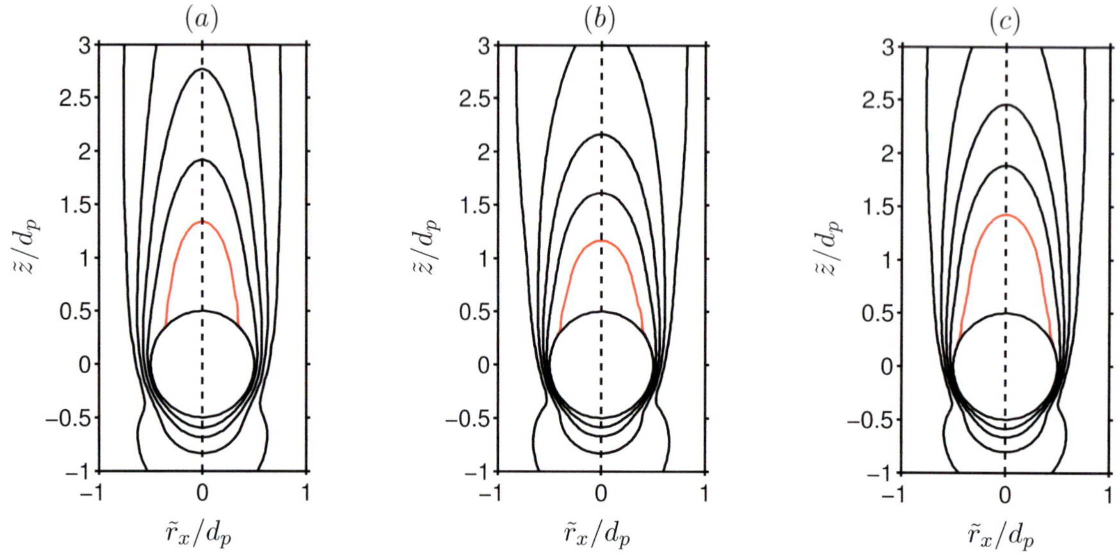

Figure 8.9: Contour plot of the relative velocity $\tilde{w}_{p,rel}/u_g$ in the plane parallel to the vertical axis passing trough the particle center $\boldsymbol{x}_p$. (a) case $M121$. (a) case $M178$. (c) case $ML178$. Contours are shown for values $(0{:}0.2{:}0.8)$. The red line (———) marks the extent of the recirculation region, i.e. $\tilde{w}_{f,rel} = 0$. The position of the z-axis is denoted by the black dashed line (- - - -). $\tilde{r}_x$ denotes the radial distance from the reference particle in the horizontal direction.

test particle aligned with the vertical direction, resembling the mean particle wake. It can be observed that the mean particle wake is in all three flow cases symmetric with respect to the vertical axis. In fact, due to the invariance of the flow in the horizontal direction, the mean particle wake is symmetric in the horizontal direction as well, this circumstance is already accounted for in figure 8.9 and is denoted by the radial distance $\tilde{r}_x$. The region encompassed by the red line represents the recirculation region of the mean particle wake, i.e. $\tilde{w}_{f,rel} \leq 0$. Due to the axi-symmetry of the mean wake, the recirculation length can be defined as in section §4.3.4 as the position on the vertical axis, where the velocity $\tilde{w}_{f,rel}$ changes sign. This can can be also observed in the inset of figure 8.10, where the velocity profile of $\tilde{w}_{f,rel}$ on the vertical axis is shown. We measured the recirculation length in case $M121$ to $l_e = 0.839 d_p$, which is approximately 12% smaller than the value measured in the case with the fixed particles. In the clustering case $M178$, the recirculation length was evaluated to $l_e = 0.663 d_p$, resulting in even large difference of 36.8%. This indicates that the particle clustering in case $ML178$ lead to a significant attenuation of the particle mean wake. The recirculation length was evaluated to $0.663 d_p$. The attenuation of the recirculation length can be mainly attributed to the orientation of the particle clusters in the vertical direction and to the increased probability of finding another particle in the wake of the reference particle. On the other hand in case $ML178$, the recirculation length of the average particle wake is measured to be substantially larger than in case $M178$, evidencing that the solid volume fraction has a significant effect upon the mean particle wake. The recirculation length in case $ML178$ is calculated to $l_e = 0.923 d_p$ being approximately 28.1% larger than on case $M178$. Although the particle in case $ML178$ agglomerate in similar clusters as in case $M178$, due to the smaller solid volume fraction, the individual particle wakes are less disturbed by the presence of neighboring particles. It is well known that the particle wakes and thus their recirculation length depend on the relative turbulence intensity of the flow field. Increase of the relative turbulence intensity causes a decrease of the particle wake recirculation length. As showed in section §7.1.3, figure 7.6, the smaller number of particles in case $ML178$ results in smaller relative turbulence intensity than in case $M178$, which also supports the above made observations.

Finally, let us mention that the value of the recirculation length calculated in case $M121$ resembles well the value observed in García-Villalba et al. (2012) for multiple finite-size particles settling in vertical channel flow. The authors reported the mean recirculation length of the particles located in the core of the channel to be $l_e = 0.86d_p$, which is approximately 2.3% larger than in case $M121$. The average Reynolds number of the particles located in the core of the channel, where the influence of the walls is the smallest, is slightly smaller (6.4%) than in case $M121$. Moreover, in the present case $M121$ nearly the same value of the relative turbulence intensity is measured as in García-Villalba et al. (2012), viz. the value of I_{r_V} in case $M121$ being 13% smaller than in García-Villalba et al. (2012). Further, comparison with the recirculation length of a single fixed particle subjected to homogeneous isotropic turbulence with similar value of the relative turbulence intensity as in Bagchi and Balachandar (2004) proved to be in good agreement with the value in case $M121$. The difference was measured to be approximately 3.5% with larger value in case $M121$. The above results imply that the average wake of the particles in case $M121$ closely resembles the wake of a single particle in otherwise similar flow conditions (w.r.t. the alues of I_{r_V}). This is in line with the observations made for case $M121$ throughout the current work, that the particles in case $M121$ settle on average as an isolated particle. At this point we would like to point out that the flow conditions and the flow structures in the above two references are significantly different from that in case $M121$. Nevertheless, the effect on the recirculation length in all these cases appear to be very similar.

On the other hand, comparison of the recirculation length of the mean particle wake in the clustering cases differ significantly from that of a single particle. We have compared the values of l_e in the cases $M178$ and $ML178$ with the results for fixed particle in homogeneous isotropic turbulence with similar relative turbulence intensity (Bagchi and Balachandar, 2004). It turned out, that due to the mobility and the clustering of the particles the recirculation length in case $M178$ and case $ML178$ is 18.1% shorter than the reported value in Bagchi and Balachandar (2004).

Concerning the profile of the vertical velocity component along the vertical axis passing through the center of the reference particle (figure 8.10), it is observed that the velocity experiences similar evolution as in the fixed particle simulations. The velocity $\tilde{w}_{f,rel}$ in all three flow cases decays quickly for small distances upstream form the test particle and experience negative values immediately behind the reference particle. The negative values of $\tilde{w}_{f,rel}$ indicate the presence of the recirculation region in the particle wake. The minimum of the velocity $\tilde{w}_{f,rel}$ is observed in all three cases at distance of approximately $0.81dp$, except in case $ML178$, where $0.86d_p$ was measured. After reaching its minimum value in the recirculation region, the velocity increases and converges steadily towards its asymptotic value. However, the recovery of the wake downstream of the reference particle appear to be in all three cases faster than in the fixed particle simulations. The rapid recovery is mainly due to the increased relative turbulence intensity of the flow field and the herewith resulting enhancement of the mixing.

Now let us turn to the velocity deficit in the average wake of the particles, which is shown for all three flow cases in figure 8.11. It can be observed that in the present flow cases, for small distances downstream from the reference particle, the fluid velocity in the particle wake

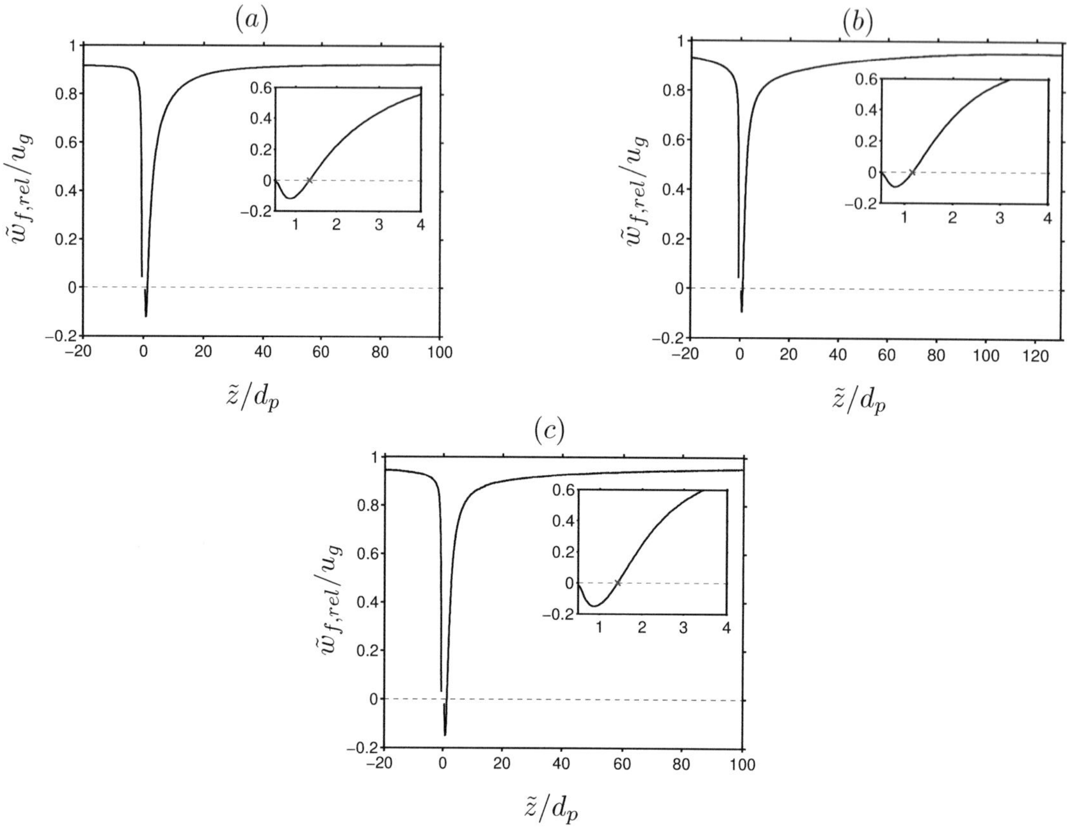

Figure 8.10: Vertical component of the relative velocity on the center line passing trough the particle center $\boldsymbol{x}_p$ and parallel to the vertical axis. (a) Case $M121$. (b) Case $M178$. (c) case $ML178$. The coordinate system is particle centered, i.e. origin at $\boldsymbol{x}_p$. The inset-figures show close-up view of the recirculation region. The blue cross $\times$ marks the position at which $\tilde{w}_{f,rel}/u_g$ changes sign. Negative values of the $\boldsymbol{x}$-axis correspond to upstream positions in front of the particle and positive values represent the domain downstream of the test particle. The distance from the particle was normalized by the particle diameter d_p. The velocity is normalized with the gravitational velocity u_g.

resembles the characteristic decay power law of a turbulent wake, i.e. it decays as $\tilde{z}^{-2/3}$. The range of distances, where this power law holds, is observed in all three flow cases for distances up to $\tilde{z} = 1.8d_p$, which is significantly smaller than for the fixed particle simulations ($\tilde{z} \approx 2.6d_p$). In case $M121$ this region is followed by a region, where the velocity deficit decays as a laminar wake, i.e. $\tilde{z}^{-1}$. This range extends up to $16d_p$ downstream from the particle. Beyond this distance a -2 power law is observed. On the other hand, the results for the clustering cases show that the $-2/3$ decay region is followed by a region, where the velocity deficit decays at rates somehow larger than -1. At large distances ($\tilde{z} > 27d_p$) from the reference particle the velocity deficit appears to decelerate and a small region with -1 power law is observed. Further downstream at distances larger than $50d_p$ a decay rate similar to -2 is observed. The decay of the velocity deficit proportional to $\tilde{z}^{-2}$ was previously reported for particles subjected to a turbulent flow environment (Wu and Faeth, 1994; Legendre et al., 2006; Amoura et al., 2010). Legendre et al. (2006) noted that the change of slope to -2 takes place at a distance downstream from the reference particle where the velocity deficit and the turbulence intensity are of the same order, i.e. $\tilde{u}_{d0} \approx \langle u'_f u'_f \rangle^{1/2}$. Closer look at our data revealed that this observation holds for the present flow cases.

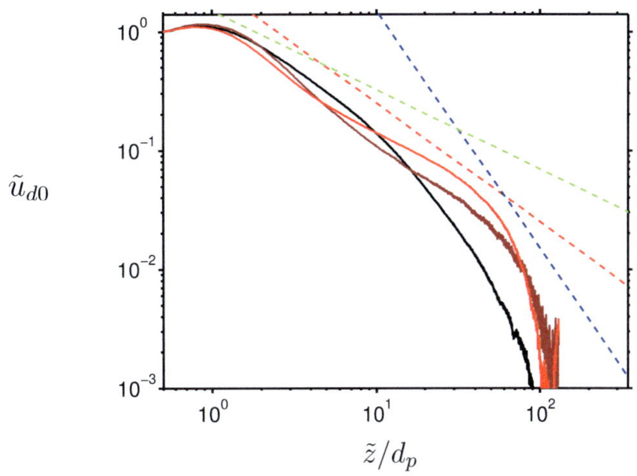

Figure 8.11: Streamwise velocity deficit $\tilde{u}_{d0}$ defined as in equation 3.11 along the vertical axis $\tilde{z}$. Case $M121$ (——), case $M178$ (——) and case $ML178$ (——). The dashed straight lines indicate decay rates proportional to $\tilde{z}^{-1}$ (----), $\tilde{z}^{-2}$ (----) and $\tilde{z}^{-2/3}$ (----).

It is interesting to note that the fluid velocity deficit for the fixed particle simulations did not exhibit a -2 power law. The present results indicate that the mobility of the particles and the resulting dynamical interaction between the wakes of the individual particles changes the decay behavior of the fluid velocity in the mean particle wake.

The spreading of the wake with the distance downstream from the reference particle can be investigated by considering the half-width (equation 3.12) of the wake for different downstream location from the test particle. Figure 8.12 depicts the evolution of the wake half-width h_{wz} with the distance from the particle for all three cases. It can be observed that, as for the fixed particles, the half-width of the wake in case $M121$ experience linear spreading. However, the observed slope of the linear function is in case $M121$ nearly three times larger than in case $F121$, indicating that the averaged wake in the present case spreads faster than in the case of fixed particles. On the other hand, in the clustering case $M178$ the wake is observed to be wider and to expands more quickly than in the non-clustering case and the fixed particle simulations. Initially for distances up to ten particle diameter the wake exhibits a linear spreading at rates significantly larger than in case $F121$. The slope of the linear function was evaluated to be six times larger than the in case $F178$. Beyond the distances of $10d_p$ the half-width of the wake appear to converge to a asymptotic value of $h_{wz}/d_p = 1.8$, as suggested by the flattening of the graph in figure 8.12. The asymptotic value of h_{wz} stems most probably from the cluster structures observed in cases $M178$ and $ML178$. As showed in section §6.3 the particle spatial distribution conditioned on the position of the particles evidenced a columnar like structure extending throughout the entire vertical length of the computational box, that was confined in the horizontal direction. Similar evolution of the wake half-width can be observed also for case $ML178$, indicating that the solid volume fraction has only a marginal effect upon the spreading of the wake. For comparison, we have additionally plotted the data from Bagchi and Balachandar (2004, p.104, figure 7) for a single fixed particle subjected to homogeneous isotropic turbulence and in a uniform flow. As can be seen, the rate of increase of the wake half-width in case $M121$ (represented by the black dashed line) resembles very closely the slope observed for the particle subjected to uniform flow. García-Villalba et al. (2012) reported also a value ($\tilde{a}$) very close to the one observed in

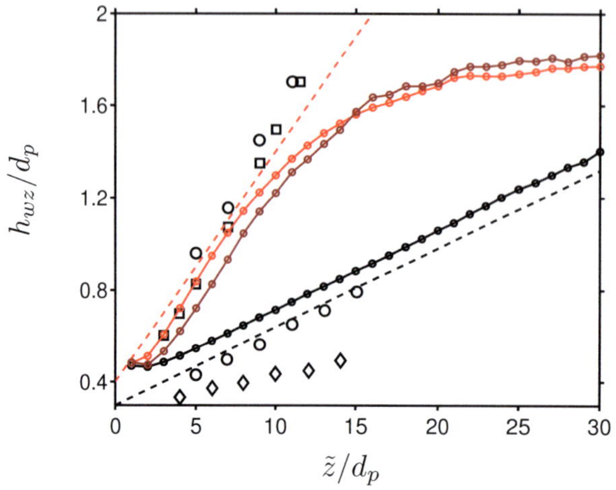

Figure 8.12: Cross-stream half-width of the average particle wake as function of the downstream distance $\tilde{z}$. Case $M121$ (—•—, - - - -). Case $M178$ (—•—, - - - -). Case $ML178$ (—•—). Dashed lines represent a linear fit of the spreading of the wake, $h_{wz}/r_p = \tilde{a}\tilde{z}/d_p + \tilde{b}$, with $\tilde{a} = 0.034$, $\tilde{b} = 0.4$ for case $M121$ and $\tilde{a} = 0.1$, $\tilde{b} = 0.4$ for case $M178$ and $ML178$. Open symbols (○, □, ◇) represent digitalized data for single fixed particle subjected to homogeneous isotropic turbulence form Bagchi and Balachandar (2004, p.104, figure 7).

the present work. In cases $M178$ and $ML178$ we found that the rate of spreading for $\tilde{z} \lesssim 10d_p$ coincide with the rate observed in Bagchi and Balachandar (2004). Incidentally, the case considered in Bagchi and Balachandar (2004) has similar particle Reynolds number as the Reynolds number in case $M178$ and case $ML178$. This observations indicate that despite the strong anisotropic clustering in cases $M178$ and $ML178$, the rate of spreading of the average particle wake for small distances downstream form the reference particle is the same as for a single fixed particle subjected to isotropic turbulence. As a consequence, the flow conditions in case $M178$ and case $ML178$ to which the particles are subjected appear to be very close to that of homogeneous isotropic turbulence and are relatively independent of the solid volume fraction and the relative turbulence intensity of the flow (at least with respect to the wake half-width). It is however surprising that the spreading of the wake for the assembly of fixed particles in case $F121$ and case $F178$ experience different behavior (note, the particle in Bagchi and Balachandar (2004) is also fixed in space). This circumstance needs to be addressed in future studies.

In a theoretical work Eames et al. (2011) suggested that the rate of spreading is comparable to the value of the relative turbulence intensity. As shown above, the results for the present flow configurations do not support this relation, especially for the clustering cases where the relative turbulence intensity of the flow vary approximately by a factor of two. The authors assumed that the particle is subdued to a homogeneous isotropic turbulence, whereas in the present work a pseudo-turbulent flow is considered. Thus, it is intriguing that our results follow the trend observed in Bagchi and Balachandar (2004).

Finally let us turn to the self similarity of the mean particle wake. To this end we consider the cross-stream profiles of the velocity $\tilde{u}_d$ in the particle wake, which is shown in figure 8.13 for all three flow case at different downstream locations from the test particle. The velocity deficit $\tilde{u}_d$ is calculated as in equations 3.11, 3.13. It can be observed that the mean particle wake experience indeed self similar behavior, the cross-stream profiles at different distance

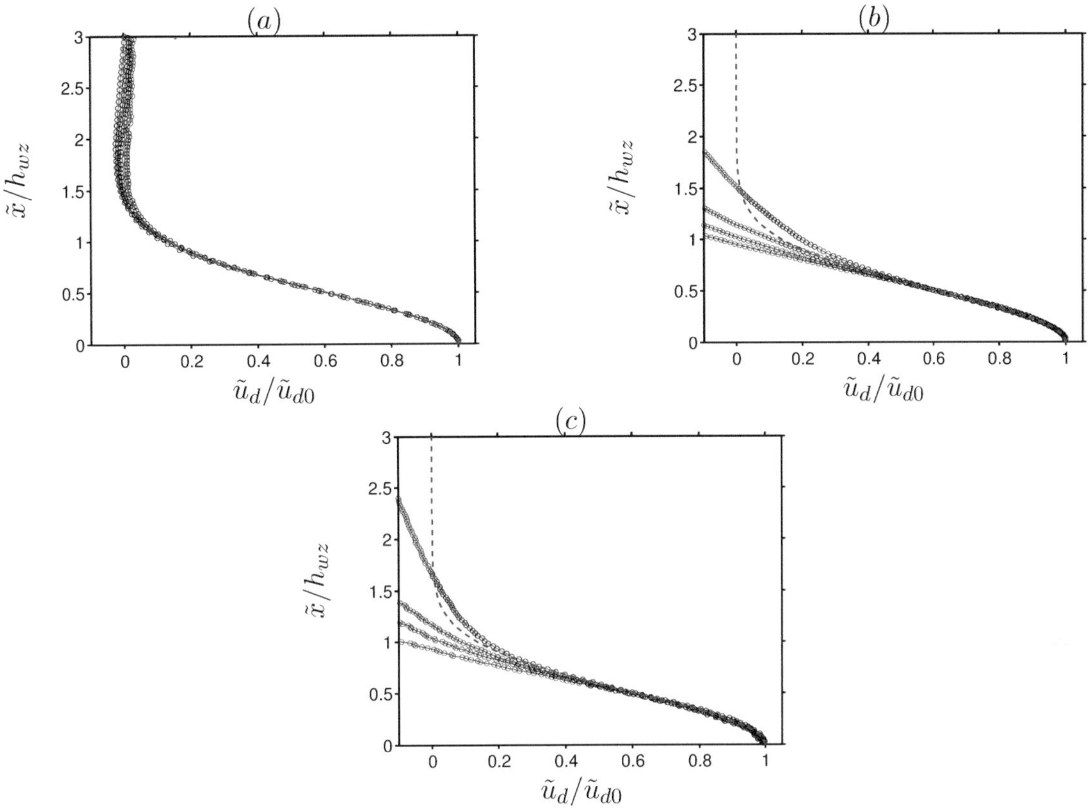

Figure 8.13: Profiles of the velocity deficit in the average particle wake as function of the cross-stream coordinate $\tilde{x}$. The profile was taken at 4 different downstream locations, $\tilde{z}/d_p = [6, 10, 13, 17]$. The blue dashed (----) line corresponds to a Gaussian function. The velocities $\tilde{u}_d$ and $\tilde{u}_{d0}$ are defined as in equations 3.11, 3.13. (a) Case $M121$. (b) Case $M178$. (c) case $ML178$. The coordinate system $(\tilde{x}, \tilde{y}, \tilde{z})$ is chosen as in figure 3.5. The distance from the particle center was normalized by the wake half-width h_{wz}.

from the reference particle collapse into a single curve following a Gaussian function. This again confirms that the mean particle wake in case $M121$ resembles very closely the wake of a isolated particle. For the clustering cases however, the velocity $\tilde{u}_d$ follows a Gaussian function only for distances up to $\tilde{x}/h_{wz} \lesssim 1$. Beyond this distance the deviation from the velocity profile deviates from the Gaussian profile, with the increasing difference for increasing downstream distances. Most probably this can be accounted to the deviation of the wake half-width from the linear rate for large downstream locations (cf. figure 8.12).

8.6 Conclusion

In this chapter we have presented and discussed results for the flow field induced by the settling of finite-size heavy particles in an initially quiescent flow field. To this end, we have considered the flow field induced by the particles from the three flow cases $M121$, $M178$ and $ML178$ introduced in chapter §5. Thus, the fluid agitations in the present work are solely due to the presence and the motion of the particles and are dominated by the particle wakes. The overall conclusion is that the fluid velocity field, as the particle velocity, is highly anisotropic in all considered flow cases, with the vertical fluid velocity component being the dominant one. The highest amplitude of the fluid velocity fluctuations was measured in the clustering case $M178$, whereas the smallest were observed in case $ML178$. It was found out that the particle induced flow field in case $M121$ was represented well by the flow field generated by the fixed assembly of particles in case $F121$. Moreover, the local flow field of the particles in case $M121$ is well resembled by the flow field of an isolated particles as in case $S121$. On the other hand, in the clustering cases the flow field is significantly affected by the dynamics of the particle clusters and the resulting hydrodynamical interactions of the individual particle wakes led to substantial attenuation of the mean particle wake. In particular we have addressed the following issues.

We have considered and discussed three widely used approaches for the prediction of the amplitude of the fluid velocity variance. Due to the high anisotropy of the flow field, a potential model for the fluid velocity fluctuations failed to provide reasonable approximations for the amplitude of the velocity fluctuations. A simple superposition of the fluid velocity fluctuations induced by the motion of an isolated particle in otherwise similar flow conditions, showed that except in case $M121$, the mutual hydrodynamic interactions of the particle wakes lead to a significant increase of the fluid velocity variance in the present cases. A detailed analysis of the local flow field in vicinity of the particles in case $M121$ revealed to be very similar to that of an isolated particle (case $S121$), leading to reasonable agreement of the predicted and observed amplitude of the fluid velocity fluctuations in case $M121$. Moreover, the fluid motion induced by the particle assembly in case $M121$ resembles closely the particle-induced agitations in the simulations with fixed assembly of particles and otherwise same solid volume fraction and particle Reynolds number $Re_{p,rel}$ (case $F121$). As a consequence, empirical models as proposed by Parthasarathy and Faeth (1990a) lead to reasonable predictions of the amplitude of the fluid velocity fluctuations. A detailed study of the different contributions to the total fluctuations in the model by Parthasarathy and Faeth (1990a), revealed that the amplitude of the fluid velocity fluctuations in case $M121$ are stemming from the mean properties of the particle wakes and are close to that observed in laminar flows. In the clustering cases however large deviations of the predicted value from the present results were observed, clearly showing the limits of such models. Nevertheless, the overall trend was reproduced by the empirical model of Parthasarathy and Faeth (1990a)

Study of the probability density functions of the fluid velocity components revealed that in all three flow cases the flow field exhibits super-Gaussian distribution. The horizontal velocity components proved to be nearly symmetric and highly intermittent, whereas the vertical component were found to be negatively skewed. Comparison with the p.d.f.s from the fixed particle simulations showed that the mobility of the particles lead to attenuation of the intermittency of the fluid velocity, regardless if the particles agglomerate into clusters or not. The value of the solid volume fraction showed to have only marginal effect upon the distribution of the horizontal fluid velocity, while the vertical component in case $M178$ evidenced smaller intermittency. The distribution of the local fluid velocity in the vicinity of the particles showed that the local flow field in the clustering cases is very similar and nearly independent of the value of the solid volume fraction. This results proved to be in line with the observations in the available literature, that the local flow field of the particles is independent from the solid volume fraction. Further, the fluid velocity variance was evidenced to evolve in a self-similar manner when scaled with the solid volume fraction ($\sim \Phi_s^{0.4}$), confirming the scaling proposed in Risso and Ellingsen (2002).

We have identified the contributions to the turbulent kinetic energy for the present flow conditions and studied their relative importance for the present flow cases. It was observed that in all cases the turbulent kinetic energy was due to the work done by the gravitational potential and was balanced by the dissipation rate. Comparison between the two clustering cases showed that the potential energy was found to depend in a linear manner from the solid volume fraction. The two-way coupling term was found to be in all three flow cases negative but very small, thus acting in the present flow conditions as a sink of turbulent kinetic energy, whereas a positive values were observed for neutrally buoyant particles in homogeneous isotropic turbulence (Lucci et al., 2010). Thus a more detailed study is required for the understanding of the exact physical mechanisms responsible for the sign of the two-way coupling term.

The energy spectrum of the fluid velocity components evidenced features characteristic to the energy decay in pseudo-turbulent flows. In the horizontal direction the fluid velocity components in all three flow cases were found to follow a decay power law proportional to -3 and -3.5, respectively. Regarding the spectra in the vertical direction, the -3 power law was found only for the clustering cases. The mobility of the particles proved to have a significant effect on the decay rate of the energy spectrum in case $M121$. A comparison with the spectra for the fixed particle assembly in case $M178$ showed that the distribution of the particles does not affect the decay rate of the fluid velocity components. The particle wakes and the hydrodynamic interactions between the wakes appeared to be the main reason for the decay rate of the fluid velocity energy spectra. Finally, in all three case the energy spectrum showed signs of a reverse cascade. We have additionally analyzed the two-point correlation function of the fluid velocity in the present cases. It was observed that the vertical velocity component in all three flow cases was positively correlated over the half-length of the computational domain. In the clustering cases the correlation was found to be more prominent than in the non-clustering case. Additionally the solid volume fraction proved to have, for the present flow conditions, significant effect upon the two-point correlation function, with larger correlations observed for the smaller values of Φ_s.

Finally, a comprehensive analysis of the mean particle wake was performed. To this end, we have calculated the particle conditioned flow field in the vicinity of the particles. We found out that the mean particle wake in all three flow cases was axi-symmetric and that the recirculation length of the wake was significantly reduced, when compared to the mean wake from the fixed particle simulations. In case $M121$, it was observed that the mean wake resembles closely the statistical characteristics (apart from the value of the recirculation length) of the wake of an isolated particle (case $S121$) and to be in good agreement with results for the mean particle wake in substantially different background flow conditions. The clustering in cases $M178$ and $ML178$ was found to have a significant effect upon the recirculation length, where the attenuation was found to be more pronounced in case $M178$. The solid volume fraction proved to have significant effect upon the length of the recirculation length as well. In contrast to the fixed particle simulations, the velocity deficit in the particle wake evidenced regions with decay rates proportional to -2. Our results confirmed that the change of the slope from -1 to -2 of the decay rate of the fluid velocity deficit occur at locations in the wake, where the velocity deficit is of the same order of magnitude as the fluid turbulence intensity.

The spreading of the mean particle wake in the present flow cases was found to be at faster rates than for the flow cases with fixed particles. With respect to the effect of the particle clusters, the mean particle wake in case $M178$ was found to be wider and to expand more quickly than in case $M121$. In the clustering cases, it was observed that for large downstream distances ($\tilde{z} \gtrsim 10d_p$) the spreading of the mean particle wake was at approximately constant rate and the solid volume fraction was found to have only marginal effect on the spreading of the mean particle wake. Comparison with the literature (Bagchi and Balachandar, 2004) revealed that the spreading of the mean particle wake in case $M121$ followed the rate for a single particle (with similar parameters and relative turbulence intensity) subjected to uniform flow. On the other hand, in the clustering cases the spreading rate for $\tilde{z} \lesssim 10d_p$ was found to coincide with that of a single particle subjected to isotropic turbulence. These observations indicated that the spreading of the wake, for small distances downstream form the particle, was independent from the particle spatial distribution in cases $M178$ and $ML178$. Despite the different statistical characteristics of the present flows and the homogeneous isotropic turbulence, the effect on the particle spreading rate appear to be very similar. This and the circumstance that the spreading of the wake for fixed particles differ from the observations in the literature (note, the particle in Bagchi and Balachandar (2004) is fixed in space as well) needs to be addressed in more detail in future studies.

The overall conclusion is that the flow field induced by the particles in the present flow cases agrees well with the general observations in the literature. The particle motion in the non-clustering case induced an agitations in the flow field similar to that induced by a fixed assembly of particles with the same solid volume fraction and similar particle Reynolds number $Re_{p,rel}$. As a consequence, empirical models as proposed by Parthasarathy and Faeth (1990a) lead to reasonable predictions of the amplitude of the fluid velocity fluctuations. On the other hand, in the clustering cases significant deviations of the predicted fluid variance from the variance observed in the simulations $M178$ and $ML178$ was evidenced. Thus, the effect of the particle clusters on the fluid velocity fluctuations needs to be addressed in future studies. In particular, it will be highly beneficial to isolate the contributions in the fluid velocity variance due to the regions of the flow associated with particle clusters (cf. large vertical streamers in figures 8.1 - 8.2). For this we plan to perform a analysis conditioned on the particle clusters and voids, respectively. Unfortunately, we were not able to calculate the temporal correlations for the flow field, which will allow us to study the temporal behavior of the flow field. The knowledge of the fluid velocity autocorrelation will allow us further to investigate the theory provided by Ferrante and Elghobashi (2003) for the sign of the two-way coupling term in the turbulent kinetic energy equation.

Chapter 9

Summary, conclusions and recommendations for future works

Particulate two-phase flows are of particular interest in large number of environmental, natural and technical processes. The fundamental understanding of the different mechanisms that take place in such flows are of particular interest. Thus, a reliable and correct prediction of the processes that take place in such flows is of great importance. Despite the importance and the wide range of applications of particulate flows, the fundamental mechanisms involved in such flows are far from being completely understood. One possible reason for this is the complex characteristics of the flows under consideration. Unfortunately, the theoretical description of the physical processes is only accessible for very limited flow conditions. Nowadays, most relations for the prediction of quantities such as the average particle settling velocity, particle drag coefficients, dispersion and distributions of the particles, amplitude of the fluid velocity fluctuations are based on empirical relations, which are derived by considering a large number of assumptions and simplifications of the problem under consideration. Therefore, the range of applicability of such predictions is limited as well.

This thesis addresses some of the topics involved in the dynamics of particles settling under the influence of gravity, as well the characteristics of the particle-induced flow field and the herewith resulting two-way coupling mechanisms. Furthermore, this work contributes to the understanding of the fundamental mechanisms involved in such flows by providing and analyzing high fidelity data for flows with large number of heavy finite-size particles settling under the influence of gravity. The problems considered in this thesis are simplified by eliminating the influences due to wall effects, while allowing to focus solely on the fundamental physics involved in such flows.

The problems in the present work are tackled by performing direct numerical simulations employing an immersed boundary method for the representation of the fluid-solid interface. Three different flow configurations have been considered in the present work: (i) the settling of a single particle in a uniform unbounded flow, (ii) flow through a stationary array of randomly distributed spheres and (iii) the settling of many particles in an initially quiescent flow field. The main focus in this work was on the characterization of the particle induced flow field, as well as the characterization of the particle velocities, hydrodynamic forces, wakes and the spatial distribution of the particles. A more detailed discussion of this aspects can be found in the conclusions of chapters §3 - §8.

9.1 Settling of a single particle in an ambient fluid

Direct numerical simulations of the sedimentation of single heavy finite-size spherical particle under the influence of gravity in an initially ambient flow field was studied. The settling of the particle in two different settling regimes was considered, namely in the steady axi-symmetric regime with Galileo number of Ga $= 121$ and in the steady oblique regime with Ga $= 178$. The particle-to-fluid density ratio in both regimes was kept at value $\rho_p/\rho_f = 1.5$. Special care was taken to ensure appropriate spatial and temporal resolutions for the particle motion.

The main objective was the characterization of the particle terminal settling velocity, hydrodynamic forces and the wake structure of the single particle for both settling regimes. It was found that the resolution of 15 and 24 grid points per particle diameter were sufficient for the correct representation of the particle wake and hydrodynamic forces in both flow cases. The results were in a very good agreement with the benchmark data from Uhlmann and Dušek (2014). The obtained results showed that the characteristic features of the particle motion and particle wake was correctly captured by the numerical method. For the axi-symmetric flow case the particle wake was found to be axi-symmetric and consists of a toroidal vortex around the particles. For the steady oblique regime the particle wake was observed to consist of a double threaded wake with counter-rotating vortices. Various statistics, such as the particle terminal settling velocity, particle drag coefficient and the oblique angle (only for the particle with Ga $= 178$) proved to be in very good agreement with the reference data.

9.2 Flow through a stationary array of randomly distributed spheres

The flow field induced by a uniform flow past an assembly of stationary randomly distributed finite-size particles was investigated. Again two direct numerical simulations were considered. The flow under consideration consisted of randomly distributed particles, with shape and size as the particles considered in chapter §3. The number of particles in both cases was chosen so that the solid volume fraction calculates to $\Phi_s = 0.005$. For this values of Φ_s the flows are considered as dilute. The spatial and temporal discretization was kept as in the simulations of the single particles in chapter §3.

It was observed that in both flow cases the fluid velocity fluctuations are highly anisotropic with the vertical component being the dominant one. This was attributed to the wake character of the fluid velocity and the levels of anisotropy were in line with the observations from the literature. The energy spectra of the fluid velocity was found to exhibit a decay power law proportional to the power of -3 of the wave number, which is known to be a characteristic feature of pseudo-turbulent flows. The fluid velocity fluctuations were decomposed in their spatial and temporal contributions. The results evidenced that the flow field is unsteady and that large-scale wake interactions are responsible for the fluctuations of the flow field. The fluctuations with respect to the volume averaged mean flow field were found to be more prominent than the fluctuations with respect to the temporal mean value, accounting for the major contribution to the total fluctuations.

The mean particle drag coefficient was observed in both cases to resemble the drag coefficient of the single particle simulations from chapter §3. Furthermore, the distribution of the hydrodynamic force with respect to the position of a test particle was studied. The results revealed that in close proximity behind the reference particle, the drag coefficient of the following particles was substantially smaller than of the leading particle. The domain of influence downstream from the test particle was found to extend over distances of several tens particle diameters from the reference particle.

Analysis of the average particle wake evidenced significant attenuation of the recirculation length due to the presence of multiple particles. The velocity deficit in the particle wake was found to decay at a rate similar as in a turbulent wake (i.e. with rate proportional to $\tilde{z}^{-2/3}$) for small distances downstream from the reference particle and as in a laminar wake (i.e. proportional to $\tilde{z}^{-1}$) further downstream. Moreover, the half-width of the particle wake was found to evolve in a linear manner with the downstream distance from the reference particle.

9.3 Settling of many particles in an initially quiescent flow field

Direct numerical simulations of heavy finite-size particles freely settling in an unbounded (initially) quiescent flow field were carried out. A total of three simulations were performed. The first two simulations correspond to the fixed particle flow cases with the difference that the particles are allowed to move freely in the computational domain. This flow cases are unambiguously characterized by the triplets ($\rho_p/\rho_f = 1.5, \mathrm{Ga} = 121, \Phi_s = 0.005$) and ($\rho_p/\rho_f = 1.5, \mathrm{Ga} = 178, \Phi_s = 0.005$). The third simulation was derived from the simulation with the Galileo number of $\mathrm{Ga} = 178$ (case $M178$) by reducing the number of particles by factor of ten leading to the triplet ($\rho_p/\rho_f = 1.5, \mathrm{Ga} = 178, \Phi_s = 0.0005$). The spatial and temporal resolutions were kept the same as in the simulations of the fixed particle assembly from chapter §4. This allowed us to study the effects of the Galileo number ($\mathrm{Ga} = 121$ and $\mathrm{Ga} = 178$) and the effect of the solid volume fraction ($\Phi_s = 0.005$, and $\Phi_s = 0.0005$).

The spatial structure of the dispersed phase in all three flow cases was investigated by means of Voronoï analysis. Concerning the effect of the Galileo number, the particles in case $M178$ sampled the flow in an inhomogeneous manner and evidenced strong clustering, while no evidence of clustering was observed for the particles with Galileo number $\mathrm{Ga} = 121$ (case $M121$), on the contrary the particles were observed to be rather organized in slightly more regular manner than a randomly distributed particles. The particle clusters in case $M178$ were observed to be highly anisotropic columnar like structures extending in the vertical direction over the entire length of the computational domain. Regarding the effect of the solid volume fraction upon the particle clusters in case $M178$, it was observed that while the overall cluster structure was maintained, the level of clustering increased with decrease of the solid volume fraction (case $ML178$).

We have preformed a comprehensive study of the particle settling velocity. It was observed that the reference fluid velocity could have a significant effect upon the mean particle settling rate. We have defined and analyzed the properties of a fluid velocity seen by the particles. It was found out, that the mean settling rate of the particles in the non-clustering case was similar to that observed for an isolated particle. On the other hand, in the clustering case the average particle settling velocity was observed to vary, depending on the definition of the reference fluid velocity. When computed with the global average of the fluid velocity, the particles were found to settle on average significantly faster than an isolated particle. However, when the local fluid velocity is used the mean particle settling rate is measured to be close to that of a single particle. The particles in the clustering cases were found to be preferentially located in regions with downward fluid motion.

We have identified and calculated the contributions to the particle velocity variance stemming from the particle motion with respect to the surrounding fluid and the fluctuations of the fluid velocity in vicinity of the particle. The results revealed that the larger fluctuations of the particle velocity component in the clustering case $M178$ is due to an increase of the fluctuating energy of the fluid velocity seen by the particles.

Study of the particle drag coefficient revealed that the non-linear drag effect lead to a net increase of the drag and that it correlates with the relative turbulence intensity, i.e. the largest effect was observed in case $M178$, where the highest value of the relative turbulence intensity is observed. Particle conditioned average showed that the particle settling velocity and drag force are distributed in an anisotropic manner, viz. in vertically oriented structures.

Similarly, the fluid velocity field is in all three flow cases distributed in an anisotropic manner. The highest amplitude of the fluid velocity fluctuations is measured in the clustering case $M178$. It was observed that the flow field induced by the particles in case $M121$ resembles well the flow field induced by the fixed particle assembly, indicating that the motion of the particles has only a marginal effect upon the flow field. Moreover, the local flow field of the particles in case $M121$ is very similar to that induced by single particle. Consequently, an empirical model as in Parthasarathy and Faeth (1990a) proved to predict reasonably the amplitude of the fluid velocity fluctuations in case $M121$. On the other hand, in the clustering cases the flow field is significantly affected by the presence of the particles.

We have studied the turbulent kinetic energy (TKE) budget for the present flow cases. It was observed that in all flow cases the TKE was due to work done by the gravitational potential and was balanced by the dissipation rate. Moreover, the two-way coupling term was found to be negative in all three flow cases, i.e. it acts as a sink of TKE.

The mean particle wake in the clustering cases was significantly attenuated due to the particle motion and clustering. The statistical description of the flow field proved to be in line with the observation in the literature.

9.4 Recommendation for future work

This thesis has covered large number of aspects involved in mono-dispersed particulate flows by performing several direct numerical simulations of heavy finite-size particles settling under the influence of gravity in an unbounded initially quiescent flow field. The simulations provided high fidelity data sets, of both the particle and the fluid motion, of the assembly of large numbers of particles at particle Reynolds numbers of the order of $\mathcal{O}(100)$. Nevertheless, review of the literature revealed that the number of experiments and direct numerical simulations of such flow configurations is clearly limited and a systematic exploration of the parameter space is missing. Additional numerical and experimental parameter studies are required in order to fully understand the effect of each control parameter of the system. It would be highly beneficial, if the future studies are performed by the application of complementary experimental and DNS studies, allowing the exploitation of the respective advantages of each approach. Most of the existing studies are limited to either small particle Reynolds numbers, small aspect ratios of the domain extensions to the particle size or sometimes even both. Ideally, any future studies should consider set-ups in a parameter range that overcomes some of the above mentioned limitations and close the existing knowledge gaps of the large number of physical mechanisms that take place in such flows. In particular, we propose that future studies should focus on the following (still not completely understood) issues.

As evidenced in this work, the particles may sample the flow field in an inhomogeneous manner and agglomerate into clusters. In our view, the exact mechanism responsible for the formation of particle clusters are still not completely understood. We suggest that future works could tackle this problem by identifying the role of the regime of settling of the individual particles and its correlation with the wake instabilities of the individual particles. Thus, simulations of particles settling in different regimes would be highly beneficial. For example, to our knowledge there is virtually no information for the clustering of particles settling in the chaotic regime (in the absence of background turbulence). In future works the identifying of the critical Galileo number demarcating the onset of clustering would allow more detailed analysis of the mechanisms of wake-induced clustering. Apart from this, the effects of the solid volume fraction, particle rotation and relative turbulence intensity needs to be addressed in detail in future studies.

With respect to the power law of the energy spectra observed in pseudo-turbulent flows and its contributions to the turbulent kinetic energy in turbulent carrier flows, a further studies are required. Especially the effect of the particle wakes and the hydrodynamic interactions between the particles in the inter-wake regions needs to be further investigated. An open question is the role of the relative turbulence intensity (or the so called bubblance parameter (Rensen et al., 2005)) or the particle distribution for the power law of the energy spectra. In the context of finite-size heavy rigid particles, relatively little is known about the turbulent kinetic energy spectra and definitely needs to be addressed in future studies.

The correlation function of the flow field in this work revealed the one of the common problems of the pre-existing studies (especially numerical simulations) is the size of the domain under consideration. Unfortunately, there is no *a-priory* approach to estimate the required size of the domain, which is further complicated by the long times the particles require to reach a statistically steady state. Thus, it will be instrumental to consider large enough domains, allowing for the correct estimation of length scales of large particle clusters and the corresponding coherent flow structures. Additionally, the study of large Reynolds numbers would contribute to the establishment of possible scaling laws that would contribute to the development of numerical models.

The present work clearly indicated the limits of the capability of the available numerical models to accurately predict the motion of the particles and the flow field over wide range of parameters. Nevertheless, some of the models showed an encouraging results as for the case of particles settling with $Ga = 121$, thus we think that the development of numerical models should be further actively pursued. The available models could be further improved, for example, by the knowledge of the local particle concentration along the particle trajectories, or typical residence time of the particles in clusters or void regions.

This thesis considered mainly pseudo-turbulent flows, i.e. fluid fluctuations are generated solely by the relative motion of the particles to the fluid. The future study of the two-way coupling mechanisms, especially the turbulence modulation by the presence of the particles for *a-priory* turbulent flows would be a logical consequence. This kind of analysis opens the way to completely new perspectives and approaches and is still considered one of the most challenging topics in particulate flows.

With respect to the present simulations, the results should be further analyzed in order to understand some of the mechanisms involved in the present flows. In particular, a more detailed analysis of the different particle and fluid statistics with respect to the particle clusters should provide some insight into the statistical consequences of the particle clustering and will contribute for the improvement of the available numerical models. The richness of the accumulated data with respect to the temporal evolution of the particles, could further contribute to the study of the three dimensional and temporal behavior of the particle clusters and provide insight into open questions as the time needed for the particle clusters to form or break. Additionally, the investigation of the spatial correlation between clusters and the so called *super-clustering* (study of the clusters of clusters) (Springel et al., 2005; Ramella et al., 2001; Obligado, 2013) can be addressed. Due to the high efficiency of the Voronoï analysis, the statistical footprint of the particles in the surrounding flow can be further investigated.

This is only a small part of the still open questions in particulate flows. The present simulations showed that with the advance of the computational resources (and the experimental techniques), the study and the understanding of the aforementioned issues and the revealing of the secrets behind these processes is in reach and needs to be further pursued. The new approaches open the way to a better understanding of the physics involved in particulate flows under similar conditions and allow the consideration of more complex flows, such as the presence of walls, higher Reynolds numbers and etc..

Appendices

Appendix A

Averaging procedures

In this section we provide a detailed definitions of the various averaging operators used in the present work.

A.1 Phase averaging

Let us first define an indicator functions for both the fluid and the dispersed phase. An indicator function for the fluid phase $\Phi_f(\boldsymbol{x}, t)$ specifies whether a given point $\boldsymbol{x}$ in the computational domain at certain time t is exclusively occupied by the fluid:

$$\Phi_f(\boldsymbol{x}, t) = \begin{cases} 1 & \text{if} \quad \boldsymbol{x} \in \Omega_f(t) \\ 0 & \text{else} \end{cases}, \tag{A.1}$$

where $\Omega_f(t)$ denotes the part of the computational domain Ω which is occupied by the fluid at time t. The corresponding indicator function for the dispersed phase $\Phi_p(\boldsymbol{x}, t)$ can be readily obtained as:

$$\Phi_p(\boldsymbol{x}, t) = 1 - \Phi_f(\boldsymbol{x}, t). \tag{A.2}$$

Based on the definitions in equations A.1 and A.2 a discrete counter of sample points occupied by the carrier phase n_f and by the dispersed phase n_p at certain time t^m can be defined as follows:

$$n_f(t^m) = \sum_{i=1}^{N_x} \sum_{j=1}^{N_y} \sum_{k=1}^{N_z} \Phi_f(\boldsymbol{x}_{ijk}, t^m), \qquad \forall\ m \in [1, N_{f,t}], \tag{A.3}$$

$$n_p(t^m) = N_x N_y N_z - n_f = \sum_{i=1}^{N_x} \sum_{j=1}^{N_y} \sum_{k=1}^{N_z} \Phi_p(\boldsymbol{x}_{ijk}, t^m), \qquad \forall\ m \in [1, N_{p,t}], \tag{A.4}$$

where N_x, N_y and N_z are the number of grid nodes in the corresponding coordinate direction and $N_{f,t}$ ($N_{p,t}$) is the number of available snapshots of the flow field (particles) in time. Please note, that during the computation the number and size of the particles does not change and thus the values of n_f and n_p remain constant during the simulation.

The fluid-phase averaging for any fluid related quantity $\psi_f(\boldsymbol{x}_{ijk}, t^m)$ can be defined as:

$$\langle \psi_f(t^m) \rangle_{\Omega_f} = \frac{1}{n_f} \sum_{i=1}^{N_x} \sum_{j=1}^{N_y} \sum_{k=1}^{N_z} \Phi_f(\boldsymbol{x}_{ijk}, t^m) \psi_f(\boldsymbol{x}_{ijk}, t^m), \qquad \forall\ m \in [1, N_{f,t}]. \tag{A.5}$$

We define the time and space averaging of $\psi_f(\boldsymbol{x}_{ijk}, t^m)$ as follows:

$$\langle \psi_f \rangle_{\Omega_f,t} = \frac{1}{N_{f,t}} \sum_{m=1}^{N_{f,t}} \langle \psi_f(t^m) \rangle_{\Omega_f} \tag{A.6}$$

Similarly, by defining a counter of fluid phase samples along the vertical direction, $n_f^{(z)}$, viz.

$$n_f^{(z)}(x_i, y_j, t^m) = \sum_{k=1}^{N_z} \Phi_f(\boldsymbol{x}_{ijk}, t^m), \qquad \forall m \in [1, N_{f,t}], i \in [1, N_x], j \in [1, N_y], \tag{A.7}$$

the spatial average of a fluid related quantity $\psi_f(\boldsymbol{x}_{ijk}, t^m)$ over the vertical direction is defined as follows:

$$\begin{aligned} \langle \psi_f \rangle_z(x_i, y_j, t^m) \;&=\; \frac{1}{n_f^{(z)}(x_i, y_j, t^m)} \sum_{k=1}^{N_z} \Phi_f(\boldsymbol{x}_{ijk}, t^m) \psi_f(\boldsymbol{x}_{ijk}, t^m), \\ &\quad \forall m \in [1, N_{f,t}], i \in [1, N_x], j \in [1, N_y]. \end{aligned} \tag{A.8}$$

The dispersed-phase averaging operator $\langle \cdot \rangle_p$ for a given particle related quantities $\psi_p^i(t^m)$ can be defined as:

$$\langle \psi_p(t^m) \rangle_p = \frac{1}{N_p} \sum_{i=1}^{N_p} \psi_p^i(t^m), \qquad \forall\ m \in [1, N_{p,t}]. \tag{A.9}$$

A time averaging operator for both the dispersed and the carrier phase can be defined as follows:

$$\langle \cdot \rangle_t = \frac{1}{N_{\alpha,t}} \sum_{m=1}^{N_{\alpha,t}} \cdot, \tag{A.10}$$

where $\alpha = f$ or $\alpha = p$. An averaging operator for a given particle related quantity $\psi_p^i(t^m)$ over all available particle samples is then easily defined as:

$$\langle \psi_p \rangle_{p,t} = \frac{1}{N_{p,t} N_p} \sum_{m=1}^{N_{p,t}} \sum_{i=1}^{N_p} \psi_p^i(t^m). \tag{A.11}$$

A.2 Particle-conditioned averaging

Let us first define a three dimensional averaging domain Ω_A with grid nodes $\boldsymbol{y}_{ijk}$ with indices i, j, k, with origin of the coordinate system at $\boldsymbol{y}_{ijk} = \mathbf{0}$. The averaging domain Ω_A is discretized in the same manner as the simulations, i.e. with an isotropic mesh with constant grid spacing in all spacial directions. The averaging domain can be easily shifted to the position of one of the particles. Then the instantaneous coordinate of a grid node $\boldsymbol{y}_{ijk} \in \Omega_A$ relative to the position of the l-th particle is defined as:

$$\tilde{\boldsymbol{x}}^l_{ijk}(t^m) = \boldsymbol{y}_{ijk} - \boldsymbol{x}^l_p(t^m), \qquad \forall\ m \in [1, N_{p,t}], \qquad l = 1 \dots N_p. \tag{A.12}$$

We can now define a discrete counter field $\tilde{n}^p_{ijk}(\tilde{\boldsymbol{x}}_{ijk})$ which holds the number of samples obtained through dispersed-phase averaging at given grid node in the averaging domain Ω_A at time t^m, i.e. $\tilde{\boldsymbol{x}}_{ijk} \in \Omega_A \bigcap \Omega_p(t^m)$:

$$\tilde{n}^p_{ijk}(\tilde{\boldsymbol{x}}_{ijk}) = \sum_{m=1}^{N_{p,t}} \sum_{l=1}^{N_p} \Phi_p(\tilde{\boldsymbol{x}}^l_{ijk}(t^m), t^m). \tag{A.13}$$

By proper normalization the quantity $\tilde{n}^p_{ijk}(\tilde{\boldsymbol{x}}_{ijk})$ defines the particle conditioned local solid volume fraction ϕ_s^{cond}:

$$\phi_s^{cond} = \frac{1}{N_{p,t} N_p} \tilde{n}^p_{ijk}(\tilde{\boldsymbol{x}}_{ijk}) \tag{A.14}$$

The particle-conditioned space-time averaging of a any particle related quantity $\psi^l_p(t^m)$ can be defined as follows:

$$\langle \tilde{\psi}^(_p \tilde{\boldsymbol{x}}_{ijk}) \rangle_{p,t} = \frac{1}{\tilde{n}^p_{ijk}(\tilde{\boldsymbol{x}}_{ijk})} \sum_{m=1}^{N_{p,t}} \sum_{l=1}^{N_p} \Phi_p(\tilde{\boldsymbol{x}}^l_{ijk}(t^m), t^m) \psi^l_p(t^m). \tag{A.15}$$

Analogously, for any fluid related quantity the $\psi_f(\boldsymbol{x}_{ijk}, t^m)$ we can define the particle conditioned space-time average as:

$$\langle \tilde{\psi}_f(\tilde{\boldsymbol{x}}_{ijk}) \rangle_{p,t} = \frac{1}{\tilde{n}^f_{ijk}(\tilde{\boldsymbol{x}}_{ijk})} \sum_{m=1}^{N_{f,t}} \sum_{l=1}^{N_p} \Phi_f(\tilde{\boldsymbol{x}}_{ijk}, t^m) \psi_f(\tilde{\boldsymbol{x}}_{ijk}, t^m). \tag{A.16}$$

By the way of definition the coordinates $\tilde{\boldsymbol{x}}_{ijk}(t^m)$, the grid nodes of the averaging domain Ω_A do not coincide with the grid nodes used for the discretization in the present simulations. This implies that in order to evaluate the quantity $\psi_f(\boldsymbol{x}_{ijk}, t^m)$ in equation A.16 at $\tilde{\boldsymbol{x}}_{ijk}(t^m)$ a spacial interpolation of $\psi_f(\boldsymbol{x}_{ijk}, t^m)$ is required. The tri-linear spatial interpolation is employed in the present work.

Appendix B

Flow field velocity decomposition

Here detailed derivation of the decomposition of the fluid velocity field into its *spatial* and *temporal fluctuations* is presented (cf. section §4.3.1). First let us recall the decomposition of the instantaneous velocity field with respect to the spatial, temporal and the space and time averages (cf. equations 4.1 - 4.3). For simplicity the definitions of these decompositions are repeated here:

$$\begin{aligned} \boldsymbol{u}_f(\boldsymbol{x},t) &= \langle \boldsymbol{u}_f \rangle_{\Omega_f}(t) + \boldsymbol{u}'_f(\boldsymbol{x},t) && \text{(B.1)} \\ \boldsymbol{u}_f(\boldsymbol{x},t) &= \langle \boldsymbol{u}_f \rangle_t(\boldsymbol{x}) \;\; + \boldsymbol{u}''_f(\boldsymbol{x},t) && \text{(B.2)} \\ \boldsymbol{u}_f(\boldsymbol{x},t) &= \langle \boldsymbol{u}_f \rangle_{\Omega_f,t} \;\; + \boldsymbol{u}'''_f(\boldsymbol{x},t). && \text{(B.3)} \end{aligned}$$

A detailed definitions of the averaging operators $\langle \cdot \rangle_{\Omega_f}$, $\langle \cdot \rangle_t$ and $\langle \cdot \rangle_{\Omega_f,t}$ can be found in appendix A. Now let us define the relation between the implicitly defined velocity fluctuations in equations B.1 - B.3. First the time average of equation B.1 is taken, which yields:

$$\langle \boldsymbol{u}_f \rangle_t(\boldsymbol{x}) = \langle \boldsymbol{u}_f \rangle_{\Omega_f,t} + \langle \boldsymbol{u}'_f \rangle_t(\boldsymbol{x}) \qquad \text{(B.4)}$$

Substituting relation B.2 into the above equation leads to:

$$\begin{aligned} & \boldsymbol{u}_f - \boldsymbol{u}''_f = \langle \boldsymbol{u}_f \rangle_{\Omega_f,t} + \langle \boldsymbol{u}'_f \rangle_t(\boldsymbol{x}) && \text{(B.5)} \\ \Rightarrow \quad & \langle \boldsymbol{u}_f \rangle_{\Omega_f,t} = \boldsymbol{u}_f - \boldsymbol{u}''_f - \langle \boldsymbol{u}'_f \rangle_t(\boldsymbol{x}). && \text{(B.6)} \end{aligned}$$

Substitution of equation B.6 into equation 4.3 yields the following relation:

$$\begin{aligned} \boldsymbol{u}'''_f(\boldsymbol{x},t) &= \boldsymbol{u}_f(\boldsymbol{x},t) - \langle \boldsymbol{u}_f \rangle_{\Omega_f,t} = && \text{(B.7)} \\ &\underbrace{=}_{eq.\ B.6} \underbrace{\boldsymbol{u}_f(\boldsymbol{x},t) - \langle \boldsymbol{u}_f \rangle_t(\boldsymbol{x})}_{=\boldsymbol{u}''_f(\boldsymbol{x},t),\ \text{(eq. 4.2)}} + \langle \boldsymbol{u}'_f \rangle_t(\boldsymbol{x}) && \text{(B.8)} \\ &= \langle \boldsymbol{u}'_f \rangle_t(\boldsymbol{x}) + \boldsymbol{u}''_f(\boldsymbol{x},t). && \text{(B.9)} \end{aligned}$$

Using the above relation B.9, the variance of the velocity field with respect to the time and space average can be expressed as:

$$\langle \boldsymbol{u}_f''' \boldsymbol{u}_f''' \rangle_{\Omega_f,t} \underbrace{=}_{eq.\ 4.3} \langle (\boldsymbol{u}_f - \langle \boldsymbol{u}_f \rangle_{\Omega_f,t})^2 \rangle_{\Omega_f,t} = \tag{B.10}$$

$$\underbrace{=}_{eq.\ B.9} \langle (\boldsymbol{u}_f'' + \langle \boldsymbol{u}_f' \rangle_t)^2 \rangle_{\Omega_f,t} = \tag{B.11}$$

$$= \langle \boldsymbol{u}_f'' \boldsymbol{u}_f'' + 2\boldsymbol{u}_f'' \langle \boldsymbol{u}_f' \rangle_t + \langle \boldsymbol{u}_f' \rangle_t \langle \boldsymbol{u}_f' \rangle_t \rangle_{\Omega_f,t} = \tag{B.12}$$

$$= \langle \boldsymbol{u}_f'' \boldsymbol{u}_f'' \rangle_{\Omega_f,t} + 2\langle \boldsymbol{u}_f'' \langle \boldsymbol{u}_f' \rangle_t \rangle_{\Omega_f,t} + \langle \langle \boldsymbol{u}_f' \rangle_t \langle \boldsymbol{u}_f' \rangle_t \rangle_{\Omega_f,t} = \tag{B.13}$$

$$= \langle \boldsymbol{u}_f'' \boldsymbol{u}_f'' \rangle_{\Omega_f,t} + 2\langle \underbrace{\langle \boldsymbol{u}_f'' \rangle_t}_{=0\ eq.\ 4.2} \langle \boldsymbol{u}_f' \rangle_t \rangle_{\Omega_f} + \langle \langle \boldsymbol{u}_f' \rangle_t \langle \boldsymbol{u}_f' \rangle_t \rangle_{\Omega_f,t} = \tag{B.14}$$

$$= \langle \boldsymbol{u}_f'' \boldsymbol{u}_f'' \rangle_{\Omega_f,t} + \langle \langle \boldsymbol{u}_f' \rangle_t \langle \boldsymbol{u}_f' \rangle_t \rangle_{\Omega_f,t} \tag{B.15}$$

Appendix C

Two-point correlation and one dimensional spectrum of the fluid velocity field

In this section detailed definitions of the two-point correlations and the one dimensional spectrum of the fluid velocity field, discussed in the present work, are introduced. First let us introduce the two-point spatial correlation of the fluid velocity as follows:

$$R_{ij}(\boldsymbol{r},t) \equiv \langle u_i'(\boldsymbol{x}+\boldsymbol{r},t)u_j'(\boldsymbol{x},t)\rangle_{\Omega_f} \tag{C.1}$$

where $\mathbf{r}$ is a separation vector and $\boldsymbol{u}$ is the fluid velocity field. The two-point auto-correlation is a good indicator for the adequacy of the box size, especially when dealing with periodic boundary conditions. Zero values of R_{ij} indicates that the domain under considerations is large enough to accommodate the largest flow structures in a particular direction. The value of R_{ij} at zero separation ($\boldsymbol{r}=0$) represents the covariance of the fluid velocity $\langle u_i'u_j'\rangle_{\Omega_f}$. Using the two-point correlation function one can define various length scales, e.g. the integral length scale:

$$L_{11} = \frac{1}{R_{11}(0,t)}\int_0^\infty R_{11}(\boldsymbol{e}_1 \boldsymbol{r},t)dr, \tag{C.2}$$

where $\boldsymbol{e}_1$ is the unit vector in the x-direction. For statistically homogeneous field the two-point correlation is a symmetric function with respect to the separation vector $\boldsymbol{r}$, i.e.

$$R_{ij}(\boldsymbol{r},t) = R_{ij}(-\boldsymbol{r},t). \tag{C.3}$$

The one-dimensional spectrum of the fluid velocity field can be defined as the twice the one-dimensional Fourier transform of R_{ij}:

$$E_{ij}(\kappa_1,t) = \frac{1}{\pi}\int_{-\infty}^{\infty} R_{ij}(r_1)\exp^{-\imath\kappa_1 r_1} dr_1, \tag{C.4}$$

where $\kappa_1 = \boldsymbol{e}_1\boldsymbol{\kappa}$ is the component of the wave number vector $\boldsymbol{\kappa}$ in direction of the unity vector $\boldsymbol{e}_1$.

With this the discrete one-dimensional spectra is computed as the Fourier transform of the two point correlation function $R_{ij}(\kappa)$.

$$E_{ij}(\kappa,t) = \sum_{\boldsymbol{\kappa}} \hat{R}_{ij}(\boldsymbol{\kappa}) = \sum_{\boldsymbol{\kappa}} \langle \hat{u}_i^*(\boldsymbol{\kappa})\hat{u}_j(\boldsymbol{\kappa})\rangle, \tag{C.5}$$

where $\hat{R}_{ij}(\kappa)$ denotes the Fourier transform of $R_{ij}(\boldsymbol{r})$, $\hat{u}(\kappa)$ the Fourier transform of the flow velocity u and $\hat{u}^*(\kappa)$ the complex conjugate of $\hat{u}(\kappa)$.

Since particles are present in the computational domain, at the position of the particles there is no fluid and therefore the flow velocity field is absent. Thus difficulties arise by the computation of the Fourier transform of $\hat{R}_{ij}(\kappa)$, since the velocity signal exhibits discontinuities at the particle positions. In order to avoid this the the velocity signal at the position of the particles is approximated by means of interpolation.

Appendix D

Derivation of the Energy equation for two-phase particulate flows

In this section we derive the turbulent kinetic energy for two-phase particulate flows. The flows under consideration are shear-free and homogeneous in all space directions, i.e. no mean shear gradients are present. The particles are settling in an quiescent fluid, i.e. the fluid was at rest, at the time the particles were released in the domain. As we will see later, this flow configuration allows us to make some simplifications of the kinetic energy equation. The only driving force of the flow is the gravity. This allow us to study the energy production by the dispersed phase, since the fluctuations of the flow are only due to the settling particles. The motion of the particles generates gradients of the flow velocity in vicinity of the particle surface, i.e. a boundary layer develops. Pressure gradient across the particles is developed as well. Depending on the particle Galileo/Reynolds number this results in a separation of the flow from the particle surface and particle wakes are formed. These wakes are the only perturbations of the flow field. Thus, this study is a good starting point for the study and the modeling of particulate flows and especially the contribution of the dispersed phase to the kinetic energy of the fluid.

This section is organized as follow. Before we start deriving and analyzing the energy equation for multiphase flows we briefly define the notation used in this section in section D.1. In section D.2 the kinetic energy equation single phase flow is derived. Than in section D.3 the kinetic energy equation two-phase flows in the entire domain is derived. Here the terms in the kinetic energy equation stemming from the momentum exchange between the two phases are presented/discussed. Finally in section D.4 the turbulent kinetic energy budget for two-phase flows is discussed.

D.1 Notation

In this section a brief description of the notation used in this section is presented. The instantaneous fluid velocity vector is noted by $\mathbf{U} = (U_1, U_2, U_3)$, the Pressure is denoted by $\mathbf{P}$ and the density of the fluid and the particles are denoted by ρ_f and ρ_p respectively. The kinematic viscosity of the fluid is denoted by ν. The domain occupied by the two phases is given by $\Omega = \Omega_f \cup \Omega_p$, where Ω_f and Ω_p are the domains occupied by the fluid and the particles respectively. Please note, that the fluid velocity $\mathbf{U}$ and pressure $\mathbf{P}$ are defined only in the domain Ω_f. An indicator function $\Phi_f(\mathbf{x},t)$ which is defined as in equation A.1 obeys the topological equation (Drew, 1983):

$$\frac{\partial \Phi_f}{\partial t} + U_j^I \frac{\partial \Phi_f}{x_j} = 0, \tag{D.1}$$

where U_j^I is the velocity of the fluid-particle interface. Further, for the gradient of Φ_f the following equality holds (Drew, 1983):

$$\frac{\partial \Phi_f}{\partial x_j} = -n_j^{(f)} \delta(\mathbf{x} - \mathbf{x}^I), \tag{D.2}$$

where $n_j^{(f)}$ is the unit normal vector pointing outward from the fluid surface into the particle and $\delta(\mathbf{x} - \mathbf{x}^I)$ is a generalized delta function at the fluid-particle surface $\mathbf{x}^I$.
A sample averaging operator conditioned on the presence of the fluid or solid phase is defined as:

$$\langle \phi_f \rangle_\Omega(\mathbf{x},t) = \frac{1}{\int_{\Omega_f} \Phi_f(\mathbf{x},t)dx} \int_\Omega \Phi_f(\mathbf{x},t)\phi_f(\mathbf{x},t)dx \quad \text{- carrier phase.} \tag{D.3}$$

Analogous for the dispersed phase the operator can be written as:

$$\langle \phi_p \rangle_\Omega(\mathbf{x},t) = \frac{1}{\int_{\Omega_p} (1-\Phi_f)(\mathbf{x},t)dx} \int_\Omega (1-\Phi_f)(\mathbf{x},t)\phi_p(\mathbf{x},t)dx \quad \text{- dispersed phase.} \tag{D.4}$$

Thus, the fluctuations of the fluid velocity, or any flow field quantity can be written as:

$$u_i = \Phi_f u_{f,i} = \Phi_f U_i - \langle \Phi_f U_i \rangle_\Omega. \tag{D.5}$$

Additionally, the following equality can be written:

$$\langle \phi \rangle_{\Omega_f} = \langle \Phi_f \phi \rangle_\Omega, \tag{D.6}$$

The rate of strain tensor is defined as:

$$S_{ij} = \frac{1}{2}\left(\frac{\partial \Phi_f U_i}{\partial x_j} + \frac{\partial \Phi_f U_j}{\partial x_i} \right). \tag{D.7}$$

The mean and fluctuating rates of strain are analogously defined as:

$$\bar{S}_{ij} = \langle S_{ij} \rangle_\Omega = \frac{1}{2}\left(\frac{\partial \langle \Phi_f U_i \rangle_\Omega}{\partial x_j} + \frac{\partial \langle \Phi_f U_j \rangle_\Omega}{\partial x_i} \right), \tag{D.8}$$

$$s_{ij} = S_{ij} - \bar{S}_{ij} = \frac{1}{2}\left(\frac{\partial \Phi_f u_i}{\partial x_j} + \frac{\partial \Phi_f u_j}{\partial x_i} \right). \tag{D.9}$$

The rates of dissipation due to the mean fluctuating flow are defined as:

$$\varepsilon = 2\nu \langle s_{ij} s_{ij} \rangle_\Omega. \tag{D.10}$$

The turbulent kinetic energy is defined as:

$$k = \frac{1}{2}\langle \Phi_f u_i u_i \rangle_\Omega. \tag{D.11}$$

The term describing the production of kinetic energy is given by:

$$P = -\langle \Phi_f u_i u_j \rangle_\Omega \frac{\partial \langle \Phi_f U_i \rangle_\Omega}{\partial x_j}. \tag{D.12}$$

The energy fluxes in the equations for the kinetic energy, kinetic energy of the mean flow and the turbulent kinetic energy (Pope, 2000) are defined as:

$$T_i = \frac{1}{\rho_f} \Phi_f U_i P - 2\nu \Phi_f U_j S_{ij} \tag{D.13}$$

$$T_i' = \frac{1}{2} \langle \Phi_f u_i u_j u_j \rangle_\Omega + \frac{1}{\rho_f} \langle \Phi_f u_i p \rangle_\Omega - 2\nu \langle \Phi_f u_j s_{ij} \rangle_\Omega \tag{D.14}$$

D.2 Derivation of the kinetic energy equation

The Navier-Stokes (NSE) equations for single phase flow is given by:

$$\underbrace{\frac{\partial U_i}{\partial t} + \frac{\partial U_i U_j}{\partial x_j}}_{\frac{DU_i}{Dt}} = -\frac{1}{\rho_f} \frac{\partial P}{\partial x_i} + 2\nu \frac{\partial S_{ij}}{\partial x_j} \tag{D.15}$$

Multiplication of D.15 with the velocity U_i results in:

$$\frac{DE}{Dt} = -\frac{1}{\rho_f} \frac{\partial U_i P}{\partial x_i} + 2\nu U_i \frac{\partial S_{ij}}{\partial x_j}, \tag{D.16}$$

$$\frac{DE}{Dt} + \frac{1}{\rho_f} \frac{\partial U_i P}{\partial x_i} = 2\nu \left(\frac{\partial U_i S_{ij}}{\partial x_j} - S_{ij} \frac{\partial U_i}{\partial x_j} \right), \tag{D.17}$$

where the first term on the left hand side is defind as:

$$\mathbf{U} \cdot \frac{D\mathbf{U}}{Dt} = \frac{DE}{Dt} = \frac{\partial E}{\partial t} + \nabla \cdot (\mathbf{U}E) \tag{D.18}$$

The expression in the brackets $(\cdots)$ in equation D.17 can be re-written as follow:

$$\frac{\partial U_i S_{ij}}{\partial x_j} - S_{ij} \frac{\partial U_i}{\partial x_j} = \tag{D.19}$$

$$\frac{\partial U_i S_{ij}}{\partial x_j} - S_{ij} \frac{\partial U_i}{\partial x_j} - S_{ij} S_{ij} + S_{ij} S_{ij} = \tag{D.20}$$

$$\frac{\partial U_i S_{ij}}{\partial x_j} - \underbrace{S_{ij} \left(\frac{\partial U_i}{\partial x_j} - S_{ij} \right)}_{\Omega_{ij}} - S_{ij} S_{ij} = \tag{D.21}$$

$$\frac{\partial U_i S_{ij}}{\partial x_j} - \underbrace{S_{ij} \Omega_{ij}}_{=0} - S_{ij} S_{ij} = \tag{D.22}$$

$$\frac{\partial U_i S_{ij}}{\partial x_j} - S_{ij} S_{ij} \underbrace{=}_{S_{ji} = S_{ij}} \frac{\partial U_j S_{ij}}{\partial x_i} - S_{ij} S_{ij} \tag{D.23}$$

Substituting this expression in equation D.17 and rearranging follows:

$$\frac{DE}{Dt} + \frac{\partial}{\partial x_i}\underbrace{\left(\frac{U_i P}{\rho_f} - 2\nu U_j S_{ij}\right)}_{T_i} = -2\nu S_{ij} S_{ij} \tag{D.24}$$

which is the equation for the kinetic energy as defined in (Pope, 2000).

D.3 Energy equation for particulate flows

The momentum equation D.15 is valid for single phase flow. In order to obtain an equation, which is valid for particulate flows, equation D.15 is multiplied by the indicator function defined in A.1

$$\Phi_f\left(\frac{\partial U_i}{\partial t} + \frac{\partial U_i U_j}{\partial x_j} = g_i - \frac{1}{\rho_f}\frac{\partial P}{\partial x_i} + 2\nu\frac{\partial S_{ij}}{\partial x_j}\right) \tag{D.25}$$

In the equation above the term g_i represents the volume force. Using the chain rule (e.g. $g\partial f/\partial x = \partial(fg)/\partial x - f\partial g/\partial x$) for differentials one obtains:

$$\frac{\partial \Phi_f U_i}{\partial t} - U_i\frac{\partial \Phi_f}{\partial t} + \frac{\partial \Phi_f U_i U_j}{\partial x_j} - U_i U_j\frac{\partial \Phi_f}{\partial x_j} = \Phi_f g_i - \frac{1}{\rho_f}\frac{\partial \Phi_f P}{\partial x_i} + \frac{P}{\rho_f}\frac{\partial \Phi_f}{\partial x_i} + 2\nu\left(\frac{\partial \Phi_f S_{ij}}{\partial x_j} - S_{ij}\frac{\partial \Phi_f}{\partial x_j}\right). \tag{D.26}$$

Substituting the property of the indicator function defined in equation D.1 and rearranging the terms on the left hand side the equation results in:

$$\frac{\partial \Phi_f U_i}{\partial t} + \frac{\partial \Phi_f U_i U_j}{\partial x_j} - U_i\left(U_j - U_j^I\right)\frac{\partial \Phi_f}{\partial x_j} = \Phi_f g_i - \frac{1}{\rho_f}\frac{\partial \Phi_f P}{\partial x_i} + \frac{P}{\rho_f}\frac{\partial \Phi_f}{\partial x_i} + 2\nu\left(\frac{\partial \Phi_f S_{ij}}{\partial x_j} - S_{ij}\frac{\partial \Phi_f}{\partial x_j}\right). \tag{D.27}$$

The third term on the left hand side vanishes, since the gradient of Φ_f exist only at the particle surface, where the fluid velocity U_i is equal to the fluid-particle surface velocity U_i^I and thus $U_i\left(U_j - U_j^I\right)\frac{\partial \Phi_f}{\partial x_j} = 0$ in the entire domain Ω. Thus:

$$\frac{\partial \Phi_f U_i}{\partial t} + \frac{\partial \Phi_f U_i U_j}{\partial x_j} = \Phi_f g_i - \frac{1}{\rho_f}\frac{\partial \Phi_f P}{\partial x_i} + \frac{P}{\rho_f}\frac{\partial \Phi_f}{\partial x_i} + 2\nu\left(\frac{\partial \Phi_f S_{ij}}{\partial x_j} - S_{ij}\frac{\partial \Phi_f}{\partial x_j}\right). \tag{D.28}$$

Combining the expression with the gradient of Φ_f on the right hand side, the momentum equation for particulate flows in the entire domain Ω reads as:

$$\frac{\partial \Phi_f U_i}{\partial t} + \frac{\partial \Phi_f U_i U_j}{\partial x_j} = \Phi_f g_i - \frac{1}{\rho_f}\frac{\partial \Phi_f P}{\partial x_i} + 2\nu\frac{\partial \Phi_f S_{ij}}{\partial x_j} + \left(P/\rho_f - 2\nu S_{ij}\right)\frac{\partial \Phi_f}{\partial x_j} \tag{D.29}$$

The last expression on the right hand side represents the momentum transfer at the fluid-particle interface due to pressure and viscous stresses at the particle surface.

Similar to the procedure in section D.2, we multiply equation D.29 with $\Phi_f U_i$ to obtain the equation for the kinetic energy for particulate flows in the domain Ω.

$$\Phi_f U_i\left(\underbrace{\frac{\partial \Phi_f U_i}{\partial t}}_{A} + \underbrace{\frac{\partial \Phi_f U_i U_j}{\partial x_j}}_{B} = \underbrace{\Phi_f g_i}_{C} \underbrace{- \frac{1}{\rho_f}\frac{\partial \Phi_f P}{\partial x_i}}_{D} + \underbrace{2\nu\frac{\partial \Phi_f S_{ij}}{\partial x_j}}_{E} + \underbrace{\left(P/\rho_f - 2\nu S_{ij}\right)\frac{\partial \Phi_f}{\partial x_j}}_{F}\right) \tag{D.30}$$

As next the expressions A - F in equation D.30 are considered individually. Please note, that once more the chain rule is applied.

$$\text{A:}\quad \Phi_f U_i \frac{\partial \Phi_f U_i}{\partial t} = \frac{\partial \frac{1}{2}\Phi_f U_i \Phi_f U_i}{\partial t} \underbrace{=}_{\Phi_f \cdot \Phi_f = \Phi_f} \frac{\partial \Phi_f \frac{1}{2} U_i U_i}{\partial t} = \frac{\partial \Phi_f E}{\partial t} \tag{D.31}$$

$$\text{B:}\quad \Phi_f U_i \frac{\partial \Phi_f U_i \Phi_f U_j}{\partial x_j} = \Phi_f U_i \left[\Phi_f U_j \frac{\partial \Phi_f U_i}{\partial x_j} + \Phi_f U_i \underbrace{\frac{\partial \Phi_f U_j}{\partial x_j}}_{=0} \right] = \Phi_f U_j \left[\Phi_f U_i \frac{\partial \Phi_f U_i}{\partial x_j} \right] = \tag{D.32}$$

$$= \Phi_f U_j \frac{\partial \frac{1}{2}\Phi_f U_i \Phi_f U_i}{\partial x_j} = \Phi_f U_j \frac{\partial E}{\partial x_j} = \frac{\partial \Phi_f U_j E}{\partial x_j} - \Phi_f E \underbrace{\frac{\partial \Phi_f U_j}{\partial x_j}}_{=0} \tag{D.33}$$

$$= \frac{\partial \Phi_f U_j E}{\partial x_j} \tag{D.34}$$

$$\text{C:}\quad \Phi_f g_i \tag{D.35}$$

$$\text{D:}\quad -\frac{1}{\rho_f} \Phi_f U_i \frac{\partial \Phi_f P}{\partial x_i} \underbrace{=}_{\frac{\partial U_i}{\partial x_i}=0} -\frac{1}{\rho_f} \frac{\partial \Phi_f U_i P}{\partial x_i} = \nabla \cdot \left(-\frac{1}{\rho_f} \Phi_f U P \right) \tag{D.36}$$

$$\text{E:}\quad \frac{\partial \Phi_f U_i \Phi_f S_{ij}}{\partial x_j} = \Phi_f U_i \frac{\partial \Phi_f S_{ij}}{\partial x_j} + \Phi_f S_{ij} \frac{\partial \Phi_f U_i}{\partial x_j} \tag{D.37}$$

$$\text{F:}\quad \Phi_f U_i \left(P/\rho_f - 2\nu S_{ij} \right) \frac{\partial \Phi_f}{\partial x_j} \tag{D.38}$$

Using the transformation steps from section D.2, equations D.19 - D.23, on expression D.37 and combining with expression D.36, equation D.30 simplifies to:

$$\frac{D\Phi_f E}{Dt} + \nabla \cdot \mathbf{T} = -2\nu \Phi_f S_{ij} S_{ij} + \Phi_f U_i g_i + \Phi_f U_i \left(P/\rho_f - 4\nu S_{ij} \right) \frac{\partial \Phi_f}{\partial x_j}, \tag{D.39}$$

where $\mathbf{T}$ is defined as:

$$\mathbf{T} = \frac{1}{\rho} \Phi_f U_i P - \Phi_f U_j S_{ij} \tag{D.40}$$

Using the property of the indicator function D.2, equation D.39 results in:

$$\frac{D\Phi_f E}{Dt} + \nabla \cdot \mathbf{T} = -2\nu \Phi_f S_{ij} S_{ij} + \Phi_f U_i g_i + \Phi_f U_i \left(P/\rho_f - 4\nu S_{ij} \right) \mathbf{n}_j^I \delta \left(\mathbf{x} - \mathbf{x}^I \right). \tag{D.41}$$

The last term on the right hand side represents the energy contribution trough the presence of the fluid-particle interface. This term can be split in a contribution due to pressure and a contribution due to viscous stresses.

D.4 Turbulent kinetic energy budget

In this section the turbulent kinetic energy (TKE) budget for the carrier phase is discussed. First the equation for the TKE is derived. To derive the equation for the TKE, first we derive the conservation equation for the turbulent velocity fluctuations by subtracting the mean momentum equation from the momentum equation D.29. Here the mean is performed by sample averaging. Thus, the fluctuations are with respect to the box averaged quantities.

With this, the mean momentum equation for particulate flows in the entire domain Ω is yield by applying the sample averaging operator D.3 on equation D.29:

$$\left\langle \frac{\partial \Phi_f U_i}{\partial t} + \frac{\partial \Phi_f U_i U_j}{\partial x_j} = \underbrace{\Phi_f g_i}_{f_{i,b}} - \frac{1}{\rho_f}\frac{\partial \Phi_f P}{\partial x_i} + 2\nu \frac{\partial \Phi_f S_{ij}}{\partial x_j} + \underbrace{(P/\rho_f - 2\nu S_{ij})\frac{\partial \Phi_f}{\partial x_j}}_{f_{i,p}} \right\rangle_\Omega . \quad \text{(D.42)}$$

By taking advantage of the commutative property of the gradient, equation D.42 can be written as:

$$\frac{\partial \langle \Phi_f U_i\rangle_\Omega}{\partial t} + \frac{\partial \langle \Phi_f U_i\rangle_\Omega \langle \Phi_f U_j\rangle_\Omega}{\partial x_j} + \frac{\partial \langle \Phi_f U_{\Phi_f} u_j\rangle_\Omega}{\partial x_j} = -\frac{1}{\rho_f}\frac{\partial \langle \Phi_f P\rangle_\Omega}{\partial x_i} + 2\nu \frac{\partial \langle \Phi_f S_{ij}\rangle_\Omega}{\partial x_j} + \langle f_{i,b}\rangle_\Omega + \langle f_{i,p}\rangle_\Omega . \quad \text{(D.43)}$$

Subtracting equation D.43 from equation D.42 yields:

$$\frac{\partial \Phi_f u_i}{\partial t} + \frac{\partial}{\partial x_j}\left[\Phi_f U_i U_j - (\langle \Phi_f U_i\rangle_\Omega \langle \Phi_f U_j\rangle_\Omega + \langle \Phi_f u_i \Phi_f u_j\rangle_\Omega)\right] = -\frac{1}{\rho_f}\frac{\partial \Phi_f p}{\partial x_i} + 2\nu \frac{\partial \Phi_f s_{ij}}{\partial x_j} + f'_{i,b} + f'_{i,p}, \quad \text{(D.44)}$$

where

$$\Phi_f S_{ij} = \langle \Phi_f S_{ij}\rangle_\Omega + \Phi_f s_{ij}, \quad \text{(D.45)}$$
$$f_{i,b} = \langle f_{i,b}\rangle_\Omega + f'_{i,b}, \quad \text{(D.46)}$$
$$f_{i,p} = \langle f_{i,p}\rangle_\Omega + f'_{i,p}. \quad \text{(D.47)}$$

Using the decomposition of the velocity defined in equation D.5, the expressin in the brackets on the left hand side can be expressed as follow:

$$\underbrace{\Phi_f U_i U_j}_{A} - \langle \Phi_f U_i\rangle_\Omega \langle \Phi_f U_j\rangle_\Omega - \langle \Phi_f u_i \Phi_f u_j\rangle_\Omega$$

$$\begin{aligned} A &= \left[\langle \Phi_f U_i\rangle_\Omega + \Phi_f u_i\right]\left[\langle \Phi_f U_j\rangle_\Omega + \Phi_f u_j\right] \\ &= \langle \Phi_f U_i\rangle_\Omega \langle \Phi_f U_j\rangle_\Omega + \langle \Phi_f U_i\rangle_\Omega \Phi_f u_j + \Phi_f u_i \langle \Phi_f U_j\rangle_\Omega + \Phi_f u_i \Phi_f u_j \end{aligned} \quad \text{(D.48)}$$

$$\begin{aligned} \Rightarrow A &- \langle \Phi_f U_i\rangle_\Omega \langle \Phi_f U_j\rangle_\Omega - \langle \Phi_f u_i \Phi_f u_j\rangle_\Omega = \\ &= \langle \Phi_f U_i\rangle_\Omega \langle \Phi_f U_j\rangle_\Omega + \langle \Phi_f U_i\rangle_\Omega \Phi_f u_j + \Phi_f u_i \langle \Phi_f U_j\rangle_\Omega + \Phi_f u_i \Phi_f u_j - \\ &- \langle \Phi_f U_i\rangle_\Omega \langle \Phi_f U_j\rangle_\Omega - \langle \Phi_f u_i \Phi_f u_j\rangle_\Omega = \\ &= \langle \Phi_f U_i\rangle_\Omega \Phi_f u_j + \Phi_f u_i \langle \Phi_f U_j\rangle_\Omega + \Phi_f u_i \Phi_f u_j - \langle \Phi_f u_i \Phi_f u_j\rangle_\Omega \end{aligned} \quad \text{(D.49)}$$

$$\Rightarrow (\Phi_f U_i U_j - \langle \Phi_f U_i\rangle_\Omega \langle \Phi_f U_j\rangle_\Omega - \langle \Phi_f u_i \Phi_f u_j\rangle_\Omega) = \quad \text{(D.50)}$$

$$= \langle \Phi_f U_i \rangle_\Omega \Phi_f u_j + \Phi_f u_i \langle \Phi_f U_j \rangle_\Omega + \Phi_f u_i \Phi_f u_j - \langle \Phi_f u_i \Phi_f u_j \rangle_\Omega$$

Hence applying the chain rule one obtains:

$$\frac{\partial}{\partial x_j} \left(\langle \Phi_f U_i \rangle_\Omega \Phi_f u_j + \Phi_f u_i \langle \Phi_f U_j \rangle_\Omega + \Phi_f u_i \Phi_f u_j - \langle \Phi_f u_i \Phi_f u_j \rangle_\Omega \right) = \tag{D.51}$$

$$= \Phi_f u_j \frac{\partial \langle \Phi_f U_i \rangle_\Omega}{\partial x_j} + \langle \Phi_f U_j \rangle_\Omega \frac{\partial \Phi_f u_i}{\partial x_j} + \Phi_f u_j \frac{\partial \Phi_f u_i}{\partial x_j} - \frac{\partial \langle \Phi_f u_i \Phi_f u_j \rangle_\Omega}{\partial x_j} \tag{D.52}$$

Substituting equation D.52 in equation D.44 and reorganizing one obtains:

$$\frac{D\Phi_f u_i}{Dt} = -\Phi_f u_j \frac{\partial \langle \Phi_f U_i \rangle_\Omega}{\partial x_j} + \frac{\partial \langle \Phi_f u_i \Phi_f u_j \rangle_\Omega}{\partial x_j} - \frac{1}{\rho_f} \frac{\partial \Phi_f p}{\partial x_i} + 2\nu \frac{\partial \Phi_f s_{ij}}{\partial x_j} + f'_{i,b} + f'_{i,p}, \tag{D.53}$$

where

$$\frac{D\Phi_f u_i}{Dt} = \frac{\partial \Phi_f u_i}{\partial t} + \left(\langle \Phi_f U_j \rangle_\Omega + \Phi_f u_j \right) \frac{\partial \Phi_f u_i}{\partial x_j} \tag{D.54}$$

Multiplication of equation D.54 with $\Phi_f u_i$ and applying the average operator D.3 one yields (for detailed derivation, please see appendix D.5):

$$\frac{\bar{D}k}{\bar{D}t} + \nabla \cdot \mathbf{T}' = -P - \varepsilon + \underbrace{\langle \Phi_f u_i f'_{i,b} \rangle_\Omega}_{\Psi_b} + \underbrace{\langle \Phi_f u_i f'_{i,p} \rangle_\Omega}_{\Psi_p}, \tag{D.55}$$

In equation D.55 $k = \langle \Phi_f u_i u_i \rangle_\Omega / 2$ is the turbulent kinetic energy (TKE) averaged over the entire domain, $\varepsilon = 2\nu \langle s_{ij} s_{ij} \rangle_\Omega$ is the average dissipation rate, P is the production of turbulent kinetic energy, Ψ_b is the buoyancy force, which represents the work done by the gravitational potential and Ψ_p is the correlation between the fluid velocity and the forcing term due to the presence of the particles. The definitions of s_{ij}, P and $\mathbf{T}'$ are given in section D.1. Since the flow under consideration here is homogeneous in all three directions, the production term and the energy flux term disappear $P = 0$, $\nabla \cdot \mathbf{T}' = 0$. Thus the final form of the turbulent kinetic energy equation is:

$$\frac{\bar{D}k}{\bar{D}t} = -\varepsilon + \Psi_b + \Psi_p \tag{D.56}$$

The buoyancy term Ψ_b is defined as follow:

$$\Psi_b = w_{rel} \Phi_s \left(\frac{\rho_p}{\rho_f} - 1 \right) g_z, \tag{D.57}$$

where w_{rel} is the averaged relative velocity between the two phases $w_{rel} = \langle w^{(p)} \rangle_\Omega - \langle w \rangle_\Omega$ and Φ_s is the solid volume fraction. The contribution to the TKE stemming from the dispersed phase is defined as follow:

$$\Psi_p = \langle \Phi_f u_i f'_{i,p} \rangle_\Omega \tag{D.58}$$

where f_i^p is the force which imposes the rigid body motion of the particles on the fluid. For statistically steady state the term on the left hand side vanishes and the equation becomes the following form:

$$\varepsilon = \Psi_b + \Psi_p \tag{D.59}$$

As can be seen the contributions to the TKE from the dispersed phase is compensated by the dissipation rate of TKE, which is expected. For heavy particles ($\rho_p/\rho_f > 1$) and gravity acting in the opposite third space direction the first term on the right hand side of equation D.59 is always positive and acts as a source of TKE. Considering the term Ψ_p no *a priory* allegation of the its sign can be made. Further analysis is required to determine whether Ψ_p acts as sink or source of TKE.

D.5 Extended derivation

Derivation of equation D.55. Equation D.54 can be simplified as follow:

$$\frac{\partial \Phi_f u_i}{\partial t} + (\langle \Phi_f U_j \rangle_\Omega + \Phi_f u_j) \frac{\partial \Phi_f u_i}{\partial x_j} = -\Phi_f u_j \frac{\partial \langle \Phi_f U_i \rangle_\Omega}{\partial x_j} + \frac{\partial \langle \Phi_f u_i u_j \rangle_\Omega}{\partial x_j} - \frac{1}{\rho_f} \frac{\partial \Phi_f p}{\partial x_i} + $$
$$= +2\nu \frac{\partial s_{ij}}{\partial x_j} + f'_{i,b} + f'_{i,p}, \quad \text{(D.60)}$$

$$\frac{\partial \Phi_f \frac{1}{2} u_i u_i}{\partial t} + \langle \Phi_f U_j \rangle_\Omega \frac{\partial \Phi_f \frac{1}{2} u_i u_i}{\partial x_j} + \frac{\partial \Phi_f \frac{1}{2} u_j u_i u_i}{\partial x_j} = -\Phi_f u_i u_j \frac{\partial \langle \Phi_f U_i \rangle_\Omega}{\partial x_j} + \Phi_f u_i \frac{\partial \langle \Phi_f u_i u_j \rangle_\Omega}{\partial x_j} \quad \text{(D.61)}$$

$$- \frac{1}{\rho_f} \frac{\partial \Phi_f u_i p}{\partial x_i} + +2\nu \Phi_f u_i \frac{\partial s_{ij}}{\partial x_j}$$
$$+ \Phi_f u_i f'_{i,b} + \Phi_f u_i f'_{i,p} \quad \text{(D.62)}$$

Taking the mean of the above equation and using $\langle u_i \rangle_\Omega = 0$ one obtains:

$$\frac{\partial k}{\partial t} + \langle \Phi_f U_j \rangle_\Omega \frac{\partial k}{\partial x_j} + \frac{\partial \Phi_f \frac{1}{2} u_j u_i u_i}{\partial x_j} = -\langle \Phi_f u_i u_j \rangle_\Omega \frac{\partial \langle \Phi_f U_i \rangle_\Omega}{\partial x_j} - \frac{1}{\rho_f} \frac{\partial \langle \Phi_f u_i p \rangle_\Omega}{\partial x_i} +$$
$$+ 2\nu \left\langle \Phi_f u_i \frac{\partial s_{ij}}{\partial x_j} \right\rangle_\Omega + \langle \Phi_f u_i f'_{i,b} \rangle_\Omega + \langle \Phi_f u_i f'_{i,p} \rangle_\Omega . \quad \text{(D.63)}$$

The third term on the right hand side can be written as:

$$2\nu \left\langle \Phi_f u_i \frac{\partial \Phi_f s_{ij}}{\partial x_j} \right\rangle_\Omega = 2\nu \left[\left\langle \frac{\partial \Phi_f u_i s_{ij}}{\partial x_j} \right\rangle_\Omega - \left\langle s_{ij} \frac{\partial \Phi_f u_i}{\partial x_j} \right\rangle \right] = \quad \text{(D.64)}$$

$$= 2\nu \left[\frac{\partial \langle \Phi_f u_i s_{ij} \rangle_\Omega}{\partial x_j} - \left\langle s_{ij} \frac{\partial \Phi_f u_i}{\partial x_j} - s_{ij} s_{ij} + s_{ij} s_{ij} \right\rangle_\Omega \right] = \quad \text{(D.65)}$$

$$= 2\nu \left[\frac{\partial \langle \Phi_f u_i s_{ij} \rangle_\Omega}{\partial x_j} - \left\langle s_{ij} \left(\frac{\partial \Phi_f u_i}{\partial x_j} - \frac{1}{2} \frac{\partial \Phi_f u_i}{\partial x_j} - \frac{1}{2} \frac{\partial \Phi_f u_i}{\partial x_i} \right) + s_{ij} s_{ij} \right\rangle_\Omega \right] = \quad \text{(D.66)}$$

$$= 2\nu \left[\frac{\partial \langle \Phi_f u_i s_{ij} \rangle_\Omega}{\partial x_j} - \left\langle s_{ij} \underbrace{\left(\frac{1}{2} \frac{\partial \Phi_f u_i}{\partial x_j} - \frac{1}{2} \frac{\partial \Phi_f u_i}{\partial x_i} \right)}_{\Omega_{ij}} + s_{ij} s_{ij} \right\rangle_\Omega \right] = \quad \text{(D.67)}$$

$$= 2\nu \left[\frac{\partial \langle \Phi_f u_i s_{ij} \rangle_\Omega}{\partial x_j} - \left\langle \underbrace{s_{ij}\Omega_{ij}}_{=0} + s_{ij}s_{ij} \right\rangle_\Omega \right] = \tag{D.68}$$

$$= 2\nu \left[\frac{\partial \langle \Phi_f u_i s_{ij} \rangle_\Omega}{\partial x_j} - \langle s_{ij}s_{ij} \rangle_\Omega \right] = \tag{D.69}$$

$$= 2\nu \frac{\partial \langle \Phi_f u_i s_{ij} \rangle_\Omega}{\partial x_j} - 2\nu \langle s_{ij}s_{ij} \rangle_\Omega = \tag{D.70}$$

$$= 2\nu \frac{\partial \langle \Phi_f u_i s_{ij} \rangle_\Omega}{\partial x_j} - \varepsilon \tag{D.71}$$

Substituting this in equation D.63 yields:

$$\underbrace{\frac{\partial k}{\partial t} + \langle \Phi_f U_j \rangle_\Omega \frac{\partial k}{\partial x_j}}_{\frac{\bar{D}k}{\bar{D}t}} + \frac{\partial}{\partial x_j} \underbrace{\left(\langle \frac{1}{2} \Phi_f u_j u_i u_i \rangle_\Omega + \frac{1}{\rho_f} \langle \Phi_f u_i p \rangle_\Omega - 2\nu \langle \Phi_f u_i s_{ij} \rangle_\Omega \right)}_{T_i'} =$$

$$\underbrace{- \langle \Phi_f u_i u_j \rangle_\Omega \frac{\partial \langle \Phi_f U_i \rangle_\Omega}{\partial x_j}}_{P} - \varepsilon + \underbrace{\langle \Phi_f u_i f_{i,b}' \rangle_\Omega}_{\Psi_b} + \underbrace{\langle \Phi_f u_i f_{i,p}' \rangle_\Omega}_{\Psi_p} \tag{D.72}$$

$$\Rightarrow \qquad \frac{\bar{D}k}{\bar{D}t} + \nabla \cdot \mathbf{T}' = P - \varepsilon + \Psi_b + \Psi_p. \tag{D.73}$$

Please note that according to definition D.5 the fluid velocity and pressure fluctuations are defined in the domain Ω and that for simplicity the indicator function is omitted.

Bibliography

Ahmed, A. and Elghobashi, S. (2001). Direct numerical simulation of particle dispersion in homogeneous turbulent shear flows. *Phys. Fluids*, 13:3346–3364.

Aliseda, A., Cartellier, A., Hainaux, F., and Lasheras, J. C. (2002). Effect of preferential concentration on the settling velocity of heavy particles in homogeneous isotropic turbulence. *J. Fluid Mech.*, 486:77–105.

Amoura, Z., Roig, V., Risso, F., and Billet, A.-M. (2010). Attenuation of the wake of a sphere in an intense incident turbulence with large length scales. *Phys. Fluids*, 22(5):055105.

Ayyalasomayajula, S., Gylfason, a., Collins, L., Bodenschatz, E., and Warhaft, Z. (2006). Lagrangian Measurements of Inertial Particle Accelerations in Grid Generated Wind Tunnel Turbulence. *Phys. Rev. Lett.*, 97(14):1–4.

Bagchi, P. and Balachandar, S. (2003). Effect of turbulence on the drag and lift of a particle. *Phys. Fluids*, 15(11):3496–3513.

Bagchi, P. and Balachandar, S. (2004). Response of the wake of an isolated particle to an isotropic turbulent flow. *J. Fluid Mech.*, 518:95–123.

Balachandar, S. and Eaton, J. K. (2010). Turbulent Dispersed Multiphase Flow. *Ann. Rev. Fluid Mech.*, 42(1):111–133.

Basset, A. (1888). On the motion of a sphere in a viscous liquid. *Phil. Trans. Roy. Soc. London*, 179(1888):43–63.

Batchelor, G. K. (1967). *An Introduction to Fluid Dynamics*. Cambridge University Press.

Batchelor, G. K. (1972). Sedimentation in a dilute dispersion of spheres. *J. Fluid Mech.*, 52(2):245–268.

Bec, J., Biferale, L., Boffetta, G., Celani, A., Cencini, M., Lanotte, a. S., Musacchio, S., and Toschi, F. (2006). Acceleration statistics of heavy particles in turbulence. *J. Fluid Mech.*, 550:349.

Bec, J., Biferale, L., Cencini, M., Lanotte, A., Musacchio, S., and Toschi, F. (2007). Heavy Particle Concentration in Turbulence at Dissipative and Inertial Scales. *Phys. Rev. Lett.*, 98(8):1–4.

Beetstra, R., Hoef, M. A. V. D., and Kuipers, J. A. M. (2007). Drag Force of Intermediate Reynolds Number Flow Past Mono- and Bidisperse Arrays of Spheres. 53(2):489–501.

Bellani, G. and Variano, E. A. (2012a). Slip velocity of large neutrally buoyant particles in turbulent flows. *New J. Phys.*, 14(12):125009.

Bellani, G. and Variano, E. a. (2012b). Slip velocity of large neutrally buoyant particles in turbulent flows. *New J. Phys.*, 14(12):125009.

Biesheuvel, A. and Wijngaarden, L. (1984). Two-phase flow equations for a dilute dispersion of gas bubbles in liquid. *J. Fluid Mech.*, 148:301–318.

Boffetta, G., De Lillo, F., and Gamba, A. (2004). Large scale inhomogeneity of inertial particles in turbulent flows. *Phys. Fluids*, 16(4):L20.

Boivin, M., Simonin, O., and Squires, K. (1998). Direct numerical simulation of turbulence modulation by particles in isotropic turbulence. *J. Fluid Mech.*, 375:235–263.

Bosse, T., Kleiser, L., and Meiburg, E. (2006). Small particles in homogeneous turbulence: Settling velocity enhancement by two-way coupling. *Phys. Fluids*, 18(2):027102.

Bouchet, G., Mebarek, M., and Dušek, J. (2006). Hydrodynamic forces acting on a rigid fixed sphere in early transitional regimes. *Eur. J. Mech. - B/Fluids*, 25(3):321–336.

Brown, P. and Lawler, D. (2003). Sphere drag and settling velocity revisited. *J. Environ. Eng.*, 129(3):222–231.

Brucato, A., Grisafi, F., and Montante, G. (1998). Particle drag coefficients in turbulent fluids. *Chem. Eng. Sci.*, 53(18):3295–3314.

Buckingham, E. (1914). On Physically Similar Systems; Illustrations of the Use of Dimensional Equations. *Phys. Rev.*, 4(4):345–376.

Bunner, B. and Tryggvason, G. (1999). Direct numerical simulations of three-dimensional bubbly flows. *Phys. Fluids*, 11(8):1967.

Bunner, B. and Tryggvason, G. (2002). Dynamics of homogeneous bubbly flows Part 2. Velocity fluctuations. *J. Fluid Mech.*, 466:53–84.

Burton, T. M. and Eaton, J. K. (2005). Fully resolved simulations of particle-turbulence interaction. Technical Report July.

Caflisch, R. E. and Luke, J. H. C. (1985). Variance in the sedimentation speed of a suspension. *Phys. Fluids*, 28(3):759.

Calzavarini, E., Volk, R., Bourgoin, M., Levêque, E., Pinton, J.-F., and Toschi, F. (2009a). Acceleration statistics of finite-sized particles in turbulent flow: the role of Faxen forces. *J. Fluid Mech.*, 630(10):179–189.

Calzavarini, E., Volk, R., Bourgoin, M., Levêque, E., Pinton, J.-F., and Toschi, F. (2009b). Acceleration statistics of finite-sized particles in turbulent flow: the role of Faxen forces. *J. Fluid Mech.*, 630:179–189.

Cartellier, A., Andreotti, M., and Sechet, P. (2009). Induced agitation in homogeneous bubbly flows at moderate particle Reynolds number. *Phys. Rev. E*, 80(6):1–4.

Cartellier, A. and Riviére, N. (2001). Bubble-induced agitation and microstructure in uniform bubbly flows at small to moderate particle Reynolds numbers. *Phys. Fluids*, 13(8):2165.

Chan-Braun, C., García-Villalba, M., and Uhlmann, M. (2011). Force and torque acting on particles in a transitionally rough open-channel flow. *J. Fluid Mech.*, 684:441–474.

Chen, J.-H. and Faeth, G. M. (2000). Interwake turbulence properties of homogeneous dilute particle-laden flows. *AIAA J.*, 38(6):995–1001.

Chen, J.-H., Wu, J., and Faeth, G. M. (2000). Turbulence generation in homogeneous particle-laden flows. *AIAA J.*, 38(4).

Chun, J., Koch, D. L., Rani, S. L., Ahluwalia, A., and Collins, L. (2005). Clustering of aerosol particles in isotropic turbulence. *J. Fluid Mech.*, 536:219–251.

Clift, R., Grace, J., and Weber, M. (1978). *Bubbles, drops, and particles.* Academic Press.

Climent, E. and Maxey, M. (2003). Numerical simulations of random suspensions at finite Reynolds numbers. *Int. J. Multiph. Flow*, 29(4):579–601.

Coleman, S. W. and Vassilicos, J. C. (2009). A unified sweep-stick mechanism to explain particle clustering in two- and three-dimensional homogeneous, isotropic turbulence. *Phys. Fluids*, 21(11):113301.

Courant, R., Friedrichs, K., and Lewy, H. (1928). Über die partiellen Differenzengleichungen der mathematischen Physik. *Math. Ann.*, 100(1):32–74.

Crowe, C. T. (2000). On models for turbulence modulation in fluid-particle flows. *Int. J. Multiph. Flow*, 26:719–727.

Crowe, C. T., Schwarzkopf, J., Sommerfeld, M., and Tsuji, Y. (2011). *Multiphase Flows with Droplets and Particles, Second Edition.* CRC Press.

Csanady, G. (1963). Turbulent diffusion of heavy particles in the atmosphere. *J. Atmos. Sci.*, 20:201–208.

Currie, I. G. (2002). *Fundamental Mechanics of Fluids, Third Edition.* Dekker Mechanical Engineering. Taylor & Francis.

De Jong, J., Salazar, J. P. L. C., Woodward, S., Collins, L., and Meng, H. (2010). Measurement of inertial particle clustering and relative velocity statistics in isotropic turbulence using holographic imaging. *Int. J. Multiph. Flow*, 36(4):324–332.

Dejoan, A. and Monchaux, R. (2013). Preferential concentration and settling of heavy particles in homogeneous turbulence. *Phys. Fluids*, 25(1):013301.

Doroodchi, E., Evans, G., Schwarz, M., Lane, G., Shah, N., and Nguyen, a. (2008). Influence of turbulence intensity on particle drag coefficients. *Chem. Eng. J.*, 135(1-2):129–134.

Drew, D. A. (1983). *Mathematical Modeling of Two-Phase Flow*, volume 15.

Druzhinin, O. A. and Elghobashi, S. (1999). On the decay rate of isotropic turbulence laden with microparticles. *Phys. Fluids*, 11:602–610.

Dušek, J., Gal, P. L., and Fraunié, P. (1994). A numerical and theoretical study of the first Hopf bifurcation in a cylinder wake. *J. Fluid Mech.*, 264:59–80.

Eames, I., Johnson, P. B. P., Roig, V., and Risso, F. (2011). Effect of turbulence on the downstream velocity deficit of a rigid sphere. *Phys. Fluids*, 23(9):095103.

Eaton, J. K. (1994). Experiments and simulations on turbulence modification by dispersed particles. *Appl. Mech. Rev.*

Eaton, J. K. (2009). Two-way coupled turbulence simulations of gas-particle flows using point-particle tracking. *Int. J. Multiph. Flow*, 35(9):792–800.

Eaton, J. K. and Fessler, J. R. (1994). Preferential concentration of particles by turbulence. *Int. J. Multiph. Flow*, 20:169–209.

Eaton, J. K. and Rouson, D. W. I. (2001). On the preferential concentration of solid particles in turbulent channel flow. *J. Fluid Mech.*, 428:149–169.

Eaton, J. K., Wood, A. M., and Hwang, W. (2005). Preferential concentration of particles in homogeneous and isotropic turbulence. *Int. J. Multiph. Flow*, 31:1220–1230.

Elghobashi, S. (1994). On Predicting Particle-Laden Turbulent Flows. *Appl. Sci. Res.*, 52:309–329.

Elghobashi, S. (2006). An updated classification map of particle-laden turbulent flows. *Proc. IUTAM Symp. Comput. Multiph. Flow*, pages 3–10.

Elghobashi, S. and Truesdell, G. (1993). On the two-way interaction between homogeneous turbulence and dispersed solid particles. I: Turbulence modification. *Phys. Fluids*, 5(7):1790–1801.

Ellingsen, K. and Risso, F. (2001). On the rise of an ellipsoidal bubble in water: oscillatory paths and liquid-induced velocity. *J. Fluid Mech.*, 440:235–268.

Ergun, S. (1952). Fluid flow through packed columns. *Chem. Eng. Prog.*, 48:89–94.

Ern, P., Risso, F., Fabre, D., and Magnaudet, J. (2012). Wake-Induced Oscillatory Paths of Bodies Freely Rising or Falling in Fluids. *Annu. Rev. Fluid Mech.*, 44(1):97–121.

Esmaeeli, A. and Tryggvason, G. (2005). A direct numerical simulation study of the buoyant rise of bubbles at O(100) Reynolds number. *Phys. Fluids*, 17(9):093303.

Faxén, H. (1922). Der Widerstand gegen die Bewegung einer starren Kugel in einer zähen Flüssigkeit, die zwischen zwei parallelen ebenen Wänden eingeschlossen ist. *Ann. Phys.*, 373(10):89–119.

Felice, R. D. (1999). The sedimentation velocity of dilute suspensions of nearly monosized spheres. *Int. J. Multiph. flow*, 25:559–574.

Felice, R. D., Gibilaro, L., and Foscolo, P. (1995). On the hindered settling velocity of spheres in the inertial flow regime. *Chem. Eng. Sci.*, 50(18):3005–3006.

Ferenc, J.-S. and Néda, Z. (2007). On the size distribution of Poisson Voronoi cells. *Phys. A Stat. Mech. its Appl.*, 385(2):518–526.

Ferrante, A. and Elghobashi, S. (2003). On the physical mechanisms of two-way coupling in particle-laden isotropic turbulence. *Phys. Fluids*, 15(2):315.

Ferziger, J. H. and Perić, M. (1996). *Computational Methods for Fluid Dynamics.* Numerical methods: Research and development. Springer-Verlag GmbH.

Fessler, J. R., Kulick, J. D., and Eaton, J. K. (1994). Preferential concentration of heavy particles in a turbulent channel flow. *Phys. Fluids*, 6(11):3742.

Fiabane, L., Volk, R., Pinton, J.-F., Monchaux, R., Cartellier, A., and Bourgoin, M. (2012). Do finite size neutrally buoyant particles cluster? *arXiv Prepr.*, pages 1–6.

Frisch, U. (1995). *Turbulence: The Legacy of A. N. Kolmogorov.* Cambridge University Press.

Galileo, G. (1612). Discourse on Bodies that Stay atop Water of Move within It.

Gao, H., Li, H., and Wang, L. (2013). Lattice Boltzmann simulation of turbulent flow laden with finite-size particles. *Comput. Math. with Appl.*, 65(2):194–210.

García-Villalba, M., Kidanemariam, A. G., and Uhlmann, M. (2012). DNS of vertical plane channel flow with finite-size particles: Voronoi analysis, acceleration statistics and particle-conditioned averaging. *Int. J. Multiph. Flow*, 46:54–74.

Gatignol, R. (1983). The Faxen formulae for a rigid particle in an unsteady non-uniform stokes flow. *J. Mec. Theor. Appl.*, 1(2):143–160.

Geiss, S., Dreizler, a., Stojanovic, Z., Chrigui, M., Sadiki, a., and Janicka, J. (2004). Investigation of turbulence modification in a non-reactive two-phase flow. *Exp. Fluids*, 36(2):344–354.

Ghidersa, B. and Dušek, J. (2000). Breaking of axisymmetry and onset of unsteadiness in the wake of a sphere. *J. Fluid Mech.*, 423:33–69.

Glowinski, R., Pan, T., Hesla, T., Joseph, D. D., and Periaux, J. (1999). A distributed Lagrange multiplier/fictitious domain method for flows around moving rigid bodies: application to particulate flow. *Int. J. Numer. Meth. Fluids*, 1066(March 1998):1043–1066.

Gore, R. A. and Crowe, C. T. (1989). Effect of particle size on modulating turbulent intensity. *Int. J. Multiph. Flow*, 15(2):279–285.

Goto, S. and Vassilicos, J. (2008). Sweep-Stick Mechanism of Heavy Particle Clustering in Fluid Turbulence. *Phys. Rev. Lett.*, 100(5):054503.

Goto, S. and Vassilicos, J. C. (2006). Self-similar clustering of inertial particles and zero-acceleration points in fully developed two-dimensional turbulence. *Phys. Fluids*, 18(11):115103.

Guala, M., Liberzon, A., Hoyer, K., Tsinober, A., and Kinzelbach, W. (2008). Experimental study on clustering of large particles in homogeneous turbulent flow. *J. Turbul.*, 9(919644653).

Guazzelli, E. and Hinch, J. (2011). Fluctuations and Instability in Sedimentation. *Annu. Rev. Fluid Mech.*, 43(1):97–116.

Hetsroni, G. (1989). Particles-turbulence interaction. *Int. J. Multiph. Flow*, 15(5):735–746.

Hill, R. J., Koch, D. L., and Ladd, A. J. C. (2001a). Moderate-Reynolds-number flows in ordered and random arrays of spheres. *J. Fluid Mech.*, 448:243–278.

Hill, R. J., Koch, D. L., and Ladd, A. J. C. (2001b). The first effects of fluid inertia on flows in ordered and random arrays of spheres. *J. Fluid Mech.*, 448(November 2001):213–241.

Hinze, J. O. (1971). Turbulent fluid and particle interaction.

Hinze, J. O. (1975). *Turbulence*. McGraw-Hill classic textbook reissue series. McGraw-Hill.

Homann, H. and Bec, J. (2009). Finite-size effects in the dynamics of neutrally buoyant particles in turbulent flow. *J. Fluid Mech.*, 651(2010):81–91.

Horowitz, M. and Williamson, C. H. K. (2010). The effect of Reynolds number on the dynamics and wakes of freely rising and falling spheres. *J. Fluid Mech.*, 651:251.

Hu, H., Fortes, A., and Joseph, D. (1993). Experiments and direct simulations of fluid particle motions. *Intern. Video J. Engng Res.*, 2:17–24.

Hu, H. H., Patankar, N. A., and Zhu, M. (2001). Direct numerical simulations of fluid-solid systems using the arbitrary Lagrangian-Eulerian technique. *J. Comput. Phys.*, 169(2):427–462.

Hussainov, M. and Kartushinsky, A. (2000). Experimental investigation of turbulence modulation by solid particles in a grid-generated vertical flow. ... *of Heat Fluid Flow*, 21:365–373.

Hwang, W. and Eaton, J. K. (2006). *Homogeneous and isotropic turbulence modulation by small heavy ($St \sim 50$) particles*, volume 564.

Iliopoulos, I., Mito, Y., and Hanratty, T. J. (2003). A stochastic model for solid particle dispersion in a nonhomogeneous turbulent field. *Int. J. Multiph. Flow*, 29(3):375–394.

Ishihara, T., Gotoh, T., and Kaneda, Y. (2009). Study of High-Reynolds Number Isotropic Turbulence by Direct Numerical Simulation. *Annu. Rev. Fluid Mech.*, 41(1):165–180.

Jenny, M., Bouchet, G., and Dušek, J. (2003). Nonvertical ascension or fall of a free sphere in a Newtonian fluid. *Phys. Fluids*, 15(1):L9.

Jenny, M., Dušek, J., and Bouchet, G. (2004). Instabilities and transition of a sphere falling or ascending freely in a Newtonian fluid. *J. Fluid Mech.*, 508:201–239.

Jeong, J. and Hussain, F. (1995). On the identification of a vortex. *J. Fluid Mech.*, 285:69–94.

Johnson, T. a. and Patel, V. C. (1999). Flow past a sphere up to a Reynolds number of 300. *J. Fluid Mech.*, 378:19–70.

Kajishima, T. (2004a). Influence of particle rotation on the interaction between particle clusters and particle-induced turbulence. *Int. J. Heat Fluid Flow*, 25(5):721–728.

Kajishima, T. (2004b). Numerical Investigation of Collective Behavior of Gravitationally-Settling Particles in a Homogeneous Field. In *5th Int. Conf. Multiph. Flow, ICMFÂ´04*.

Kajishima, T. and Takiguchi, S. (2002). Interaction between particle clusters and particle-induced turbulence. *Int. J. Heat Fluid Flow*, 23(5):639–646.

Kaneda, Y. (1986). The drag on a sparse random array of fixed spheres in flow at small but finite Reynolds number. *J. Fluid Mech.*, 167:455–463.

Kidanemariam, A. G., Chan-Braun, C., Doychev, T., and Uhlmann, M. (2013). DNS of horizontal open channel flow with finite-size, heavy particles at low solid volume fraction. *New J. Phys.*, 15(2):025031.

Kim, I., Elghobashi, S., and Sirignano, W. (1993). Three-dimensional flow over two spheres placed side by side. *J. Fluid Mech.*, 246:465–488.

Kim, J. and Balachandar, S. (2011). Mean and fluctuating components of drag and lift forces on an isolated finite-sized particle in turbulence. *Theor. Comput. Fluid Dyn.*, 26(1-4):185–204.

Koch, D. L. (1993). Hydrodynamic diffusion in dilute sedimenting suspensions at moderate Reynolds numbers. *Phys. Fluids A Fluid Dyn.*, 5(5):1141.

Koch, D. L. and Hill, R. (2001). Inertial effects in suspension and porous-media flows. *Annu. Rev. Fluid Mech.*, 33:619–647.

Koch, D. L. and Shaqfeh, E. S. G. (1991). Screening in sedimenting suspensions. *J. Fluid Mech.*, 224:275–303.

Kolmogorov, A. N. (1941). The Local Structure of Turbulence in Incompressible Viscous Fluid for Very Large Reynolds' Numbers. In *Dokl. Akad. Nauk SSSR*, volume 30, pages 301–305.

Lance, M. and Bataille, J. (1991). Turbulence in the liquid phase of a uniform bubbly air-water flow. *J. Fluid Mech.*, 222:95–118.

Lee, H. and Balachandar, S. (2010). Drag and lift forces on a spherical particle moving on a wall in a shear flow at finite Re. *J. Fluid Mech.*, 657:89–125.

Lee, K., Faeth, G. M., and Chen, J.-H. (2003). Properties of particle-generated turbulence in the final-decay period. *AIAA J.*, 41(7):1332–1340.

Legendre, D., Merle, A., and Magnaudet, J. (2006). of a spherical bubble or a solid sphere set fixed in a turbulent environment. *Phys. Fluids*, 18:48102–48104.

Lovalenti, P. M. and Brady, J. F. (1993). The hydrodynamic force on a rigid particle undergoing arbitrary time-dependent motion at small Reynolds number. *J. Fluid Mech.*, 256:561–605.

Lucci, F., Ferrante, A., and Elghobashi, S. (2010). Modulation of isotropic turbulence by particles of Taylor length-scale size. *J. Fluid Mech.*, 650:5.

Magarvey, R. H. and Bishop, R. L. (1961). Wakes in Liquid-Liquid Systems. *Phys. Fluids*, 4(7):800.

Martínez-Mercado, J., Chehata Gomez, D., Van Gils, D., Sun, C., and Lohse, D. (2010). On bubble clustering and energy spectra in pseudo-turbulence. *J. Fluid Mech.*, 650(2010):287.

Martínez-Mercado, J., Palacios-Morales, C. a., and Zenit, R. (2007). Measurement of pseudoturbulence intensity in monodispersed bubbly liquids for $10 < Re < 500$. *Phys. Fluids*, 19(10):103302.

Martínez-Mercado, J., Prakash, V. N., Tagawa, Y., Sun, C., and Lohse, D. (2012). Lagrangian statistics of light particles in turbulence. *Phys. Fluids*, 24(5):055106.

Maxey, M. (1987). The gravitational settling of aerosol particles in homogeneous turbulence and random flow fields. *J. Fluid Mech.*, 174:441–465.

Maxey, M. and Riley, J. (1983). Equation of motion for a small rigid sphere in a nonuniform flow. *Phys. Fluids*, 26(4):883–889.

Mazzitelli, I. and Lohse, D. (2009). Evolution of energy in flow driven by rising bubbles. *Phys. Rev. E*, 79(6):066317.

Mazzitelli, I., Lohse, D., and Toschi, F. (2003). The effect of microbubbles on developed turbulence. *Phys. Fluids*, 15(1):L5.

Melheim, J. a. (2005). Cluster integration method in Lagrangian particle dynamics. *Comput. Phys. Commun.*, 171(3):155–161.

Mendez-Diaz, S., Serrano-García, J. C., Zenit, R., and Hernández-Cordero, J. a. (2013). Power spectral distributions of pseudo-turbulent bubbly flows. *Phys. Fluids*, 25(4):043303.

Merle, A., Legendre, D., and Magnaudet, J. (2005). Forces on a high-Reynolds-number spherical bubble in a turbulent flow. *J. Fluid Mech.*, 532:53–62.

Michaelides, E. (2006). *Particles, Bubbles & Drops: Their Motion, Heat And Mass Transfer*. World Scientific Publ.

Mittal, R. (2000). Response of the Sphere Wake to Freestream Fluctuations. *Theor. Comput. Fluid Dyn.*, 13:397–419.

Mittal, R. and Iaccarino, G. (2005). Immersed boundary methods. *Annu. Rev. Fluid Mech.*, 37(1):239–261.

Mizukami, M., Parthasarathy, R., and Faeth, G. M. (1992). Particle-generated turbulence in homogeneous dilute dispersed flows. *Int. J. Multiph. Flow*, 18(3):397–412.

Moin, P. and Mahesh, K. (1998). DIRECT NUMERICAL SIMULATION: A Tool in Turbulence Research. *Annu. Rev. Fluid Mech.*, 30(1):539–578.

Monchaux, R., Bourgoin, M., and Cartellier, A. (2010). Preferential concentration of heavy particles: A Voronoï analysis. *Phys. Fluids*, 22(10):103304.

Monchaux, R., Bourgoin, M., and Cartellier, A. (2012). Analyzing preferential concentration and clustering of inertial particles in turbulence. *Int. J. Multiph. Flow*, 40:1–18.

Mordant, N., Crawford A. M., and Bodenschatz, E. (2004). Three-Dimensional Structure of the Lagrangian Acceleration in Turbulent Flows. *Phys. Rev. Lett.*, 93(21):214501.

Mudde, R., Groen, J., and Van Den Akker, H. (1997). Liquid velocity field in a bubble column: LDA experiments. *Chem. Eng. Sci.*, 52(21-22):4217–4224.

Natarajan, R. and Acrivos, A. (1993). The instability of the steady flow past spheres and disks. *J. Fluid Mech.*, 254:323–344.

Newton, I. (1687). *Philosophiae Naturalis Principia Mathematica*. Berkeley: Univ. Calif. Press, 1999 edition.

Nicolai, H. and Guazzelli, E. (1995). Effect of the vessel size on the hydrodynamic diffusion of sedimenting spheres. *Phys. Fluids*, 7(1):3.

Nielsen, P. (1993). Turbulence Effects on the Settling of Suspended Particles. *SEPM J. Sediment. Res.*, Vol. 63(5):835–838.

Nishino, K. and Matsushita, H. (2004). Columnar particle accumulation in homogeneous turbulence. *Proc. ICMF04 (5th Int. Conf. Multiph. Flow)*, (248):1–12.

Nyquist, H. (2002). Certain topics in telegraph transmission theory. *Proc. IEEE*, 90(2):280–305.

Obligado, M. (2013). *Fluid-particle interactions :*. PhD thesis, Université de Grenoble.

Okabe, A., Boots, B., Sugihara, K., and Chiu, S. (1992). *Spatial tessellations: concepts and applications of Voronoi diagrams*.

Oseen, C. (1911). *Über die Stoke'sche Formel und über eine verwandte Aufgabe in der Hydrodynamik*. Almqvist & Wiksell.

Oseen, C. (1913). *Ueber den Gültigkeitsbereich der Stokesschen Widerstandsformel*. Arkiv för Matematik, Astronomi och Fysik. Friedländer.

Pai, M. G. and Subramaniam, S. (2012). Two-way coupled stochastic model for dispersion of inertial particles in turbulence. *J. Fluid Mech.*, 700:29–62.

Parthasarathy, R. and Faeth, G. M. (1990a). Turbulence modulation in homogeneous dilute particle-laden flow. *J. Fluid Mech.*, 220:485–514.

Parthasarathy, R. and Faeth, G. M. (1990b). Turbulent dispersion of particles in self-generated homogeneous turbulence. *J. Fluid Mech.*, 220:515–537.

Peskin, C. S. (1972). Flow patterns around heart valves: A digital computar method for solving the equations of motions. Albert Einstein College of Medicine.

Poelma, C. and Ooms, G. (2006). Particle-Turbulence Interaction in a Homogeneous, Isotropic Turbulent Suspension. *Appl. Mech. Rev.*, 59(2):78–90.

Poelma, C., Westerweel, J., and Ooms, G. (2007). Particle-fluid interaction in grid generated turbulence. *J. Fluid Mech.*, 589:315–351.

Poisson, S. (1831). Memoire sur les Mouvements Simultanes d' un Pendule et de L' air Environnemant. *Mem. V Acad. des Sci. Paris*, 9:521–523.

Pope, S. (2000). *Turbulent Flows*. Cambridge university press.

Pozorski, J. and Minier, J.-P. (1998). On the Lagrangian turbulent dispersion models based on the Langevin equation. *Int. J. Multiph. Flow*, 24(6):913–945.

Prakash, V. N., Tagawa, Y., Calzavarini, E., Martínez-Mercado, J., Toschi, F., Lohse, D., and Sun, C. (2012). How gravity and size affect the acceleration statistics of bubbles in turbulence. *New J. Phys.*, 14(10):105017.

Prosperetti, A. and Tryggvason, G. (2007). *Computational Methods for Multiphase Flow*. Cambridge books online. Cambridge University Press.

Qureshi, M. N., Arrieta, U., Baudet, C., Cartellier, A., Gagne, Y., and Bourgoin, M. (2008). Acceleration statistics of inertial particles in turbulent flow. *Eur. Phys. J. B*, 66(4):531–536.

Qureshi, M. N., Bourgoin, M., and Baudet, C. (2007). Experimental investigation of turbulent transport of material particles. In *Adv. Turbul. XI*.

Ramella, M., Boschin, W., Fadda, D., and Nonino, M. (2001). Finding Galaxy Clusters using Voronoi Tessellations. *arXiv Prepr.*, 03:11.

Reade, W. and Collins, L. (2000). Effect of preferential concentration on turbulent collision rates. *Phys. Fluids*, 12(10):2530–2540.

Rensen, J., Luther, S., and Lohse, D. (2005). The effect of bubbles on developed turbulence. *J. Fluid Mech.*, 538(-1):153.

Riboux, G., Legendre, D., and Risso, F. (2013). A model of bubble-induced turbulence based on large-scale wake interactions. *J. Fluid Mech.*, 719:362–387.

Riboux, G., Risso, F., and Legendre, D. (2010). Experimental characterization of the agitation generated by bubbles rising at high Reynolds number. *J. Fluid Mech.*, 643(2010):509–539.

Richardson, J. and Zaki, W. (1954). Sedimentation and fluidisation: Part I. *Chem. Eng. Res. Des.*, 32A:35–53.

Richardson, L. (1922). *Weather Prediction by Numerical Process*. Cambridge Mathematical Library. Cambridge University Press.

Risso, F. (2011). Theoretical model for k[sup âˆ'3] spectra in dispersed multiphase flows. *Phys. Fluids*, 23(1):011701.

Risso, F. and Ellingsen, K. (2002). Velocity fluctuations in a homogeneous dilute dispersion of high-Reynolds-number rising bubbles. *J. Fluid Mech.*, 453:395–410.

Risso, F., Roig, V., Amoura, Z., Riboux, G., and Billet, A.-M. (2008). Wake attenuation in large Reynolds number dispersed two-phase flows. *Philos. Trans. A. Math. Phys. Eng. Sci.*, 366(1873):2177–90.

Roig, V. and De Tournemine, a. L. (2007). Measurement of interstitial velocity of homogeneous bubbly flows at low to moderate void fraction. *J. Fluid Mech.*, 572:87.

Roma, A. M., Peskin, C. S., and Berger, J. (1999). An Adaptive Version of the Immersed Boundary Method. *J. Comp. Phys.*, 153:509–534.

Rouson, D. W. I. and Eaton, J. K. (2001). On the preferential concentration of solid particles in turbulent channel flow. *J. Fluid Mech.*, 428:149–169.

Rudolff, R. and Bachalo, W. (1988). Measurement of droplet drag coefficients in polydispersed turbulent flow field. *AIAA Pap.*, 88:0235.

Salazar, J. P. L. C., De Jong, J., Cao, L., Woodward, S. H., Meng, H., and Collins, L. (2008). Experimental and numerical investigation of inertial particle clustering in isotropic turbulence. *J. Fluid Mech.*, 600:245–256.

Sangani, A. S. and Didwania, A. K. (1993). Dynamic simulations of flows of bubbly liquids at large Reynolds numbers. *J. Fluid Mech.*, 250(January):307–337.

Sardina, G., Schlatter, P., Brandt, L., Picano, F., and Casciola, C. M. (2012). Wall accumulation and spatial localization in particle-laden wall flows. *J. Fluid Mech.*, 699:50–78.

Saw, E. and Salazar, J. P. L. C. (2012). Spatial clustering of polydisperse inertial particles in turbulence: I. Comparing simulation with theory. *New J. Phys.*, 14(10):105030.

Saw, E. and Shaw, R. (2012). Spatial clustering of polydisperse inertial particles in turbulence: II. Comparing simulation with experiment. *New J. Phys.*, 14(10):105031.

Saw, E., Shaw, R. A., Ayyalasomayajula, S., Chuang, P., and Gylfason, A. (2008). Inertial Clustering of Particles in High-Reynolds-Number Turbulence. *Phys. Rev. Lett.*, 100(21):1–4.

Schiller, L. and Naumann, A. (1933). Uber die grundlegenden Berechnungen bei der Schwerkraftaufbereitung. *Zeitschrift des Vereines Dtsch. Ingenieure*, 77(12):318–320.

Schlichting, H. (1979). *Boundary-layer theory*. McGraw-Hill series in mechanical engineering. McGraw-Hill.

Schreck, S. and Kleis, S. (1993). Modification of grid-generated turbulence by solid particles. *J. Fluid Mech.*, 249:665–688.

Shannon, C. (1998). Communication In The Presence Of Noise. *Proc. IEEE*, 86(2):447–457.

Shotorban, B. and Balachandar, S. (2006). Particle concentration in homogeneous shear turbulence simulated via Lagrangian and equilibrium Eulerian approaches. *Phys. Fluids*, 18(6):065105.

Smereka, A. S. (1993). On the motion of bubbles in a peridic box. *J. Fluid Mech.*, 254(79-112).

Smith, T. N. (1998). A model of settling velocity. *Chem. Eng. Sci.*, 53(2):315–323.

Snyder, M. R., Knio, O. M., Katz, J., and Le Maitre, O. P. (2008). Numerical study on the motion of microscopic oil droplets in high intensity isotropic turbulence. *Phys. Fluids*, 20(7):073301.

Snyder, W. and Lumely, J. L. (1971). Some measurements of particle velocity autocorrelation functions in a turbulent flow. *J. Fluid Mech.*, 48:41–71.

Springel, V., White, S., Jenkins, A., Frenk, C., Yoshida, N., Gao, L., Navarro, J., Thacker, R., Croton, D., Helly, J., Peacock, J., Cole, S., Thomas, P., Couchman, H., Evrard, A., Colberg, J., and Pearce, F. (2005). Simulations of the formation, evolution and clustering of galaxies and quasars. *Nature*, 435(7042):629–36.

Squires, K. and Eaton, J. K. (1990). Particle response and turbulence modification in isotropic turbulence. *Phys. Fluids*, 2(7):1191–1203.

Squires, K. and Eaton, J. K. (1991). Preferential concentration of particles by turbulence. *Phys. Fluids A Fluid Dyn.*, 3(May 2011):1169–1178.

Stokes, G. (1845). On the Theories of Internal Friction of the Fluids in Motion. *Trans. Cambridge Philos. Soc.*, 8:287–319.

Sundaram, S. and Collins, L. (1997). Collision statistics in an isotropic particle-laden turbulent suspension. Part 1. Direct numerical simulations. *J. Fluid Mech.*, 335:75–109.

Sundaram, S. and Collins, L. (1999). A numerical study of the modulation of isotropic turbulence by suspended particles. *J. Fluid Mech.*, 379:105–143.

Tagawa, Y., Martínez-Mercado, J., Prakash, V. N., Calzavarini, E., Sun, C., and Lohse, D. (2012). Three-dimensional Lagrangian Voronoi analysis for clustering of particles and bubbles in turbulence. *J. Fluid Mech.*, 693(2009):201–215.

Taylor, G. (1921). Diffusion by continuous movements. *Proc. Lond. Math. Soc.*, 20:196–212.

Ten Cate, A., Derksen, J. J., Portela, L. M., and Van Den Akker, H. (2004). Fully resolved simulations of colliding monodisperse spheres in forced isotropic turbulence. *J. Fluid Mech.*, 519:233–271.

Tennekes, H. and Lumley, J. (1972). *A first Course in Turbulence*. MIT Press Cambridge, MA.

Tenneti, S., Garg, R., and Subramaniam, S. (2011). Drag law for monodisperse gas-solid systems using particle-resolved direct numerical simulation of flow past fixed assemblies of spheres. *Int. J. Multiph. Flow*, 37(9):1072–1092.

Tomboulides, A. and Orszag, S. (2000). Numerical investigation of transitional and weak turbulent flow past a sphere. *J. Fluid Mech.*, 416:45–73.

Tooby, P. F., Wick, G. L., and Isaacs, J. D. (1977). The motion of a small sphere in a rotating velocity field: A possible mechanism for suspending particles in turbulence. *J. Geophys. Res.*, 82(15):2096–2100.

Toschi, F. and Bodenschatz, E. (2009). Lagrangian Properties of Particles in Turbulence. *Annu. Rev. Fluid Mech.*, 41(1):375–404.

Tryggvason, G., Scardovelli, R., and Zaleski, S. (2011). *Direct Numerical Simulations of Gas-Liquid Multiphase Flows*. Cambridge University Press.

Tsuji, Y., Kato, N., and Tanaka, T. (1991). Experiments on the unsteady drag and wake of a sphere at high Reynolds numbers. *Int. J. Multiph. flow*, 17(3):343–354.

Uhlherr, P. and Sinclair, C. (1970). The effect of free-stream turbulence on the drag coefficients of spheres. *Proc. Chem.*, 1:1–13.

Uhlmann, M. (2003a). First experiments with the simulation of particulate flows. Technical Report No. 1020, CIEMAT, Madrid, Spain.

Uhlmann, M. (2003b). Simulation of particulate flows on multi-processor machines with distributed memory. Technical Report No. 1039, CIEMAT, Madrid, Spain.

Uhlmann, M. (2004). New Results on the Simulation of Particulate Flows. Technical Report March, CIEMAT, Madrid, Spain.

Uhlmann, M. (2005). An immersed boundary method with direct forcing for the simulation of particulate flows. *J. Comput. Phys.*, 209(2):448–476.

Uhlmann, M. (2008). Interface-resolved direct numerical simulation of vertical particulate channel flow in the turbulent regime. *Phys. Fluids*, 20(5):053305.

Uhlmann, M. (2010). Hydrodynamic turbulence induced by sedimenting particles. In Mohr, B. and Frings, W., editors, *Jülich Blue Gene/P Extrem. Scaling Work. 2010*, Technical Report FZJ-JSC-IB-2010-03, pages 35–37.

Uhlmann, M. and Doychev, T. (2012). Finite size particles in homogeneous turbulence. In Binder, K., Münster, G., and Kremer, M., editors, *NIC Symposium 2012*, volume 45 of *Publication series of the John von Neumann Institute for Computing*, pages 377–384, Jülich (Germany).

Uhlmann, M. and Doychev, T. (2014). Sedimentation of a dilute suspension of rigid spheres at intermediate Galileo numbers: particle motion and clustering. *Accept. publ. J. Fluid Mech.*

Uhlmann, M. and Dušek, J. (2014). The motion of a single heavy sphere in ambient fluid: A benchmark for interface-resolved particulate flow simulations with significant relative velocities. *Int. J. Multiph. Flow*, 59:221–243.

van Wijngaarden, L. (1998). On Pseudo Turbulence. *Theor. Comput. Fluid Dyn.*, 10(1-4):449–458.

Veldhuis, C., Biesheuvel, A., and Lohse, D. (2009). Freely rising light solid spheres. *Int. J. Multiph. Flow*, 35(4):312–322.

Veldhuis, C., Biesheuvel, A., Wijngaarden, L. V., and Lohse, D. (2005). Motion and wake structure of spherical particles. *Nonlinearity*, 18(1):C1–C8.

Verzicco, R. and Orlandi, P. (1996). A finite-difference scheme for three-dimensional incompressible flows in cylindrical coordinates. *J. Comput. Phys.*, 123:402–414.

Volk, R., Calzavarini, E., Lévêque, E., and Pinton, J.-F. (2011). Dynamics of inertial particles in a turbulent von Kármán flow. *J. Fluid Mech.*, 668:223–235.

Volk, R., Calzavarini, E., Verhille, G., Lohse, D., Mordant, N., Pinton, J.-F., and Toschi, F. (2008). Acceleration of heavy and light particles in turbulence: Comparison between experiments and direct numerical simulations. *Phys. D Nonlinear Phenom.*, 237(14-17):2084–2089.

Voth, G. A., La Porta, A., Crawford A. M., Alexander, J., and Bodenschatz, E. (2002). Measurement of particle accelerations in fully developed turbulence. *J. Fluid Mech.*, 469:121–160.

Voth, G. a., Satyanarayan, K., and Bodenschatz, E. (1998). Lagrangian acceleration measurements at large Reynolds numbers. *Phys. Fluids*, 10(9):2268.

Wang, L. and Maxey, M. (1993). Settling velocity and concentration distribution of heavy particles in homogeneous isotropic turbulence. *J. Fluid Mech.*, 256(27):68.

White, B. L. and Nepf, H. M. (2003). Scalar transport in random cylinder arrays at moderate Reynolds number. *J. Fluid Mech.*, 487:43–79.

Wood, A. M., Hwang, W., and Eaton, J. K. (2005). Preferential concentration of particles in homogeneous and isotropic turbulence. *Int. J. Multiph. Flow*, 31(10-11):1220–1230.

Wu, J. and Faeth, G. M. (1993). Sphere wakes in still surroundings at intermediate Reynolds numbers. *AIAA J.*, 31(8):1448–1455.

Wu, J. and Faeth, G. M. (1994). Sphere wakes at moderate Reynolds numbers in a turbulent environment. *AIAA J.*, 32(3):535–541.

Wu, J. and Faeth, G. M. (1995). Effect of ambient turbulence intensity on sphere wakes at intermediate Reynolds numbers. *AIAA J.*, 33(1):171–173.

Xu, H. and Bodenschatz, E. (2008). Motion of inertial particles with size larger than Kolmogorov scale in turbulent flows. *Phys. D Nonlinear Phenom.*, 237(14-17):2095–2100.

Yang, C. Y. and Lei, U. (1998). The role of the turbulent scales in the settling velocity of heavy particles in homogeneous isotropic turbulence. *J. Fluid Mech.*, 371:179–205.

Yang, T. S. and Shy, S. S. (2003). The settling velocity of heavy particles in an aqueous near-isotropic turbulence. *Phys. Fluids*, 15(4):868.

Yang, T. S. and Shy, S. S. (2005). Two-way interaction between solid particles and homogeneous air turbulence: particle settling rate and turbulence modification measurements. *J. Fluid Mech.*, 526:171–216.

Yarin, L. and Hetsroni, G. (1994). Turbulence intensity in dilute two-phase flows-3 The particles-turbulence interaction in dilute two-phase flow. *Int. J. Multiph. Flow*, 20(1):27–44.

Yeo, K., Dong, S., Climent, E., and Maxey, M. (2010). Modulation of homogeneous turbulence seeded with finite size bubbles or particles. *Int. J. Multiph. Flow*, 36(3):221–233.

Yeung, P. and Pope, S. (1989). Lagrangian statistics from direct numerical simulations of isotropic turbulence. *J. Fluid Mech.*, 207:531–586.

Yin, X. and Koch, D. L. (2007). Hindered settling velocity and microstructure in suspensions of solid spheres with moderate Reynolds numbers. *Phys. Fluids*, 19(9):093302.

Yin, X. and Koch, D. L. (2008). Velocity fluctuations and hydrodynamic diffusion in finite-Reynolds-number sedimenting suspensions. *Phys. Fluids*, 20(4):043305.

Yoshimoto, H. and Goto, S. (2007). Self-similar clustering of inertial particles in homogeneous turbulence. *J. Fluid Mech.*, 577:275–286.

Yuan, Z. and Michaelides, E. (1992). Turbulence modulation in particulate flows-A theoretical approach. *Int. J. Multiph. Flow*, 18:779–785.

Yudine, M. (1959). Physical Considerations on Heavy-Particle Diffusion. *Adv. Geophys.*, 6:185–191.

Zeng, L., Balachandar, S., Fischer, P., Najjar, F. M., and Fisher, P. (2008). Interactions of a stationary finite-sized particle with wall turbulence. *J. Fluid Mech.*, 594:271–305.

Zeng, L., Balachandar, S., and Najjar, F. M. (2010). Wake response of a stationary finite-sized particle in a turbulent channel flow. *Int. J. Multiph. Flow*, 36(5):406–422.

Zenit, R., Koch, D. L., and Sangani, A. S. (2001). Measurements of the average properties of a suspension of bubbles rising in a vertical channel. *J. Fluid Mech.*, 429(January):307–342.

Zimmermann, R., Gasteuil, Y., and Bourgoin, M. (2011). Rotational intermittency and turbulence induced lift experienced by large particles in a turbulent flow. *Phys. Rev. Lett.*, 2:4.

List of Tables

2.1 Physical properties for (turbulent) particulate flow. Additionally, various length, time and velocity scales introduced in section §2.1.3 are presented. Dimensionless groups involved in particulate flows are also listed. 26

3.1 Physical parameters of the, in the current work, performed simulations with single settling particle. Case $S121$ and case $S178$. 42
3.2 Numerical parameters of the performed simulations with single settling particle. Case $S121$ and case $S178$. 42
3.3 Comparison between various time scales for case $S121$ and $S178$. 43
3.4 Various statistical results for the direct numerical simulations of case $S121$ and $S178$. 54

4.1 Physical parameters fused in case $F121$ and case $F178$. Solid volume fraction Φ_s and Reynolds number Re_∞, lengths of the computational box L_x, L_y, L_z and number of particles N_p. 60
4.2 Numerical parameters use in case $F121$ and case $F178$. Particle resolution $d_p/\Delta x$, where Δx denotes the grid spacing. Number of grid nodes N_i in the i-th coordinate direction. Number of Lagrangian marker points used for the discretization of the fluid-particle interface, N_l. Number of processors used for the present simulations, N_{proc}. 61
4.3 Skewness and kurtosis for case $F121$ of the probability distribution functions shown in figure 4.5 (a,b,c). 67
4.4 Skewness and kurtosis for case $F178$ of the probability distribution functions shown in figure 4.5 (d,e,f). 67

5.1 Physical parameters used in case $M121$, case $M178$ and case $ML178$. Solid volume fraction Φ_s, density ratio ρ_p/ρ_f, Galileo number Ga, Reynolds number $Re_\infty = U_\infty d_p/\nu$ based on the imposed mean fluid velocity in the vertical direction, the particle diameter d_p and the fluid viscosity ν, extensions of the computational box L_x, L_y, L_z and number of particles N_p. 92
5.2 Numerical parameters use in case $M121$, case $M178$ and case $ML178$. Particle resolution $d_p/\Delta x$, where Δx denotes the grid spacing. Number of grid nodes N_i in the i-th coordinate direction. Number of Lagrangian marker points used for the discretization of the fluid-particle interface, N_l. Number of processors used for the present simulations, N_{proc}. 92

7.1 Comparison of the first and second moments (mean value and standard deviation) of the vertical component of the particle conditioned fluid velocity $w_f^S(i,t)$ ($\forall i = 1 \ldots N_p$) and the unconditioned velocity field $w_f(\boldsymbol{x},t)$ for all three flow cases $M121$, $M178$, $ML178$. 119
7.2 Comparison of various definitions of the drag coefficient for the mean particle drag in all three cases $M121$, $M178$ and $ML178$. 130
7.3 Difference between the time average of the drag coefficients defined in equations 7.7 and 7.10 (equations 7.8 and 7.11). 132
7.4 Skewness of the probability density functions of the particle velocity, force, angular velocity and torque components for all three cases $M121$, $M178$, $ML178$.148
7.5 Kurtosis of the probability density functions of the particle velocity, force, angular velocity and torque components for all three cases $M121$, $M178$, $ML178$. 148
7.6 Standard deviation, skewness and kurtosis of the probability density functions of the particle acceleration for all three cases $M121$, $M178$, $ML178$. 148
7.7 Taylor micro-scale, τ_λ calculated from the autocorrelation functions of the particle linear and angular velocity components as well as for the particle hydrodynamic force and torque components for all three flow cases $M121$, $M178$ and $ML178$. 158
7.8 Integral time scale, T_{int}, for the autocorrelation functions of the particle linear and angular velocity components as well as for the particle hydrodynamic force and torque components for all three flow cases $M121$, $M178$ and $ML178$. Additionally the integral time scale as estimated by a Langevin model is given as well. 158

8.1 Skewness of the probability density functions of the components of the fluid velocities $\boldsymbol{u}_f$ and $\boldsymbol{u}_f^S$ for all three cases $M121$, $M178$, $ML178$. 172
8.2 Kurtosis of the probability density functions of the components of the fluid velocities $\boldsymbol{u}_f$ and $\boldsymbol{u}_f^S$ for all three cases $M121$, $M178$, $ML178$. 173

List of Figures

1.1 Overview of some possible flow configurations for dispersed phase flows with selected papers from sections §1.2.1 - §1.2.3 13

2.1 A schematic diagram of the energy cascade and energy spectra at very high Reynolds number Pope (2000) 21

3.1 Diagram of single particle settling regimes digitalized from Jenny et al. (2004) . 40
3.2 Map of single particle settling regimes digitalized from Horowitz and Williamson (2010) 41
3.3 Temporal evolution of the particle settling velocity in case $S121$ 44
3.4 Temporal evolution of (a) the particle horizontal, (b) vertical velocity components and (c) the horizontal angular velocity component in case $S178$. 46
3.5 Sketch of the coordinate systems used for the evaluation of the flow field conditioned on the particle position 48
3.6 Iso-contours of the particle wake in case $S121$. (a) Iso-contour of the relative relative velocity $\tilde{w}_{p,rel}$. (b) Iso-contour of the vortical structure around the particle represented by λ_2 criterion. (c) Contour plot of the projected relative velocity $\tilde{w}_{p,rel}$ in the plane parallel to the vertical axis passing trough the particle center $\boldsymbol{x}_p$. 49
3.7 Iso-contours of the particle wake in case $S178$. (a) Iso-contour of the relative relative velocity $\tilde{w}_{p,rel}$. (b) Iso-contour of the vortical structure around the particle represented by λ_2 criterion. (c) Contour plot of the projected relative velocity $\tilde{w}_{p,rel}$ in the plane parallel to the vertical axis passing trough the particle center $\boldsymbol{x}_p$. 50
3.8 Profile of the vertical component of the conditionally averaged fluid velocity $\tilde{w}_{p,rel}$ on the center line passing trough the reference particle center for (a) case $S121$ and (b) case $S178$ 51
3.9 Comparison of the wake recirculation length l_e measured in cases $S121$ and $S178$ with results from previous studies 52
3.10 Fluid velocity deficit $\hat{u}_{d0}$ in the averaged particle wake in case $S121$ and case $S178$. 53
3.11 Cross-profiles of the conditionally averaged vertical fluid velocity component $\hat{w}_{p,rel}$ in the particle wake for (a) cases $S121$ and (b,c) $S178$. 53
3.12 Cross-profiles of the velocity deficit $\hat{u}_d/\hat{u}_{d0}$ in the particle wake as function of the cross-stream coordinate for (a) case $S121$ and (b) case $S178$. 54

4.1 Dimensions of the computational domain with the initial particle distribution for (a) case $F121$ and (b) case $F178$. 62
4.2 Intensity of the fluid velocity fluctuations for case $F121$ and case $F178$ 63
4.3 Probability density function of the flow velocity field for (a) case $F121$ and (b) case $F178$. 64
4.4 Fluid velocity fluctuations of the vertical velocity component for (a,b,c) case $F121$ and (d,e,f) case $F178$. Total instantaneous fluctuations $\boldsymbol{u}_f'''(\boldsymbol{x},t)$, (a,d). Spatial fluctuations $\langle \boldsymbol{u}_f' \rangle_t(\boldsymbol{x})$, (b,e). Temporal fluctuations $\boldsymbol{u}_f''(\boldsymbol{x},t)$, (c,f). 65
4.5 Probability density function of the fluid velocity field for (a,b,c) case $F121$ and (d,e,f) case $F178$. Total instantaneous fluctuations $\boldsymbol{u}_f'''(\boldsymbol{x},t)$ (a,d). Time average of the spatial fluctuations $\langle \boldsymbol{u}_f' \rangle_t(\boldsymbol{x})$ (b,e). Temporal fluctuations $\boldsymbol{u}_f''(\boldsymbol{x},t)$ (c,f). . 66
4.6 One dimensional energy spectrum $E_{\alpha\alpha}$ of the flow field for (a,c) case $F121$ and (b,d) case $F178$. 68
4.7 Two-point auto-correlation function and corresponding pre-multiplied spectra of the fluid velocity fluctuations in case $F121$. (a) for separations in x-direction and (b) for separations in z-direction. 69
4.8 Two-point auto-correlation function and corresponding pre-multiplied spectra of the fluid velocity fluctuations in case $F178$. (a) for separations in x-direction and (b) for separations in z-direction. 70
4.9 Time evolution of (a) the vertical component of the mean hydrodynamic force $\langle f_z \rangle_p$ and (b) the Variance of the horizontal and the vertical components $\langle f_\alpha' f_\alpha' \rangle_p^{1/2}$ for case $F121$ and case $F178$. 71
4.10 Map of the particle-conditioned average value of the vertical hydrodynamic force component $\langle \tilde{f}_z \rangle_{p,t}$ for (a) case $F121$ and (b) case $F178$. 73
4.11 Variation of the particle conditioned average value of $\tilde{f}_z$ from figure 4.10 along the horizontal and vertical axis passing through the center of the reference particle for case $F121$ and case $F178$. (a) in the horizontal direction and (b) in the vertical direction 74
4.12 Spatial map of the particle-conditioned average value of the resultant hydrodynamic force $\tilde{f}_{ij}$ as function of the position $(\tilde{x}, \tilde{z})$ relative to the reference particle for case $F121$. 75
4.13 Spatial map of the particle-conditioned average value of the resultant hydrodynamic force $\tilde{f}_{ij}$ as function of the position $(\tilde{x}, \tilde{z})$ relative to the reference particle for case $F178$. 75
4.14 Variation of the particle conditioned average values of $\langle \tilde{f}_{ij} \rangle_{p,t}$ from figures 4.12 - 4.13 along the two axis passing through the center of the reference particle for case $F121$ and case $F178$. (a) in the horizontal direction and (b) in the vertical direction. 76
4.15 Contour plot of the projected relative velocity $\tilde{w}_{p,rel}/U_\infty$ in the plane parallel to the vertical axis passing trough the particle center for (a) case $F121$ and (b) case $F178$. 77
4.16 Vertical component of the relative velocity on the center line passing trough the center of the reference particle and parallel to the vertical axis for (a) case $F121$ and (b) case $F178$. 78

4.17 Streamwise velocity deficit $\tilde{u}_{d0}$ defined as in equation 3.11 along the vertical axis $\tilde{z}$ as function of the downstream distance $\tilde{r}_z$ from the reference particle surface for case $F121$ and case $F178$. 79
4.18 Profiles of the velocity deficit in the average particle wake as function of the cross-stream coordinate $\tilde{x}$ for (a) case $F121$ and (b) case $F178$. 80
4.19 Cross-stream half-width of the average particle wake as function of the downstream distance $\tilde{z}$ for case $F121$ and case $F178$. 81
4.20 Evolution of the energy $\tilde{k}_f$ as function of the downstream distance $\tilde{z}$ form the reference particle for case $F121$ and case $F178$. 82

5.1 Dimensions of the computational domain with the initial particle distribution for case $ML178$ at the begin of the simulation. a) side view (b) top view. 93
5.2 Average of the relative particle settling velocity, w_s, for case $M121$, case $M178$ and case $ML178$. 94
5.3 Intensity of the velocity fluctuations of (a) the fluid phase and (b) dispersed phase as function of time for case $M121$, case $M178$ and case $ML178$. 95

6.1 Instantaneous particle position in case $M121$ at (a) initial and (b) statistically stationary state. 100
6.2 Instantaneous particle position in case $M178$ at (a) initial and (b) statistically stationary state. 100
6.3 Two dimensional Voronoï diagram for case $M178$. 102
6.4 Probability density function of (a) the normalized Voronoï cell volumes and (b) the aspect ratios $A_\mathcal{V}$ for case $M121$. 103
6.5 Probability density function of (a) the normalized Voronoï cell volumes and (b) the aspect ratios $A_\mathcal{V}$ for case $M178$. 104
6.6 Standard deviation of Voronoï cell volumes as function of time for case $M121$ and case $M178$. 105
6.7 Time evolution of the number of particles with high probability to be in clusters and in voids for case $M178$. 106
6.8 Lagrangian autocorrelation of Voronoï volumes and Voronoï cell aspect ratio $A_\mathcal{V}$ for (a) case $M121$ and (b) case $M178$. 106
6.9 Temporal evolution of the average distance to the nearest neighbor for cases $M121$ and $M178$. 107
6.10 Averaged local solid volume fraction ϕ_s^{cond} conditioned on the particle positions for case $M121$. 109
6.11 Averaged local solid volume fraction ϕ_s^{cond} conditioned on the particle positions for case $M178$. 109
6.12 Evolution of ϕ_s^{cond} along two perpendicular axis through the reference particle for case $M121$ and case $M178$. (a) along the horizontal axis and (b) along the vertical axis. 110
6.13 Instantaneous particle position in case $ML178$ at (a) initial and (b) statistically stationary state. 111
6.14 Probability density function of (a) the normalized Voronoï cell volumes and (b) the aspect ratios $A_\mathcal{V}$ for case $ML178$. 112

6.15 Standard deviation of Voronoï cell volumes as function of time for case $M178$ and case $ML178$. 113
6.16 Averaged local solid volume fraction ϕ_s^{cond} conditioned on the particle positions for case $ML178$. 114
6.17 Evolution of ϕ_s^{cond} along two perpendicular axis through the reference particle for case $M178$ and case $ML178$. (a) along the horizontal axis and (b) along the vertical axis. 114

7.1 Average of the relative particle settling velocity, w_s (equation 5.1) and w_s^S (equation 7.2), for case $M121$, case $M178$ and case $ML178$. 121
7.2 Joint p.d.f.s of the vertical component of the relative particle velocity fluctuations $w'_{p,rel} = w_{p,rel} - \langle w_{p,rel} \rangle_{p,t}$ and the fluctuations of the Voronoï cell volumes $V'_i = V_i - \langle V \rangle_{p,t}$ for (a) case $M121$, (b) case $M178$ and (c) case $ML178$. 122
7.3 (a,c,e) Top view of the particle locations at one instant in time. The particles are colored according to the fluctuation of their settling velocity for case $M121$, case $M178$ and case $ML178$. (b,d,f) Instantaneous, vertically-averaged fluid velocity fluctuations normalized with the instantaneous box-average for case $M121$, case $M178$ and case $ML178$. 124
7.4 Spatial map of the particle-conditioned average value of the settling velocity, $\tilde{w}_{p,rel}$ for (a) case $M121$, (b) case $M178$ and (c) case $ML178$. 125
7.5 Variation of the particle conditioned average value $\tilde{w}_{p,rel}$ from figure 7.4 along the two axis passing through the center of the reference particle for case $M121$, (a) case $M178$ and (b) case $ML178$. 126
7.6 Temporal evolution of the relative turbulence intensity $I_r = \sqrt{\langle u'_{i,f} u'_{i,f} \rangle_{\Omega_f} / 3} / w_s$ and $I_{r_V} = \sqrt{\langle w'_f w'_f \rangle_{\Omega_f}} / w_s$ as function of time for case $M121$, case $M178$ and case $ML178$. 127
7.7 (a) Time evolution of the particle ensemble average of the vertical component of the hydrodynamic force and (b) mean particle drag coefficient for case $M121$, case $M178$ and case $ML178$. 129
7.8 Joint p.d.f.s of the drag coefficient fluctuations C'_d and the fluctuations of the Voronoï cell volumes V'_i for (a) case $M121$, (b) case $M178$ and (c) case $ML178$. 131
7.9 Spatial map of the particle-conditioned average value of the vertical hydrodynamic force, $\langle \tilde{f}_z \rangle_p$ for (a) case $M121$, (b) case $M178$ and case (c) $ML178$. . . . 133
7.10 Variation of the particle conditioned average value of $\tilde{f}_z$ from figure 7.9 along the two axis passing through the center of the reference particle for case $M121$, case $M178$ and case $ML178$. (a) in the horizontal direction and (b) in the vertical direction 134
7.11 Variance of the particle vertical velocity component, $\langle w'_p w'_p \rangle_p$, and its three contributions (cf. equation 7.13): $\langle w_f^{S\prime} w_f^{S\prime} \rangle_p$, $\langle w_{p,rel}^{S\prime} w_{p,rel}^{S\prime} \rangle_p$, $2\langle w_{p,rel}^{S\prime} w_f^{S\prime} \rangle_p$ for (a) case $M121$, (b) case $M178$ and (c) case $ML178$. 138
7.12 Probability density function of the angle α between the relative particle velocity vector and the vertical axis for case $M121$, case $M178$ and case $ML178$. 140
7.13 Temporal evolution of the fluctuating energy of the particle angular velocity components for case $M121$, case $M178$ and case $ML178$. 141

7.14 Probability density functions of the particle velocity components for (a) case $M121$, (b) case $M178$ and (c) case $ML178$. 142
7.15 Probability density functions of the particle angular velocity components for (a) case $M121$, (b) case $M178$ and (c) case $ML178$. 143
7.16 Probability density functions of the particle force components for (a) case $M121$, (b) case $M178$ and (c) case $ML178$. 144
7.17 Probability density functions of the particle acceleration ($\boldsymbol{a}_p$) components for case $M121$, case $M178$ and case $ML178$. (a) horizontal component and (b) vertical component. 146
7.18 Probability density functions of the particle torque components for (a) case $M121$, (b) case $M178$ and (c) case $ML178$. 147
7.19 Lagrangian autocorrelation of the particle velocity components as function of time lag τ_{sep} for (a) case $M121$, (b) case $M178$ and (c) case $ML178$. 150
7.20 Lagrangian autocorrelation of the particle angular velocity components as function of time lag τ_{sep} for (a) case $M121$, (b) case $M178$ and (c) case $ML178$. . . 153
7.21 Lagrangian autocorrelation of the particle hydrodynamic force components as function of time lag τ_{sep} for (a) case $M121$, (b) case $M178$ and (c) case $ML178$. 155
7.22 Lagrangian autocorrelation of the particle torque components as function of time lag τ_{sep} for (a) case $M121$, (b) case $M178$ and (c) case $ML178$. 157

8.1 (a) Iso-surfaces of the vertical fluid velocity fluctuations w'_f (with respect to the box-averaged value) in case $M178$ at one instant in the statistically steady regime. (b) The same graph along with the particle positions, but viewed from the top. 162
8.2 (a) Iso-surfaces of the vertical fluid velocity fluctuations w'_f (with respect to the box-averaged value) in case $ML178$ at one instant in the statistically steady regime. (b) The same graph along with the particle positions, but viewed from the top. 163
8.3 Iso-surfaces of the vortex eduction criterion proposed by Jeong and Hussain (1995) at the same instant as in figures 8.1 - 8.2. (a) case $M178$ and (b) case $ML178$. 164
8.4 Fluid velocity fluctuations of the (a) horizontal and (b) vertical velocity components obtained from the simulations, case $M121$, case $M178$ and case $ML178$. The data is plotted along with the results from Parthasarathy and Faeth (1990a) and Mizukami et al. (1992). 170
8.5 Probability density function of the horizontal and vertical components of the flow velocities $\boldsymbol{u}_f^S$ and $\boldsymbol{u}_f$ for (a) case $M121$, (b) case $M178$ and (c) case $ML178$.171
8.6 he four terms in the average kinetic energy equation 8.6: rate of change term $d_t E_k$, rate of dissipation ε, average work due to potential energy $w_{p,rel}\phi_s(\rho_p/\rho_f - 1)g_z$ and correlation between the fluctuating velocity ad particle force $\langle u'_{f,i} f_{p,i}\rangle_\Omega$ for (a) case $M121$, (b) case $M178$ and (c) case $ML178$. 175
8.7 Energy spectra of fluid velocity fluctuations in case $M121$ (a,b), case $M178$ (c,d) and case $ML178$ (e,f). Horizontal direction (a,c,e) and in the vertical directions (b,d,f). 177

8.8 Two-point spatial autocorrelation function of fluid velocity fluctuations in case $M121$ (a,b), case $M178$ (c,d) and case $ML178$ (e,f) for separations in horizontal direction (a,c,e) and in the vertical directions (b,d,f). 179
8.9 Contour plot of the relative velocity $\tilde{w}_{p,rel}/u_g$ in the plane parallel to the vertical axis passing trough the particle center $\boldsymbol{x}_p$. (a) case $M121$. (a) case $M178$. (c) case $ML178$. 180
8.10 Vertical component of the relative velocity on the center line passing trough the particle center $\boldsymbol{x}_p$ and parallel to the vertical axis. (a) Case $M121$. (b) Case $M178$. (c) case $ML178$. 182
8.11 Streamwise velocity deficit $\tilde{u}_{d0}$ defined as in equation 3.11 along the vertical axis $\tilde{z}$. Case $M121$, case $M178$ and case $ML178$. 183
8.12 Cross-stream half-width of the average particle wake as function of the downstream distance $\tilde{z}$ for case $M121$, case $M178$ and case $ML178$. 184
8.13 Profiles of the velocity deficit in the average particle wake as function of the cross-stream coordinate $\tilde{x}$ for (a) case $M121$. (b) case $M178$ and (c) case $ML178$.185